Theoretical Challenges in the Dynamics of Complex Fluids

NATO ASI Series

Advanced Science Institutes Series

A Series presenting the results of activities sponsored by the NATO Science Committee, which aims at the dissemination of advanced scientific and technological knowledge, with a view to strengthening links between scientific communities.

The Series is published by an international board of publishers in conjunction with the NATO Scientific Affairs Division

A Life Sciences **B Physics**	Plenum Publishing Corporation London and New York
C Mathematical and Physical Sciences **D Behavioural and Social Sciences** **E Applied Sciences**	Kluwer Academic Publishers Dordrecht, Boston and London
F Computer and Systems Sciences **G Ecological Sciences** **H Cell Biology** **I Global Environmental Change**	Springer-Verlag Berlin, Heidelberg, New York, London, Paris and Tokyo

PARTNERSHIP SUB-SERIES

1. Disarmament Technologies	Kluwer Academic Publishers
2. Environment	Springer-Verlag / Kluwer Academic Publishers
3. High Technology	Kluwer Academic Publishers
4. Science and Technology Policy	Kluwer Academic Publishers
5. Computer Networking	Kluwer Academic Publishers

The Partnership Sub-Series incorporates activities undertaken in collaboration with NATO's Cooperation Partners, the countries of the CIS and Central and Eastern Europe, in Priority Areas of concern to those countries.

NATO-PCO-DATA BASE

The electronic index to the NATO ASI Series provides full bibliographical references (with keywords and/or abstracts) to more than 50000 contributions from international scientists published in all sections of the NATO ASI Series.
Access to the NATO-PCO-DATA BASE is possible in two ways:

- via online FILE 128 (NATO-PCO-DATA BASE) hosted by ESRIN, Via Galileo Galilei, I-00044 Frascati, Italy.

- via CD-ROM "NATO-PCO-DATA BASE" with user-friendly retrieval software in English, French and German (© WTV GmbH and DATAWARE Technologies Inc. 1989).

The CD-ROM can be ordered through any member of the Board of Publishers or through NATO-PCO, Overijse, Belgium.

Series E: Applied Sciences - Vol. 339

Theoretical Challenges in the Dynamics of Complex Fluids

edited by

Tom McLeish
Department of Physics,
University of Leeds,
Leeds, U.K.

Springer Science+Business Media, B.V.

Proceedings of the NATO Advanced Study Institute on
Theoretical Challenges in the Dynamics of Complex Fluids
Cambridge, U.K.
24 March – 4 April 1996

A C.I.P. Catalogue record for this book is available from the Library of Congress

ISBN 978-94-010-6305-0 ISBN 978-94-011-5480-2 (eBook)
DOI 10.1007/978-94-011-5480-2

Printed on acid-free paper

TABLE OF CONTENTS

PREFACE

No-one who took part in the NATO Advanced Studies Institute from which this book emerges will have forgotten the experience. True, the necessary conditions for a very successful workshop were satisfied: a field of physics bursting with new power and new puzzles, a matchless team of lecturers, an international gathering of students many of whom had themselves contributed at the forefront of their subject, an admirable overlap of experiment and theory, a good mix of experimenters and theorists, an enviable environment. But who could have foreseen the way the workshop became a focus for future directions, how fresh scientific ideas tumbled out of the discussion periods, how the context of *teaching* the field produced such fruitfulness of *research* at the highest level?

The organisers did have some specific aims in mind. Perhaps foremost was the desire to compare notes among different areas within the subfield of soft condensed matter physics fast becoming known as "complex fluids". For readers seeking a definition, the prosaic "fluids with bits in" can be passed rapidly over in favour of the elegant discussion of slow variables by Scott Milner in his chapter. The uniting goals of the subject are to model the essential molecular or mesoscopic structure theoretically, and to probe this structure as well as the bulk response of the system experimentally. Our famous examples were: colloids, polymers, liquid crystals, block co-polymers and self-assembling surfactant systems. Each has largely spawned its own research community and although there is some overlap it is not uniform. It was especially memorable to follow the formulations of rheological theory in colloids and polymers, which included learning to "translate" across the fields - when does a "Brownian stress" become an "entropic stress"? There was a deliberate desire to include experimentalists at each stage in the programme - the book follows suit by posing the "experimental challenges" before the "theoretical" ones in each case.

It is quite possible to start anywhere in the book, as many ideas develop in parallel. Indeed it was not possible to ensure that all the introductory material lay in the first few chapters because by necessity it often accompanies advanced new work. For the clearest possible introduction to the kinetics of phase separation in fluids, for example, the reader is advised to turn straight to the chapter by Masao Doi, although the new work on a rheological model for two-phase fluids presented there belongs after the earlier chapters on single-phase fluids.

The beginning of the book is also a good place to start, where a summary of recent dynamic light scattering and rheology on colloids (Peter Pusey) and polymers (Jean Candau) is admixed with an account of the key theoretical ideas and methods (Sriram Ramaswamy, Mike Rubinstein and Sasha Semenov). An approach to non-linear response and topology in polymers follows (Tom McLeish) before considering the coupling of flow to concentration-fluctuation in both colloids and polymers (Sriram Ramaswamy and Scott Milner).

Leaving the two archetypes behind we explore polymeric liquid crystal dynamics with Pino Marrucci and their development of textures in flow (Ron Larson). The ingredients of polymer connectivity, persistence length, and the competition between liquid randomness and crystalline order all arise in the suite of chapters on surfactant systems. These (mostly but not exclusively) experimental viewpoints look at wormlike micelles (François Lequeux and Jean Candau), foams (Lequeux with Boltenhagen), shear-induced structures including the famous "onion phases" by Didier Roux and the wonderful surfactant cubic phases (Chris Toprak). Elegant theoretical ideas which tackle worms and vesicles are then presented by Mike Cates and a possible way forward to the slow modes in disordered smectics by Sriram Ramaswamy. We finish with the theory of two-phase fluid rheology by Masao Doi and the surprising subject of reactions at interfaces in complex fluids by Glenn Fredrickson, which both gave rise to prolonged debate in Cambridge, even after nearly two weeks of the school.

It is never possible to cover all the material in a school like this one in the succeeding book, if only because some is published largely intact elsewhere. This is true of the masterly review of block co-polymer fluids by Glenn Fredrickson (which brings together ideas and phenomena from the smectic-phases of Roux, the polymer dynamics of Rubinstein and Semenov and the two-phase flow of Doi). Readers who refer to his recent published review with Frank Bates (*Annu. Rev. Mater. Sci. 1996, 26:501-550*) as an accompaniment to this volume will recover something of the flavour of the ASI material. The same is true of the important material on the theory of colloidal dynamics and computational approaches by John Brady. Where referred-to in the text of other contributors, the reader should consult (*Curr. Opin. Colloid Interface Sci. 1996, 1, 472-480*). I can likewise do little justice here to the sparking introduction given to the ASI by Sam Edwards, who reminded us of the necessity to focus on the essential physics of a problem, and of the magnitude of difficulty in the formal treatments of some of the phenomena which arise repeatedly, especially that of topological interactions.

The compensation of a book published some months after the school is of course that some of the issues raised in discussion at the ASI have been incorporated into the chapters. A glance at the similar (but not yet identical!) structures for self-diffusion constants in colloids by Ramaswamy and Pusey or of the shear-history integrals of Ramaswamy and Milner will illustrate the point. The problem of the concentration-scaling of the polymer plateau modulus in theta-conditions raised by Rubinstein was the subject of a discussion question by Scott Milner to Tom Mcleish in the context of star polymer dynamics, and became the subject of an ensuing calculation by us which has found its way into the chapter on branched polymers. In the months since the ASI there has been fresh experimental and theoretical work on the topic of flow-instabilities and banded flows of complex fluids, which was much discussed following the student's poster and oral presentations. Although this discussion could not be pursued in this volume, many readers will want to follow this up, and at the terrible risk of the accusation of over-selectivity, let me point them to publications by Callaghan and co-workers in New Zealand, Palierne and colleagues in France and Olmsted and Lu in the UK.

Many people are due thanks for their efforts in making this volume, and the meeting that generated it, possible. Without the support of the NATO Science Committee nothing would have happened, and I am very grateful for the assistance of the ASI committee, Mike Cates, Michael Rubinstein, François Lequeux and Chris Toprak. The hospitality and organisation of the Isaac Newton Institute for the Mathematical Sciences was matchless. Production of the final text was greatly assisted by Mrs. Maureen

Thompson and special thanks are due to Dr. Nigel Clarke, who produced the excellent index and made valuable comments on the manuscript. I am indebted to them and to the superb team of authors. All remaining textual errors are of course my own responsibility.

We all look forward to discovering over the next few years who will have taken up these theoretical challenges in the dynamics of complex fluids, and what they will have discovered.

Tom McLeish

Leeds 1997

S. Ramaswamy

P. Pusey

G. Marrucci

M. Rubinstein

A. Semenov

M.Doi.

S. Milner

T. McLeish

M. Cates

R. Larson

M. Doi and S.F. Edwards

C. Toprokcioglu

D. Roux

G. Fredrickson

EXPERIMENTAL CHALLENGES IN COLLOIDS

P.N. PUSEY, P.N. SEGRÈ, O.P. BEHREND,
S.P. MEEKER and W.C.K. POON
Department of Physics and Astronomy
The University of Edinburgh
Mayfield Road
Edinburgh
EH9 3JZ
UK

1. Introduction

Recent experimental studies, largely by dynamic light scattering (DLS), of suspensions in a liquid of colloidal particles which interact like hard spheres have yielded two unexpected findings whose interpretation challenges theory. First, the rate of structural relaxation (defined below) of the suspensions shows the same dependence on suspension concentration as the inverse of their zero-shear-rate viscosity. Second, the intermediate scattering functions, measured by DLS, show an interesting scaling property which suggests that structural relaxation is controlled by self diffusion of the particles.

Detailed descriptions of this work have been published recently. Thus this paper will give only a very brief summary which directs the reader to the literature. A more detailed, but still concise, summary is given in Ref. [1].

2. Background

Dynamic light scattering measures the normalised intermediate scattering function $f(k,t)$ of the suspension (see chapters by Candau, McLeish and Semenov, this volume),

$$f(k,t) \equiv \frac{S(k,t)}{S(k,0)}, \tag{1}$$

where the intermediate scattering function $S(k,t)$ is given by

T McLeish (ed), Theoretical Challenges in the Dynamics of Complex Fluids, 1-6

$$S(k,t)=\frac{1}{N}\left\langle\sum_{j=1}^{N}\sum_{k=1}^{N}\exp\left\{i\boldsymbol{k}\cdot\left[\boldsymbol{r}_j(0)-\boldsymbol{r}_k(t)\right]\right\}\right\rangle, \tag{2}$$

and

$$S(k,0)=S(k) \tag{3}$$

where $S(k)$ is the static structure factor. Here N is the number of particles, $\boldsymbol{k}$ is the scattering vector and $\boldsymbol{r}_j(t)$ the position of particle j at time t. As can be seen from its definition, in general the intermediate scattering function measures a collective motion of the particles.

In concentrated suspensions, where the fraction ϕ of the suspension's volume which is occupied by the particles may be 0.5 or larger, the static structure factor resembles that of simple atomic liquids, showing a pronounced diffraction peak at $2\pi/k\approx 2R$, where R is the particles' radius. The dominant structure in the suspension, which gives rise to this peak, is the short-ranged ordering, or cage, of particles surrounding a given particle.

In a dilute suspension, where interactions between the particles can be neglected, the intermediate scattering function takes the simple form $f(k,t)=\exp\left(-D_0k^2t\right)$, where D_0 is the free-particle (Stokes-Einstein) diffusion coefficient. In a concentrated suspension, due to both direct and hydrodynamic interactions between the particles, $f(k,t)$ has a more complicated dependence on k^2t, the slowest decay being found at the peak of $S(k)$. Furthermore $f(k,t)$ decays via a two-stage process: an initial exponential decay,

$$f(k,t)=\exp\left[-D_S(k)k^2t\right], \qquad t<<\tau_R, \tag{4}$$

where τ_R is the "structural relaxation time", essentially the lifetime of a particle's cage of neighbours; and a second, slower, approximately exponential decay at long times,

$$f(k,t)\propto\exp\left[-D_L(k)k^2\tau\right], \qquad t>>\tau_R, \tag{5}$$

with $D_L(k)<D_S(k)$.

3. Experimental

The suspensions consisted of sterically-stabilised particles of poly-methylmethacrylate in cis-decalin [see e.g. 2]. Due to a slight difference between the refractive indices of the particles and liquid, these sample were slightly opaque. The "two-colour dynamic light scattering" technique [3] was used to suppress multiple scattering, and thereby select only the strong single scattering.

4. Short-time diffusion

The short-time diffusion coefficients $D_S(k)$ describe the average motions of the particles over distances small compared to their radius and reflect both direct and hydrodynamic interactions between the particles. Extensive measurements were made of $D_S(k)$ as functions of both scattering vector k and suspension concentration ϕ [2]. These were compared with the predictions of theory and computer simulation, good agreement being found with the latter [2].

5. Long-time diffusion

The long-time diffusion coefficients $D_L(k)$ describe motions of the particles over distances comparable to, or larger than, their radius. There is no satisfactory theory to date of long-time diffusion (but see the first chapter by Ramaswamy, this volume, for a connection between self diffusion and the dynamics of concentration fluctuations). The new results outlined below may provide insights which will stimulate theoretical developments.

As noted above, in a dense fluidlike assembly of hard spheres the dominant structure, which gives rise to the main peak in $S(k)$ at $k = k_m$, is the cage of particles surrounding a given particle. Thus it can be argued that the long-time decay of $f(k_m,t)$, the intermediate scattering function measured at $k = k_m$, reflects the dominant structural relaxation of the system so that $D_L(k_m)$ is a measure of the rate of structural relaxation. By comparing measurements of $D_L(k_m)$ with measurements of the zero-shear-rate viscosity η of the suspensions [4] we have found that the rate of structural relaxation shows the same dependence on suspension concentration as the inverse of the viscosity over the whole range $0 < \phi < 0.50$, i.e. that

$$\frac{D_L(k_m)}{D_0} = \frac{\eta_0}{\eta}, \tag{6}$$

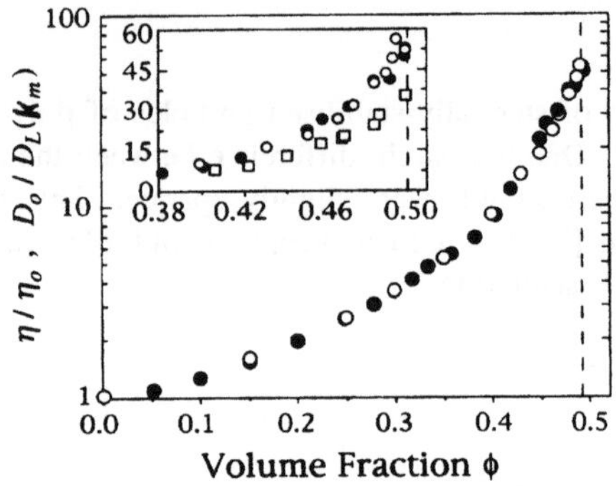

Figure 1. Relative viscosity η/η_0 (open circles) and inverse rate of structural relaxation $D_0/D_L(k_m)$ (filled circles) versus volume fraction ϕ of suspensions of PMMA spheres (from [4], q.v. for an explanation of the data points represented by squares in the inset).

where η_0 is the viscosity of the liquid in which the particles are suspended (see Fig. 1). While one would certainly expect these two quantities to show similar dependences on concentration - the processes of simple shear flow and structural rearrangement both involve the relative motions of neighbouring particles - the apparent identity found experimentally is surprising and remains to be explained by theory.

6. Scaling of the intermediate scattering functions

During an attempt to understand better the mechanism of structural relaxation we made a second surprising discovery [5]. This was that for $kR > 2.7$, a range of scattering vector k which encompasses most of the strong variation of structure factor $S(k)$ including the main peak, plots of $\ln f(k,t)/D_S(k)k^2$ against t, measured at different values of k, lay on a master curve. As can be seen from Eq. (4), this way of plotting the data ensures that they superimpose at short times (since $\ln f(k,t)/D_S(k)k^2 = -t$, for $t \ll \tau_R$). What is surprising is the additional superimposition of data at intermediate and long times, $t \geq \tau_R$ (see Fig. 2). As noted in [5], this finding implies that, for $kR > 2.7$, the intermediate structure factor can be written

$$f(k,t) \approx \exp\left(-\frac{D_S(k)}{D_S(\infty)} k^2 \langle \Delta r^2(t) \rangle / 6\right), \tag{7}$$

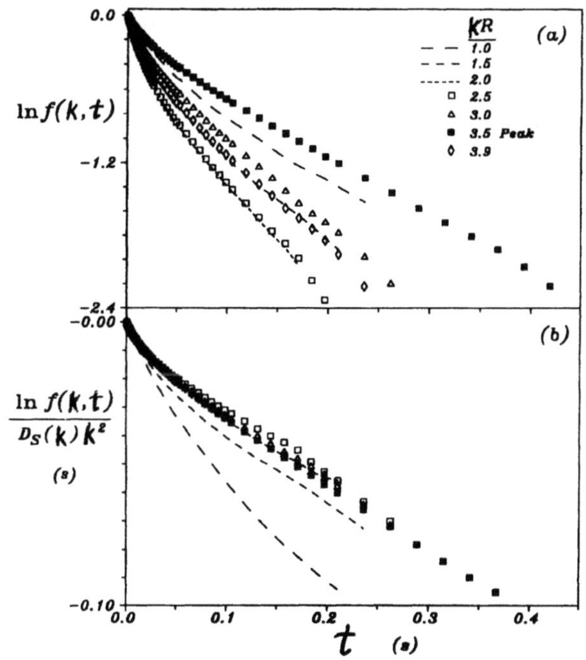

Figure 2. **(a) Logarithm of normalised intermediate scattering functions $\ln f(k,t)$ versus time t** for a PMMA suspension of volume fraction ϕ = **0.465 for different** values of kR as indicated. (b) Same data plotted as $\ln f(k,t)/D_S(k)k^2$ versus t, showing scaling for $kR > 2.5$ (from [5]).

where $\langle \Delta r^2(t) \rangle$ is the mean-square displacement of a single particle, suggesting that structural relaxation is controlled by self diffusion. Although previous work [6] has suggested a connection between structural relaxation and self diffusion, the detailed scaling implied by Eq. (7) awaits a full theoretical explanation.

7. References

1. Pusey, P.N., Segrè, P.N., Behrend, O.P., Meeker, S.P. and Poon, W.C.K. (1996) Dynamics of concentrated colloidal suspensions, *Physica A*, in press.

2. Segrè, P.N., Behrend, O.P. and Pusey, P.N. (1995) Short-time Brownian motion in colloidal suspensions: Experiment and simulation, *Phys. Rev. E* **52**, 5070-83.

3. Segrè, P.N., van Megen, W., Pusey, P.N., Schätzel, K. and Peters, W. (1995) Two-colour dynamic light scattering, *J. Mod. Opt.* **42**, 1929-52.

4. Segrè, P.N., Meeker, S.P., Pusey, P.N. and Poon, W.C.K. (1995) Viscosity and structural relaxation in suspensions of hard-sphere colloids, *Phys. Rev. Lett.* **75**, 958-61.

5. Segrè, P.N. and Pusey, P.N. (1996) Scaling of the dynamic scattering function of concentrated colloidal suspensions, *Phys. Rev. Lett.* **77**, 771-4.

6. de Schepper, I.M., Cohen, E.G.D., Pusey, P.N. and Lekkerkerker, H.N.W. (1990) Analogies between the dyanamics of concentrated charged colloidal suspensions and dense atomic liquids, *Physica A* **164**, 12-27.

SELF-DIFFUSION OF COLLOIDS AT FREEZING

SRIRAM RAMASWAMY
Department of Physics
Indian Institute of Science, Bangalore 560 012, India
sriram@physics.iisc.ernet.in

Abstract. This is an introduction to the theory of interacting Brownian particles, as applied to charge-stabilised colloidal suspensions near their equilibrium liquid-solid transition. The density functional approach to the statics of the transition is reviewed briefly, and the generalised Langevin equation method for the dynamics presented in detail. Work with A.V. Indrani [1] on a self-consistent approach for calculating the excess single-particle friction is presented, which explains the observed [2] "universal" suppression of self-diffusion at freezing as a consequence of the universal structure-factor height at this transition. Criticisms, open questions, and challenges to theory are discussed.

1. Introduction

1.1. BACKGROUND AND EXPERIMENTS

Artificial colloidal suspensions [3] of synthetic spherical particles suspended in solvents such as water are widely studied as scale models of conventional atomic liquids or crystals. The relative ease with which interparticle interactions can be controlled in these systems have made them a natural model system for the study of the phases and metastable states [3] of classical simple liquids. The large natural timescales ($\tau \sim 1$ ms) and length scales ($\sim 10^3 \mathring{A}$) and small elastic moduli ($G \sim 10^2$ dyn cm^{-2}) of these materials allow ready access to structural and dynamical information through light scattering [4] (see Peter Pusey's lecture in this volume) and make it possible as well to drive the system into strongly nonequilibrium and nonlinear regimes (shear-rates $\sim \tau^{-1}$ and stresses $\sim G$). Shear-induced distortion of spatial correlations in the fluid phase [5], enhancement of self-diffusion un-

T McLeish (ed.), Theoretical Challenges in the Dynamics of Complex Fluids, 7-20.

der shear [6] and shear-induced melting of colloidal crystals [7] are among the more dramatic nonequilibrium and nonlinear effects observed in these systems.

While my second lecture deals with nonequilibrium aspects, the present lecture is on a surprising feature of the dynamics of colloidal suspensions at *equilibrium* at the liquid-solid transition. A piece of background information about this transition is useful at this point. It has long been known empirically [8] and more recently understood theoretically [9] that the static structure factor (the fourier-transform of the pair correlation function, defined in detail below) has a nearly universal height, of about 2.85, at the equilibrium liquid-solid crystal transition for a wide variety of simple liquids, such as liquid alkali metals, liquid argon, computer simulations of soft and hard sphere liquids, and so on. The novel observation of Löwen, Palberg and Simon [2] from dynamic light scattering experiments on colloids as well as Brownian dynamics simulations of particles interacting with model potentials was that the ratio $R = D_l/D_s$ of the long-time (D_l) and short-time (D_s) self-diffusion coefficients of the particles was also "universal" at freezing with a value close to 0.1. I shall show in this lecture that this "dynamic universality" is a direct consequence of the universal structure factor discussed above, not an independent effect. The work described in this lecture and the one that follows was done in collaboration with A.V. Indrani [1, 10]. Many details can be found in [11].

It is necessary to say what is meant here by "short-time". I will always be talking about times longer than the very short timescale ($\sim 10^{-8}\,s$ for a colloidal particle of diameter 0.1 μm) over which friction with the *solvent* causes the particle to lose the memory of its initial velocity. Physically, the most important timescale in the problem is the time τ for a given particle to move one nearest-neighbour distance. For times $\ll \tau$ the friction on the particle is entirely due to the solvent, and the diffusivity measured will be D_s, while for times $\gg \tau$ the interactions with the other colloidal particles are effective in enhancing the friction, thus slowing down the particle and leading to a diffusivity $D_l < D_s$. In very dilute suspensions $D_s \simeq D_0 = k_BT/6\pi\eta a$, the familiar Stokes-Einstein diffusivity. An important difference between colloidal suspensions and single-component liquids becomes evident at this point: the molecules of the latter feel no friction at short times, so that their motion is *ballistic*, not diffusive, and the ratio R cannot be defined.

I shall deal mainly with charge-stabilised colloids such as polystyrene spheres ("polyballs") suspended in water. Acid groups on the surface of the polyballs dissociate in water, sending positively charged "counterions" into solution and leaving the polyballs with a negative charge of a few hundred electrons. The resulting repulsion between the polyballs stabilises them from aggregating due to van der Waals forces. The repulsion is however partly screened by the counterions themselves as well as by additional ionic impurities which the water may contain. The pair potential between the particles is thus approximately of the Yukawa form $e^{-\kappa r}/\epsilon r$ where r is the interparticle separation, ϵ the dielectric constant of water, and κ the inverse screening length. These systems are studied not by varying the temperature but instead by adding or removing ionic impurities to change the screening length. I shall be concerned here with systems in which the mean interparticle separation l is comparable in magnitude to the screening length κ^{-1} and much larger than the particle radius a. For the systems of interest here, $a \approx 0.1\mu$m, $\kappa \sim \ell \sim 0.5\mu$m.

Real colloidal particles in a suspension interact hydrodynamically as well (see John Brady's article in this volume): the motion of a particle induces velocity fields around it which move other particles. The simplest model Brownian systems which are often used to simulate their properties on a computer do not. For the study of *equilibrium, static* properties, this difference is immaterial: any algorithm which respects detailed balance will produce the same phase diagram and equal-time correlations. Unfortunately the same does not hold for the *dynamics*, even at equilibrium, nor for equal-time correlations away from equilibrium, e.g., in a shear flow: a complete dynamical theory of suspensions must in principle include the hydrodynamic interaction. For the case we are concerned with here however, since $\kappa a \gg 1$, the effect of the hydrodynamic interaction can simply be absorbed into a redefinition of D_s [12]. The problem then reduces to that of a collection of interacting Brownian particles and the ratio $R = D_l/D_s$ is determined exclusively by the electrostatic interactions.

1.2. SUMMARY OF THEORETICAL APPROACH

The mechanism underlying the suppression of self-diffusion (the diffusion of a specific "tagged" particle) is physically clear. Each particle moves in a background of all the other particles. The manner in which these particles slow down our tagged particle depends on how they're distributed in space (the structure factor), how their distribution relaxes in time (the diffusivity itself), and how they interact with the tagged particle (the pair potential). It might therefore seem that the degree to which self-diffusion is suppressed at freezing should depend both on the structure factor at freezing and on the pair potential, which would leave unanswered the question why the suppression at freezing is the same for widely different potentials. However, because the system is at thermal equilibrium, the effective potential between the particles, in a coarse-grained description, is also characterised by the structure factor, as will be shown below in detail. Thus both factors which govern the enhanced friction on a tagged particle in a suspension are seen to be determined by the structure factor of the fluid. Since the latter is universal at freezing, one should expect the diffusivity to be suppressed by a universal factor as well.

Note that in the argument above the mechanism for reducing the diffusivity depended on the diffusivity itself: the lower the diffusivity, the more slowly the background particle distribution changes, and the more effective the slowing down! There is clearly an element of feedback [13] in this process which should lead to substantial reductions in the diffusivity but to evaluate it quantitatively will require, as is common in "self-consistent" problems, the solution of an integral equation.

I will take a continuum, coarse-grained approach, in which the detailed particle configurations and microscopic pair-potential are replaced by concentration fields and a direct pair correlation function. The dynamics of the concentration and momentum density fields of the particles are described by generalised Langevin equations. These we solve in a perturbative [14] self-consistent [13] approximation generally known as "mode-coupling". Although the calculations I shall describe are fairly technical and involved, they are really just an enactment in detail of the physically appealing ideas presented in the paragraphs above.

1.3. RESULTS

Here are the main results of this work. (i) The ratio $R = D_l/D_s$ of long–time to short–time self–diffusivities at the freezing transition is shown to be"universal" i.e. the same for several different pair potentials, as an inevitable consequence of the "universality" of the static structure factor at this transition. (ii) The theory predicts $R \simeq 0.05$, which is rather smaller than the experimental value of 0.1, and plausible reasons are offered for this discrepancy. (iii) the *shape* of the plot (Fig. 1) of mean–square displacement versus time (scaled by D_l in a sense explained below) is shown to be in good agreement with experiment. I should add that the technique and an understanding of its strengths and limitations form the main message of the lecture, and the particular application presented here should serve as an illustration.

The rest of this paper is organised as follows. In the next section, I summarise the density functional approach which makes possible a description of interparticle interactions entirely in terms of correlations in the liquid state without reference to a pair potential, which of course is essential if we are to understand the"dynamic" universality [2]. Section III constructs the model in detail. The self-consistent "mode-coupling" calculation of the self–diffusivity is described in Section IV, and the paper closes in Section

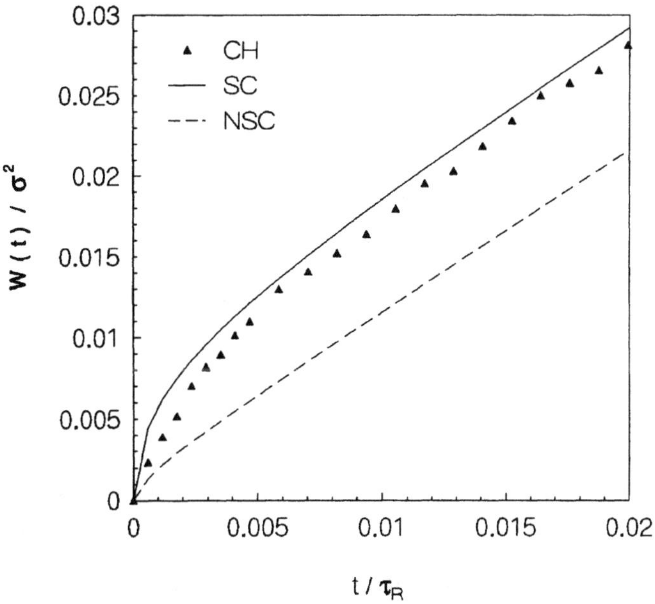

Figure 1. Mean square displacement versus scaled time: CH is data read off visually from Cichocki and Hinsen, 1992, SC is the present calculation, and NSC is from a non self-consistent calculation.

V with a discussion of the results and the approximations made.

2. The Density–Functional Approach

Since our theory of the dynamics of colloids is going to be formulated in terms of concentration fields rather than particle coordinates, we need a description of the free-energetics of concentration fluctuations at thermal equilibrium. This is done through the *density functional free-energy* [9]. The idea is to rewrite the classical canonical partition function $Z \propto \int d^{3N}\mathbf{q}^{\mathbf{N}} e^{-\beta V(q_1,\ldots,q_N}$ where the $\{\mathbf{q} \propto\}$ are the coordinates of the colloidal particles, V their total potential energy, and $\beta = T^{-1}$ the inverse temperature, as a functional integral over configurations of the number density field $n(\mathbf{r}) = \sum_{\alpha=1}^{N} \delta(\mathbf{r}-\mathbf{q}_\alpha)$ as

$$Z \propto \int \Pi_r dn(\mathbf{r}) e^{-\beta F[n]}. \tag{1}$$

Here $F[n]$ is known as the density–functional free–energy, and is simply the free–energy cost of producing a number density fluctuation $\delta n(\mathbf{r}) = n(\mathbf{r}) - n_0$ about the uniform liquid density n_0. Instead of presenting a formal calculation of $F[n]$ let us understand it by a brief physical argument. Apply an external potential $U(\mathbf{r})$ to the system. If the particles did *not* interact with each other the density field would change from the uniform n_0 to $n(\mathbf{r}) = n_0 e^{-\beta U(\mathbf{r})}$. Interactions modify this: let us still write the result as $n(\mathbf{r}) = n_0 e^{-\beta U_{eff}(\mathbf{r})}$ and expand the effective potential or molecular field in a functional Taylor series in $\delta n(\mathbf{r})$: $U_{eff}(\mathbf{r}) = U(\mathbf{r}) + \int_{r'} c(\mathbf{r}-\mathbf{r}')\delta n(\mathbf{r}') + \ldots$, and truncate this series at first order. Note the"expansion coefficient" $c(\mathbf{r}-\mathbf{r}')$, which we will relate shortly to directly measurable liquid–state properties, is independent of $U(\mathbf{r})$, so that in the limit of a vanishingly small external potential we can write

$$n(\mathbf{r}) = n_0 e^{\int_{r'} c(\mathbf{r}-\mathbf{r}')\delta n(\mathbf{r})} \tag{2}$$

or

$$\int_{r'} c(r-r')\delta n(r') = ln\frac{n(\mathbf{r})}{n_0}. \tag{3}$$

The highly nonlinear integral equation (2) or (3) can in fact be obtained as the stationary condition $\delta F/\delta n(\mathbf{r}) = 0$ for the functional

$$\beta F[n] = -\frac{1}{2}\int_{r'} \delta n(r)c(r-r')\delta n(r') + \int_r [n(r)ln\frac{n(r)}{n_0} - \delta n(r)]. \tag{4}$$

Since the average density field $n(r)$ is obtained by extremising $F[n]$, it is quite plausible that F *is* the free-energy functional of Eqn.(1), with the first term on the right-hand side of (4) representing interactions and the second the ideal-gas entropy. If we expand F to quadratic order in δn then, in terms of the fourier transformed number density $\delta n_q = \int_r \delta n(r) e^{-i\mathbf{q}.\mathbf{r}}$, $\beta F[n] \simeq \frac{1}{2n_o}\int_q (1-n_0 c_q)|\delta n_q|^2$ we see immediately by equipartition on (1) that the static structure factor of the colloidal liquid is

$$S_q \equiv \int_r e^{-i\mathbf{q}.\mathbf{r}} \frac{< \delta n(0)\delta n(r) >}{n_0} \simeq \frac{1}{1-n_0 c_q} \tag{5}$$

This tells us that c_q can be obtained from scattering experiments that measure the structure factor S_q. In fact c_q is known in liquid theory [15] as the fourier–transform of the direct pair correlation function. Equations (2) to (4) form the basis of the modern theory of freezing [9] Equation (2) can self-consistently stabilise spatially periodic density modulations if $c(r)$ is large enough, and (4) tells you whether such modulations have a lower free-energy than, and are hence thermodynamically stable relative to, the uniform liquid. This programme, if carried out for a wide variety of simple fluids, yields the universal height of about 2.85 for the structure factor at freezing [9].

Now recall that we are setting up to calculate the diffusivity of a tagged particle. We will therefore find it useful to define a "tagged" or "self" density field, $n_s(\mathbf{r},t) = \delta(\mathbf{r} - \mathbf{q}_1(t))$, which simply keeps track of a specific particle (in this case particle number 1). More precisely, imagine tagging a negligibly small fraction of particles, say $\alpha = 1$ to $N_1 \ll N$, and defining $n_s(\mathbf{r},t) = \sum_{\alpha=1}^{N_1} \delta((\mathbf{r} - \mathbf{q}_\alpha(t))$. Then for $N_1 \ll N$ the free-energy functional (scaled by the temperature) for fluctuations in n_s about the mean n_{s0} by precisely the same arguments as those above for n, is

$$F_s[n_s] = -\int_{r,r'} \delta n_s(r) c(r-r') \delta n(r') + \int_r [n_s(r) ln \frac{n_s(r)}{n_{s0}} - \delta n_s]. \quad (6)$$

Note the asymmetry of the interaction term, a result of our distinguishing a subset of the particles as "tagged." Interactions among the tagged particles have been left out since they are negligible for $N_1 \ll N$.

We are now equipped to begin the dynamical calculation. The main input from the statics is that the local tagged particle chemical potential is

$$\frac{\delta F_s}{\delta n_s(\mathbf{r},t)} = -T \int_{r'} c(r-r') \delta n(r't) + T ln n_s(r,t), \quad (7)$$

so that the local density of thermodynamic force [16, 17] is

$$\mathbf{\Phi}_s \equiv -n_s \nabla \frac{\delta F_s}{\delta n_s} = -T n_s(rt) \nabla \int_{r'} c(r-r') \delta n(r',t) + T \nabla n_s \quad (8)$$

The physics of the two terms on the right-hand side of (8) is clear. The second term is simply the gradient of the ideal-gas osmotic pressure, while the first says that particles at $\mathbf{r}$ feel a force, through the interactions, here subsumed in $c(\mathbf{r} - \mathbf{r}')$, from all neighboring points $\mathbf{r}'$. In the next section, the expressions (8) will be used to construct a dynamical description.

3. The Model

Our detailed calculation begins with the continuity equation for the self-number density $n_s(\mathbf{r},t)$ and a generalised Langevin equation for the self-momentum density $\mathbf{g}_s(\mathbf{r},t)$ (or self–velocity field $\mathbf{v}_s = \mathbf{g}_s/mn_s$) of a *tagged particle* [16] of mass m in a colloidal suspension with a fluid velocity field $\mathbf{v}_f$.

$$\frac{\partial n_s}{\partial t} + \nabla.(n_s \mathbf{v}_s) = 0 \tag{9}$$

$$\frac{\partial \mathbf{g}_s}{\partial t} + \gamma_o \mathbf{v}_s + n_s \nabla \frac{\delta F_s}{\delta n_s} = \mathbf{f}(\mathbf{r},t) \tag{10}$$

The physical content of these equations is clear. (9) says simply that any given "tagged" particle is separately conserved and cannot vanish. In (10), the second term on the left attempts to dissipate any difference between the velocities of the particle and the solvent. Since we are neglecting the hydrodynamic interaction, we have set the fluid velocity to its thermal equilibrium value, namely zero. Observe that this term contains the bare damping constant γ_o, which should be taken as equal to $\Gamma_0 n_{s0}$. The third term on the left in (10) is the density of thermodynamic force (8) constructed at the end of section II. The right–hand–side of (10) is the usual delta–correlated Gaussian thermal noise source with variance proportional to γ_o times the temperature $T = \beta^{-1}$. Inertia plays no role for the slow flows considered here, so it suffices to look at (10) with the $\partial g_s/\partial t$ term set to zero, so that the velocity autocorrelation function (VACF) in the absence of interactions is delta–correlated in time, with the same *value* of the bare friction. If we linearise (9) and (10), and consider timescales $\gg m/\Gamma_0$ or, equivalently, pass to the limit of no inertia, then (10) simply becomes a constitutive relation: $n_s \mathbf{v}_s = -(T/\Gamma_0)\nabla n_s$. In this approximation, (9) reduces to the fluctuating–diffusion equation, with a diffusivity D_0:

$$\frac{\partial n_{s\mathbf{q}}(t)}{\partial t} + D_0 q^2 n_{s\mathbf{q}} = \zeta_{\mathbf{q}}(t) \tag{11}$$

Here $\zeta_{\mathbf{q}}$ is a Gaussian thermal noise source with variance proportional to $D_0 q^2$, where $D_0 = T/\Gamma_0$ is the bare self–diffusivity. Similar arguments applied to the Fourier–transform $n_{\mathbf{q}}$ of the *collective* density yield the (linearised) equation

$$\frac{\partial n_{\mathbf{q}}(t)}{\partial t} + \tau_q^{-1} n_{\mathbf{q}} = \zeta_{\mathbf{q}}(t) \tag{12}$$

where

$$\tau_q^{-1} = D_0 q^2 / S(q) \tag{13}$$

reflecting the fact that concentration fluctuations relax slowly for wavenumbers near the maximum of the structure factor $S(q) = [1 - n_0 c(q)]^{-1}$. Of course, this linearised treatment greatly underestimates the slowing down: the maximum height of $S(q)$, as we said, is about 2.85 at freezing, while the observed slowing-down is by a factor of ten. In the next section we will understand why there is such a large effect.

4. Mode-Coupling Analysis

We would like to calculate the correlation functions of n_s and n as functions of frequency and wavenumber. If all nonlinear terms in our dynamical equations were neglected, this could be done by fourier-transforming (11) and (12) in space and time. The resulting correlation functions would have peaks at zero frequency with widths governed by the *bare* diffusivities, i.e. ultimately by Γ_0. The self-density (C_s) and collective density (C_c) correlation functions in this linearised theory, as functions of (Laplace) frequency z and wavenumber k are

$$C_s(k,z) = \frac{1}{z + ik^2T/\Gamma_0};$$

$$C_c(k,z) = \frac{S(k)}{z + \frac{ik^2T/\Gamma_0}{S(k)}} \tag{14}$$

When nonlinearities are included, we are free to write the correlation functions in a form as close to (14) as possible, provided we replace Γ_0 by a wavenumber- and frequency-dependent damping-rate

$$\Gamma(k,z) \equiv \Gamma_0\,(1 + \Sigma(k,z)), \tag{15}$$

The short- and long-time self-diffusivities are respectively $D_s = T/\Gamma_0$ and $D_l = T/\Gamma(k=0, z=0)$, so that the ratio $R = D_l/D_s = 1/[1+\Sigma(k = 0, z = 0)]$.

It is convenient to measure all lengths and times in units of the mean interparticle distance $\sigma \equiv n_0^{-\frac{1}{3}}$ and the bare interparticle diffusion time σ^2/D_0 respectively. Together with (15), this means that the correlation functions now take the form

$$C_s(k,z) = \frac{1}{z + \frac{i\,k^2}{1+\Sigma(k,z)}};$$

$$C_c(k,z) = \frac{S(k)}{z + \frac{i\,k^2\,S^{-1}(k)}{1+\Sigma(k,z)}} \tag{16}$$

The problem then reduces to calculating $\Sigma(k,z)$ from (9) and (10).

Before embarking on this stage of the calculation, a warning is in order. Strictly speaking, we should introduce separate "renormalised frictions" $\Gamma_c(k,z)$ and $\Gamma_s(k,z)$ for the collective and self motions. In fact, at *zero* wavenumber, the collective force density arising from interparticle interactions (and hence its contribution to the collective friction) must vanish since it is simply equal to the total force on the system (Newton's Third Law!), while the self-force density remains nonzero. We shall nonetheless pretend here that the two frictions are the same. The error we are making is not grave, since the important contribution to the excess friction comes from wavevectors around the peak in $S(k)$ rather than near $k = 0$. Indeed, measurements on colloids show a surprising quantitative similarity between self- and collective frictions on short timescales [18]. Further discussion of the approximation made can be found in [19, 20] (see also [21]) and alternative calculational schemes in [22, 23]. I shall remark briefly on the latter at the end of this lecture.

Let us return now to the calculation of the excess friction. The starting point is largely standard and widely used [14, 16, 22, 24], and rests on a simple application of a fluctuation-dissipation theorem: just as the *bare* friction is proportional to the autocorrelation of the random force term in (10), the excess friction or self-energy Σ in (15) is proportional to the autocorrelation of the excess thermodynamic force $\mathbf{\Phi_s}$. Since the latter [see (8)] contains one self- and one collective-density field, the relevant autocorrelation is qualitatively of the form $< [n_s n](0)[n_s n](t) >$. The major approximation lies in replacing the average of the product of four fields by the product of two 2-point correlation functions. This uncontrolled approximation has the virtue of truncating in the simplest possible way an otherwise endless hierarchy, and still yielding nontrivial answers. In this Gaussian decoupling approximation, known as "mode-coupling" because it relates the kinetic coefficient at a given wavevector and frequency to a convolution of correlation functions of modes with other wavevectors and frequencies, the autocorrelation $< \mathbf{\Phi}_s(\mathbf{0},0)\mathbf{\Phi}_s(\mathbf{r},t) >$, Laplace-transformed in time, Fourier-transformed in space, and divided by Γ_0, becomes

$$\Sigma(k,z) = \frac{1}{3}\int d^3q \int dz_1\, v(q)\, C_s(|\mathbf{k}-\mathbf{q}|, z - z_1)\, C_c(q, z_1), \qquad (17)$$

which is bilinear in the self-density and collective-density correlations. With (17) and (16) in this dimensionless form, the fact that Σ depends only on $S(q)$, and is thus likely to be "universal" at freezing, is very clear. In (17)

$$v(q) = q^2[n_0 c(q)]^2 \equiv q^2[1 - S^{-1}(q)]^2 \qquad (18)$$

is the interaction vertex arising from (8), $c(q)$ being the Fourier transform of the direct correlation function. It is clear from (17) and (16) that we have to solve an integral equation in four dimensions is order to evaluate $\Sigma(k,z)$ self–consistently.

We can get a quick insight into just how robust or "universal" the ratio R really is, from a preliminary non-self-consistent (NSC) calculation– that is, one in which the *bare* ($\Sigma = 0$) self and collective density correlation functions of (14) are used on the right hand-side of (17). In that case the integral equation (17) reduces to an integral, and $\Sigma(0,0)$ is trivial to evaluate. It is most convenient to do the calculation for systems for which tabulated data or closed-form expressions are available for the structure factor. The freezing line data tabulated in Hansen and Schiff [25] for the static structure factors of soft sphere potentials of the form $V(r) = \epsilon\left(\frac{\sigma}{r}\right)^n$ ($n = 1, 4, 6, 9, 12$), and the Percus–Yevick expression for the structure factor of hard spheres (at a density at which the peak height was 2.85) [15] are useful sources. After the rather lengthy discussion above, the reader will not be too surprised to learn that the ratio $R = D_l/D_s$, in this NSC approach, is remarkably robust, $\simeq 0.35$ in all these cases [1]. Even without a more elaborate calculation, and despite the quantitative disagreement with the observed value of about 0.1, this confirms that the origin of the universality reported by [2] lies in the structure factor.

Let us see next how the full self-consistent calculation is done. Equation (17) appears to be a four-dimensional integral equation and hence numerically somewhat involved. However, we do not attempt to eliminate the important wave–number dependence in (17), as is sometimes done [26], by assuming *ad hoc* that modes near the peak position q_0 of $S(q)$ dominate the integral, to replace the problem by a one-dimensional integral equation. Instead we use the spatial isotropy of the correlation functions (16) to render the problem effectively *two-dimensional*, and evaluate (17) iteratively in a "pseudospectral" manner. To see what this means, note that the right-hand side of (17) is the convolution of the functions $C_s(k,z)$ and $\tilde{C}_c(k,z) \equiv v(k)C_c(k,z)$ which are themselves isotropic in k–space, so that in real space-time, the self-energy kernel is simply a *product* of two spherically symmetric functions: $\Sigma(r,t) = C_s(r,t)\tilde{C}_c(r,t))$. We therefore begin by obtaining the real-space C_s and $\tilde{C}_c$ from (18) and (16) with $\Sigma = 0$, and evaluate (17) the first time around, using only two-dimensional transforms and avoiding convolutions completely. The resulting $\Sigma(k,z)$ is fed into (16) to give the new values for the correlation functions and the process iterated until $\Sigma(k,z)$ converges to within less than a percent over the entire range of (k,z). We make no assumptions whatsoever about the (k,z) dependence of Σ, C_s or C_c at any stage in the iteration scheme.

We carried out the self-consistent calculation described above for the n = 6 soft sphere case and the Percus–Yevick hard–sphere case. The ratio R was found to be the same for these two potentials, and there is no reason to believe that other potentials, such as those for which we obtained a constant ratio from the non-self-consistent calculation, will give different results here. The physics of the liquid-solid transition is essentially the same for any situation where strong short-range repulsion dominates, which is ultimately why the structure factors at freezing are the same for widely different potentials. This will inevitably lead to the "universal" suppression of self-diffusion at freezing. However, the numerical value from the self-consistent calculation does not agree with experiment: it is now about 0.05. Thus while the non-self-consistent calculation underestimated the slowing down by a factor of three, the self-consistent approach overestimates it by a factor of two.

5. Conclusion

5.1. QUESTIONS AND CRITICISM

Two questions then arise: (i) Are there other useful results can we extract from the self-consistent calculation, especially general trends and shapes rather than precise numerical values? (ii) Is there a way of improving the calculation to get better agreement? The answer to (i) is yes, while opinions on (ii) differ. Let me take up each of these in turn.

The most useful additional information that this mode-coupling calculation gives us is the structure (curvature, crossovers) of the mean-square displacement $W(t)$ versus time. Figure 1 displays Brownian dynamics simulation data for hard spheres at freezing [27] and the predictions of [1] for $W(t)$ in units of the mean interparticle distance σ, plotted against time in units of the *renormalised* diffusion time $D_l t/\sigma^2$, where D_l is the *measured* (for the simulation) and *calculated* (for the theory) long-time diffusivity. Such a scaled plot allows a detailed comparison of the predicted and measured time-dependence irrespective of the actual values of the diffusivities. For contrast, the result of the non-self-consistent (NSC) calculation described a few paragraphs above is also plotted. The satisfactory agreement between the self-consistent and simulation results is evident, as is the substantial deviation of the NSC calculation from the observed behaviour. The shortcoming of the NSC calculation is simply that the long-time behaviour of $W(t)$ begins to set in on a *short* timescale determined by the *bare* diffusivity, which means that the curve bends over much too early in our scaled plot. The main virtue of the self-consistent approach thus lies in its faithful reproduction of the *shape* of the time-dependence of $W(t)$, something which the NSC approach, which is just a lowest-order perturbative calculation, is unable to do. Other predictions such as detailed analysis of the crossover from short to long times via a subdiffusive regime, and a study of non-Gaussian features in the statistics of the particle displacements, which can be compared to light-scattering experiments or Brownian dynamics simulations can be found in [1].

Let us now return to the fact that the calculation in its present form gives too small a value for R. It has long been known that uncritical application of one-loop self-consistent methods to the dynamics of liquids or colloids predicts a "mode-coupling glass transition" [28, 13] at a critical temperature T_* or density n_*, at which viscosities diverge and diffusivities vanish, in a power-law fashion. An unbiased way to check this against experiment is to compare the predicted T_* or n_* with the extrapolated point at which the experimental data indicates a vanishing diffusivity or divergent viscosity. If this is done, the predicted T_* is found to be much higher (n_* much lower) than the experimental value. Such a method would naturally also give too small a diffusivity at freezing. Note that if the theory is taken completely literally, it contains ***no*** fitting parameter, and hence offers no escape. A given density or temperature implies a particular static structure factor which is the sole input to the calculation of R. Presumably the theory errs in replacing four-point correlations by pairs of two-point correlations. However, a method that has found favour with some workers [22, 23], but with no clear justification, is to treat the location of the apparent glass transition as a fitting parameter (in a theory which has none). This forces the data and theory to match in the region of most rapid decrease of diffusivity, giving good agreement at freezing, at the cost of very poor agreement at lower densities where one might expect perturbative theories to work well. A very recent alternative approach [29] works with the dynamics of a single-component *liquid* rather than a suspension, separates out a short-time contribution to the diffusivity, arising from uncorrelated collisions, carries out a *non*-self-consistent calculation, and finds a long-time to short-time ratio of about 0.1. However, the separation of timescales in a single-component liquid is probably not sharp enough to yield two distinct diffusive regimes, since the true short-time dynamics of a one-component liquid is *ballistic*. In any case the applicability of this model to the case of a *suspension* is rather doubtful.

5.2. CHALLENGES

Let me close by listing some challenges to theory. First, a direct numerical solution *à la* [30] of the stochastic partial differential equations (9) and (10), rather than further attempts at approximate closures, is the best way to see whether the disagreement between the experimental and theoretical values for R is a consequence of the mode-coupling approximation or a real failure of the model. A generalisation of this treatment to include rotational diffusion for anisotropic colloidal particles could yield interesting predictions near the liquid-solid and isotropic-nematic transitions as well as near orientational freezing transitions in plastic crystalline phases.

References

1. A.V. Indrani and S. Ramaswamy, Phys. Rev. Lett. **73**, 360 (1994).
2. H.Löwen, T.Palberg and R.Simon, Phys. Rev. Lett. **70** 1557 (1993).
3. A.K. Sood, in *Solid State Physics* (eds. H. Ehrenreich and D. Turnbull) **45**, 1, Academic Press, New York (1991).
4. B.J. Berne and R. Pecora, *Dynamic Light Scattering*, Wiley, New York (1976).
5. B.J. Ackerson and N.A. Clark, Physica **118A**, 221 (1983).
6. X. Qiu, H.D. Ou-Yang, D.J. Pine, and P.M. Chaikin, Phys. Rev. Lett. **61**, 2554 (1988).
7. N.A. Clark and B.J. Ackerson, Phys. Rev. Lett. **44**, 1005 (1980).
8. J.-P. Hansen and L. Verlet, Phys. Rev. **184**, 151 (1969).
9. T.V. Ramakrishnan and M. Yussouff, Phys. Rev. B **19**, 2775 (1979).
10. A.V. Indrani and S. Ramaswamy, Phys. Rev. **E52**, 6492 (1995).
11. A.V. Indrani, Ph.D. thesis, Indian Institute of Science (1995).
12. M. Medina-Noyola, Phys. Rev. Lett. **60** 2705 (1988).
13. W. Götze in *Liquids, Freezing and the Glass Transition*, Eds. D. Levesque, J.P. Hansen and J. Zinn-Justin, Elsevier, Amsterdam (1990).
14. W. Hess and R. Klein, Adv. in Phys. **32**, 173 (1983).
15. See, e.g. J.-P. Hansen and I.R. MacDonald, *Theory of Simple Liquids*, Academic Press, London (1986).
16. T.R.Kirkpatrick and J.C.Nieuwoudt, Phys. Rev. A **33**, 2651 (1986); *ibid.*, 2658 (1986).
17. B. Bagchi, J. Chem. Phys. **95**, 467 (1991).
18. T.G. Mason and D.A. Weitz, Phys. Rev. Lett. **74**, 1250 (1995).
19. M. Fuchs, Phys. Rev. Lett. **74**, C1490 (1995).
20. A.V. Indrani and S. Ramaswamy, Phys. Rev. Lett. **74**, C1491 (1995).
21. B. Cichocki and K. Hinsen, Physica **141A**, 475 (1987).
22. W. Götze and L. Sjögren, Rep. Prog. Phys. **55** 241 (1992).
23. M. Fuchs *et al.*, Phys. Rev. A **45** 898 (1992).
24. R. Biswas, S. Roy, and B. Bagchi, Phys. Rev. Lett. **75**, 1098 (1995).
25. J.P. Hansen and D. Schiff, Mol. Phys. **25**, 1281 (1973).
26. T.R.Kirkpatrick, Phys. Rev. A **32**, 3130 (1985).
27. B. Cichocki and K. Hinsen, Physica **187A**, 133 (1992).
28. B. Kim and G.F. Mazenko, Adv. Chem. Phys. LXXVIII, 129 (1990).
29. R. Ahluwalia and S.P. Das, Phys. Rev. E **54**, 463 (1996).
30. L.M. Lust, O.T. Valls, and C. Dasgupta, Phys. Rev. E **48**, 1787 (1993).

THEORETICAL CHALLENGES IN POLYMER DYNAMICS

MICHAEL RUBINSTEIN
Department of Chemistry
University of North Carolina
Chapel Hill, N.C. 27599-3290, USA

1. Introduction

Polymer dynamics is one of the most interesting areas of polymer physics. It has witnessed significant progress during the last three decades, but a number of major questions still remain unanswered (including the one on the nature of entanglements in polymer solutions and melts).

There are many excellent reviews of polymer dynamics [1-5]. The present one sketches a biased view of the author on some aspects of this very broad subject, emphasizing open questions and theoretical challenges. In Section 2 the two main models of the dynamics of unentangled polymers (Rouse [6] and Zimm [7]) are outlined and the cross-over between them is discussed. In Section 3 the question of the nature of entanglements is introduced and several predictions of the dependence of the molecular weight between entanglements on concentration are reviewed. The models of polymers in arrays of fixed obstacles are presented in Section 4, including linear chains, stars and rings. The many-chain effects are discussed in Section 5. The main theoretical challenges are summarized in Section 6.

2. Rouse and Zimm Models of Unentangled Polymers

Consider a N-mer moving through a media. The size of this polymer is

$$R \approx b\, N^{\nu} \tag{1}$$

where b is the monomer size and $\nu = 1/2$ for Gaussian configuration (screened excluded volume interaction) and $\nu \approx 3/5$ for chains swollen due to excluded volume repulsion.

The Rouse model [1,2,6] assumes that each monomer has a friction coefficient ζ_1 and there is no hydrodynamic coupling between different monomers of the chain. The

T McLeish (ed), *Theoretical Challenges in the Dynamics of Complex Fluids*, 21-51 friction coefficient of the whole chain in the Rouse model (free draining limit) is proportional to the number of monomers

$$\zeta_R \approx N\,\zeta_1 \tag{2}$$

In the Zimm model [1,2,7] the monomers are hydrodynamically coupled to each other and to the solvent inside the chain. The friction coefficient of the polymer is the Stokes friction of a sphere of size R (non-draining limit)

$$\zeta_Z \approx \eta_s R \tag{3}$$

where η_s is the viscosity of surrounding media. Note that in this section we present only the scaling relations and drop all numerical coefficients. The diffusion coefficient of the polymer can be determined from the fluctuation-dissipation theorem [2]

$$D_R \approx kT/\zeta_R \approx kT/(N\,\zeta_1) \tag{4a}$$

$$D_Z \approx kT/\zeta_Z \approx kT/(\eta_s R) \tag{4b}$$

The relaxation time of polymer is the time it takes the chain to diffuse its size

$$\tau_R \approx R^2/D_R \approx \tau_1 N^{2\nu+1} \tag{5a}$$

$$\tau_Z \approx R^2/D_Z \approx (\eta_s/kT)\,R^3 \tag{5b}$$

where τ_1 is the monomeric time (the time it takes a monomer to diffuse its own diameter).

In dilute polymer solutions the motion of the chain is hydrodynamicaly coupled and is best described by the Zimm model. By combining Eqs. (5a) and (5b) with $\tau_1 \approx \eta_s b^3/(kT)$ we note that the Zimm time is shorter than the Rouse time

$$\tau_Z \approx \tau_1 N^{3\nu} < \tau_R \approx \tau_1 N^{2\nu+1} \tag{6}$$

because $\nu < 1$. Thus, if the chain finds a faster relaxation mode that requires less dissipation (lower friction), it prefers it over a slower (higher friction) mode with the same wavelength.

In semidilute solutions the hydrodynamic interactions are screened at distance scales larger than correlation length [1,2]

$$\xi \approx b\,\phi^{-\nu/(3\nu-1)} \tag{7}$$

where ϕ is the polymer volume fraction. At length scales longer than correlation length ξ the polymer moves as a chain of N/g correlation volumes (each containing $g \approx (\xi/b)^{1/\nu} \approx \phi^{-1/(3\nu-1)}$ monomers). The relaxation time of an unentangled chain is

$$\tau \approx \tau_\xi (N/g)^2 \tag{8}$$

where

$$\tau_\xi \approx (\eta_s/kT)\, \xi^3 \tag{9}$$

is the Zimm relaxation time of a strand of g monomers with size ξ (correlation length). The viscosity is $\eta = \int G(t)\, dt$, where $G(t)$ is the stress relaxation function, defined as the ratio of the stress $\sigma(t)$ after the step strain and the value of the strain γ (see Eq. 61). The viscosity of this solution can be estimated by the product of terminal modulus $G(\tau)$ and relaxation time τ

$$\eta \approx G(\tau)\, \tau \approx \eta_s\, N/g \tag{10a}$$

Here the terminal modulus $G(\tau)$ is the energy density stored in the system at relaxation time τ. It is of the order of the thermal energy kT per chain times the density of chains

$$G(\tau) \approx kT\, \phi/(b^3 N) \tag{10b}$$

If one puts a long probe chain (degree of polymerization P) into an unentangled solution or melt of N-mers, it can relax as a Rouse chain of P/g Zimm segments with relaxation time (see Eq. 8)

$$\tau_R \approx \tau_\xi (P/g)^2 \tag{11}$$

But if the probe chain is very long, it may find it easier (and faster) to relax by Zimm motion

$$\tau_Z \approx (\eta/kT)\, R_P^{\,3} \approx \tau_\xi (N/g)\, (P/g)^{3/2} \tag{12}$$

where η is the solution viscosity (Eq. 10a) and $R_P \approx \xi\, (P/g)^{1/2}$ is the chain size. The Rouse time for not very long P - chains $[P/g < (N/g)^2]$ (Eq. 11) is shorter than Zimm time (Eq. 12) and the chains move by Rouse modes on the longest length scale. The Zimm time of very long P - chains $[P/g > (N/g)^2]$ (Eq. 12) is shorter than their Rouse time (Eq. 11). These very long chains move by the Zimm modes on the longest length scale, by Rouse modes on intermediate length scales, and by Zimm modes on the shortest length scales (smaller than correlation blob $r < \xi$) [8].

Note that this cross-over between Rouse and Zimm modes occurs at the same chain length as the swelling of long linear chains [9]. Therefore one needs to correct the

Zimm time (Eq. 12) due to this swelling. It is important to stress that the fact that both static and dynamic cross-overs overlap is a *coincidence* that occurs only for linear chains. If one puts a flexible fractal in a melt, the cross-over from Rouse to Zimm modes occurs at lower molecular weights than the cross-over from Gaussian to swollen conformations of this fractal.

3. Entanglements

3.1 THE MAJOR THEORETICAL CHALLENGE

Polymer solutions and melts exhibit qualitative change in their dynamics at higher solution concentrations and/or polymer molecular weights [10]. This qualitative change is traditionally attributed to the onset of entanglements [1-5,10]. But there is still no clear picture of what an entanglement in polymer solutions and melts really is.

There are sophisticated theories of knots and links using various topological invariants (such as Alexander polynomials) [11-13].

One invariant, known as the Gauss linking number [14]

$$m = \frac{1}{4\pi}\oint\oint d\vec{r}_1 \times d\vec{r}_2 \frac{\vec{r}_1 - \vec{r}_2}{\left|\vec{r}_1 - \vec{r}_2\right|^3}$$

was used to characterize the winding number for a pair of polymeric rings (Fig. 1) [11,15-

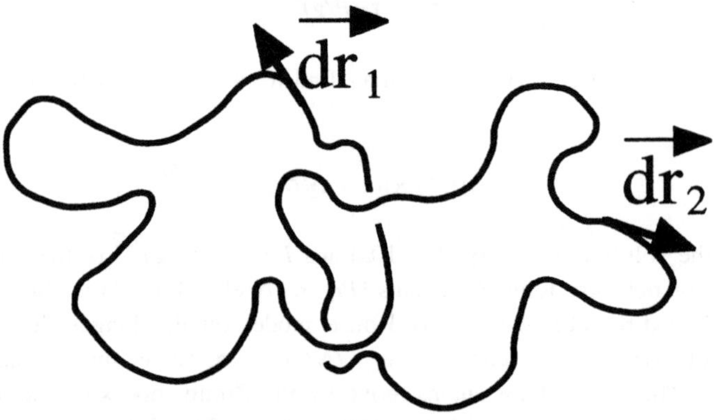

Figure 1. A pair of entangled rings. Tangent vectors $d\vec{r}_1$ and $d\vec{r}_2$ are used in the definition of the Gauss linking number.

17]. Even though it is not an exact topological invariant (there are known examples of topologically connected pairs of rings with zero Gauss linking number [17]), it gives useful information about the way two polymers wind around each other. It was demonstrated that the sections of chains that are far from each other do not significantly contribute to the linking number. The mean square Gauss linking number is proportional to the number of close encounters (contacts) between two polymers [16,17].

Gauss linking number and other topological invariants (such as Alexander polynomials) have been used to characterize entanglements between few chains, but have not yet led to a comprehensive picture of topological constraints in melts and solutions of many overlapping polymers. The current understanding of this complicated problem is based on the phenomenological models.

3.2 CONCENTRATION DEPENDENCE OF THE DEGREE OF POLYMERIZATION BETWEEN ENTANGLEMENTS

Since there in no satisfactory theory of entanglements in polymer solutions and melts, other questions, related to this major one, are also open and present a theoretical challenge. These include the dependence of the mesh size a of an entanglement net (if such a net can be defined) on polymer concentration. There have been numerous attempts to predict the size of this entanglement mesh from "first principles" [18-22]. Some of them were more successful than others, but the problem still remains unsolved.

The experimental values of this mesh size a (and of the degree of polymerization between entanglements N_e) are usually obtained from the measurements of plateau modulus G_0. The stress relaxation function $G(t)$ of entangled solutions and melts exhibits a rubber-like plateau with modulus G_0 at times shorter than terminal time (Fig. 2) [10]. Rubber elasticity theory [23] is employed to relate this plateau modulus G_0 at polymer concentration c to the degree of polymerization between entanglements N_e

$$G_0 \approx kT\, c/N_e \qquad (13)$$

Concentration c is defined as the number of monomers per unit volume. This relation (Eq. 13) assumes that modulus is equal to the thermal energy kT per entanglement strand. This expression can be rewritten [21] using $c \approx g/\xi^3$ and $a \approx \xi\ (N_e/g)^{1/2}$, where ξ is the correlation length (see Eq. 7) and g is the number of monomers in a strand of that size.

$$G_0 \approx kT/(a^2\xi) \qquad (14)$$

Most experiments are consistent with the plateau modulus increasing with concentration as

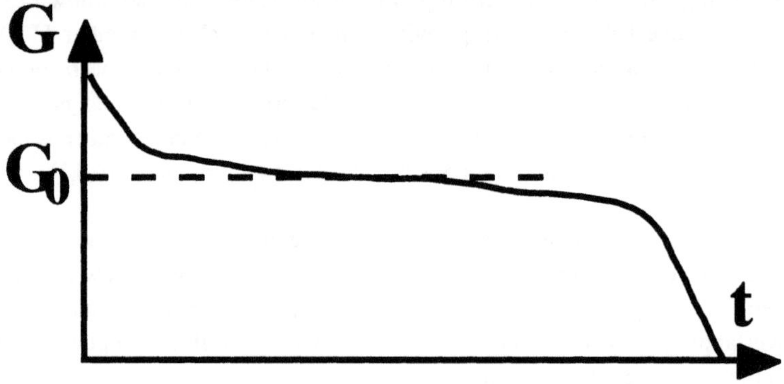

Figure 2. Sketch in logarithmic scales of a stress relaxation function G(t) of entangled polymers. It exhibits a rubber-like plateau with modulus G_0.

$$G_0 \sim c^{\alpha} \tag{15}$$

with exponent $\alpha \approx 2.3 \pm 0.2$ both in good and θ-solvents [24-27].

In good solvents most models predict the entanglement spacing a to be proportional to the correlation length ξ (see Eq. 7), leading to the plateau modulus G_0 proportional to the osmotic pressure Π [28]

$$G_0 \approx kT/(a^2\xi) \approx \Pi \approx kT/\xi^3 \sim c^{3\nu/(3\nu-1)} \tag{16}$$

For Flory exponent $\nu \approx 3/5$ equation 16 predicts $\alpha \approx 9/4$ in good agreement with experiments. However, in θ-solvents $\nu = 1/2$ leading to $\alpha = 3$ in disagreement with most experiments, which indicare a value closer to 2.

We will briefly review three models [19-22], each based on a different assumption (conjecture) about the nature of entanglements.

The first model [19] is based on the conjecture that there is a constant number n_{strand} of entanglement strands (containing $\sim N_e$ monomers each) per entanglement volume. This can be justified by saying that this fixed number of strands is needed to build an entanglement cage. This number n_{strand} is assumed to be of order 10. The number density of entanglement strands is c/N_e and the number of strands per mesh volume a^3 is

$$n_{strand} \approx a^3 c/N_e \tag{17}$$

In θ-solvent the entanglement mesh size is $a \approx b N_e^{1/2}$ and therefore

$$n_{strand} \approx c\, b^3 N_e^{1/2} \quad (18)$$

The conjecture $n_{strand} = const$ [19] leads to the following concentration dependence of the degree of polymerization between entanglements $N_e \sim 1/c^2$ and the plateau modulus $G_0 \sim c^3$, that is to $\alpha = 3$ (see Eq. 15 for the definition of α).

The second model [20] assumes that there is a constant number of binary contacts per entanglement strand. This assumption can be justified by the results on the Gauss linking number, that suggest that only close encounters (contacts) between chains lead to significant change in winding number [16,17]. The probability of a given monomer (of size b) to be in contact with any other monomer is proportional to the monomer concentration cb^3. The number of such binary contacts for a strand of N_e monomers is

$$cb^3 N_e = const \quad (19)$$

This second model predicts the degree of polymerization between entanglements reciprocally proportional to the concentration

$$N_e \approx 1/(cb^3) \quad (20)$$

and the plateau modulus

$$G_0 \sim c^2 \quad (21)$$

leading to $\alpha = 2$ (see Eq. 15).

The third model [21] assumes that there is a constant number of binary contacts per entanglement volume a^3. This recognises that both binary contacts and the presence of many chains generate entanglements. The number of monomers in this volume is ca^3 and the probability of a binary contact per monomer is $\sim cb^3$. Therefore the average number of binary contacts in volume a^3 is

$$c^2 a^3 b^3 = const \quad (22)$$

This model predicts the value of the exponent $\alpha = 7/3$ and

$$N_e \sim c^{-4/3} \quad (23)$$

$$a \sim c^{-2/3} \quad (24)$$

$$G_0 \sim c^{7/3} \quad (25)$$

The third model seems to be in a slightly better agreement with experiments [21,22] than the first two models. But there is no a priori reason to prefer one model over the other because they are all based on uncontrolled assumptions. In fact, it is not obvious that the entanglement mesh has to scale as a power law of concentration [29,30]. If it does, one can argue that the exponent α at very low concentrations cannot be smaller than 3. Otherwise there will only be monomers belonging to a single chain inside the entanglement mesh volume a^3, which would lead to many self-knots rather than to a confinement of a chain by its neighbors. Such self-knots exist in a high molecular weight chain in dilute θ-solvent solutions, but do not lead to any rheologically measurable effects. Therefore it is not clear that they should be important in semidilute solutions. Entanglements (rather than self-knots) are believed to be due to collective topological confinements of a given polymer strand by several neighboring ones.

The first model [19] (constant number of strands per entanglement volume) leads to good estimates of entanglement molecular weight in different polymer melts [31-35].

Thus there are many more questions than answers in the area of polymer entanglements. This field is awaiting another breakthrough similar to the one it witnessed 30 years ago with the introduction of the tube model [36].

4. Polymers in the Presence of Fixed Obstacles

As we discussed in the previous section, the many-chain entanglement problem still remains unsolved and presents a major theoretical challenge. In order to develop a quantitative picture for the dynamics of entangled polymers, single chain models have been introduced. The most successful ones are those of the confining tube and of the chain in an array of fixed obstacles [1, 2 36, 37].

In these models the surrounding chains are replaced by the topological constraints they impose on a given one. It was proposed by Edwards [36] that these constraints confine a given chain to a tube-like region. Displacements of the chain in the directions, transverse to its confining tube, are severely restricted by the surrounding chains. The only allowed motion is that along the contour of the tube. This longitudinal motion, called reptation, was proposed by de Gennes [1, 2, 37] as the main mechanism of the dynamics of entangled linear polymers.

4.1 REPTATION

"Pure" reptation model, as was originally proposed by de Gennes [37], describes the motion of the chain, as a whole (or of its center of mass) along the contour of its confining tube (Fig. 3). This motion is not restricted by topological constraints. The

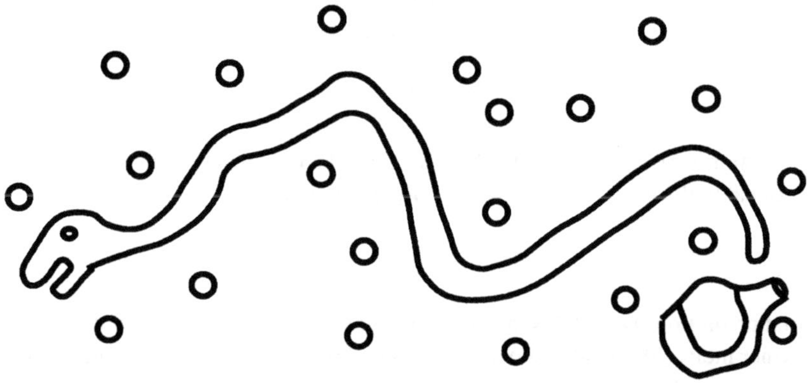

Figure 3. Drunken snake in a forest (a chain in an array of obstacles) as a model of polymer reptation.

corresponding friction coefficient for this degree of freedom in the entangled melt (in the absence of hydrodynamic coupling) is of the Rouse type (Eq. 2). The curvilinear diffusion coefficient of a chain along the contour of its confining tube is therefore Rouse-like (Eq. 4a). The curvilinear distance the chain has to diffuse in order to get out of the tube is the length L of the tube. The width a of the confining tube (the distance between entanglements) is independent of the polymer molecular weight. In the melt (or in θ-solution) it is

$$a \approx b\, N_e^{1/2} \tag{26}$$

where N_e is the degree of polymerization between entanglements (see section 3.2). The tube is a random walk of N/N_e steps of size a. It has the same root-mean-square end-to-end distance as the chain it confines $L\,a = b^2 N$, where L is the tube length

$$L \approx a\, N/N_e \tag{27}$$

The time it takes the chain to diffuse out of the tube (distance L) with curvilinear diffusion coefficient D_R (Eq. 4a) is

$$\tau_{rep} \approx L^2/D_R \approx \tau_1 N^3/N_e \tag{28}$$

During this escape time the chain diffuses distance of the order of its size $R = b\ N^{1/2}$. Therefore the three-dimensional diffusion coefficient is

$$D_{3D} \approx R^2/\tau \approx (b^2/\tau_1)(N_e/N^2) \qquad (29)$$

Thus the reptation model predicts the self-diffusion coefficient D_{3D} reciprocally proportional to the square of the degree of polymerization

$$D_{3D} \sim N^{-2} \qquad (29a)$$

and relaxation time τ (the time it takes the chain to completely renew its configuration) proportional to the cube of the degree of polymerization

$$\tau_{rep} \sim N^3 \qquad (28a)$$

The first prediction (Eq. 29a) is in very good agreement with self-diffusion measurements in polymer melts [38], while relaxation times and related viscosities of entangled systems have stronger molecular weight dependence [10]

$$\tau \sim \eta \sim N^{3.4} \qquad (30)$$

This disagreement between theory (Eq. 28a) and experiments (Eq. 30) has been a major theoretical challenge in polymer physics for the last 25 years.

4.2 TUBE LENGTH FLUCTUATIONS

The first and one of the most successful attempt to explain the disagreement between "pure" reptation theory (Eq. 28a) and experiments (Eq.30) was made by Doi [39, 40].

It is important to realize that reptation model is supposed to represent Rouse motion of a chain in a tube. But the "pure" reptation model considers only one mode of polymer motion - that of the center of mass of the chain along the contour of its tube. Other Rouse modes of the chain in the tube could also contribute to polymer dynamics. Indeed the reptation process is defined as the result of the cooperative motion of kinks ('defects") along the tube [1, 2, 37]. These kinks appear and disappear at the ends of the tube, leading to fluctuations of its length. The tube length decreases as a kink is created and increases when it is destroyed, leading to fluctuations of its length.

To estimate the magnitude and time scale of tube length fluctuations, consider a section of g monomers at one end of the tube (Fig. 4). The Rouse relaxation time of this section (Eqs. 5a or 8)

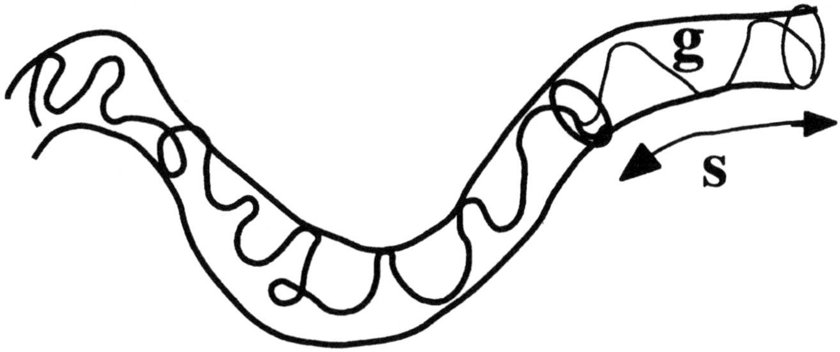

Figure 4. Tube length fluctuations.

$$t(g) \approx \tau_1 \, g^2 \tag{31}$$

is the time at which it begins to move coherently along the tube. The root mean square displacement s during this time $t(g)$ as the result of this coherent motion along the contour of the tube is the end-to-end distance of this section

$$s \approx b \, g^{1/2} \tag{32}$$

This root-mean-square displacement s of the end of the tube along its contour describes the tube length fluctuations. Substituting the time dependence of g from equation 31, $g(t) \approx (t/\tau_1)^{1/2}$, into equation 32, we get the time dependence of the tube length fluctuations

$$s(t) \approx b \, (t/\tau_1)^{1/4} \tag{33}$$

These typical tube length fluctuations grow up to the Rouse time of the whole chain (Eq. 5a or 8)

$$\tau_R \approx \tau_1 \, N^2 \tag{34}$$

with the magnitude of these fluctuations at the Rouse time of the order of the end-to-end distance of the whole chain

$$s(\tau_R) \approx b \, N^{1/2} \approx R_N \tag{35}$$

Even though this magnitude seems large, it is a small fraction of the tube length (Eq. 27)

$$R_N/L \approx (N_e/N)^{1/2} \tag{36}$$

Thus the fraction $(N_e/N)^{1/2}$ of the tube is vacated, and therefore relaxed, at the Rouse time before reptation itself has had a significant effect on stress relaxation. Thus the chain has to reptate a curvilinear distance $L[1-\kappa(N_e/N)^{1/2}]$ with the same curvilinear diffusion coefficient D_R (Eq. 4a), where κ is a numerical constant of order unity [40]. The relaxation time is therefore shorter than in "pure" reptation model (Eq. 28) by the factor $[1-\kappa(N_e/N)^{1/2}]^2$

$$\tau_{rep} \approx (\tau_1 N^3/N_e)[1-\kappa(N_e/N)^{1/2}]^2 \qquad (37)$$

The stress, relaxed at reptation time, is lower than in "pure" reptation model by the factor $[1-\kappa(N_e/N)^{1/2}]$, because the end sections of the tube relax stress by the tube length fluctuations at Rouse time. The plateau modulus (Eq. 13) is therefore

$$G_0 \approx [kT/(b^3 N_e)]\,[1-\kappa(N_e/N)^{1/2}] \qquad (38)$$

The viscosity, which is the integral of the stress relaxation function,

$$\eta = \int G(t)\, dt \qquad (39)$$

can be estimated by the product of the plateau modulus and reptation time,

$$\eta \approx G_0\,\tau \approx (\eta_1\, N^3/N_e^2)\,[1-\kappa(N_e/N)^{1/2}]^3 \qquad (40)$$

where $\eta_1 \approx \tau_1\, kT/b^3$.

This estimate of the effect of tube length fluctuations (Eq. 40), proposed by Doi [39, 40], predicts a molecular weight dependence which is well modeled by $\eta \sim N^{3.4}$ (Eq. 30) over a narrow range of molecular weights. But this range is narrower than that for the experimentally observed power law (Eq. 30).

More recent studies of the tube length fluctuations in single chain models [41 - 47] indicate that the cross-over to the asymptotic "pure" reptation behavior $\tau \sim \eta \sim N^3$ (Eq. 28a) is wider than originally predicted by Doi [39, 40]. Both the repton model [41 - 43] and the Evans-Edwards model [43 - 45] simulations observe the anomalous exponent $\tau \sim \eta \sim N^{3.4}$ over the entire range of chain sizes studied.

4.3 REPTON MODEL

Several models were developed [41 - 45] in order to find a numerical solution to the problem of "a single chain in an array of fixed obstacles". The repton model [41 - 43] maps the confining tube onto a one-dimensional lattice and represents sections of the chain by walkers (reptons) on this lattice (Fig. 5). The presence of more than one repton

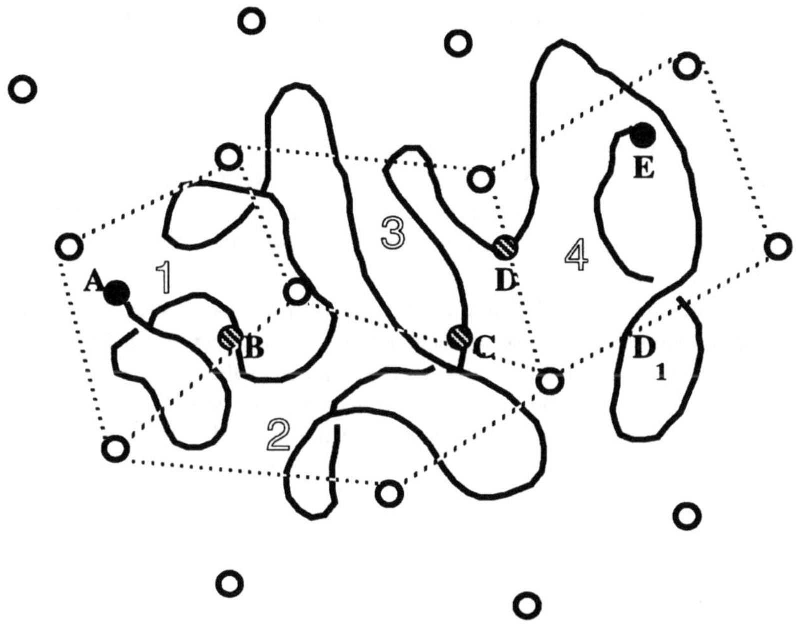

Figure 5a. A polymer AE in an entanglement net, represented by open circles, is confined to a tube consisting of four cells, numbered 1-4 and represented by dotted lines. Unentangled loops, assigned to each cell, are AB, BC, CD and DE respectively.

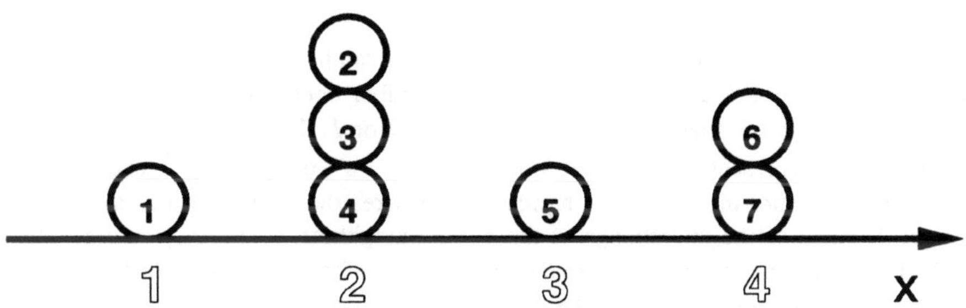

Figure 5b. Repton model representation of the configuration of Fig. 5a. Cells of the entanglement net along the confining tube are represented by a one-dimentional lattice. Unentangled loops, corresponding to these cells are modeled by walkers (reptons) on the lattice sites.

on a site (e. g. site 2 in Fig. 5b) indicates the presence of a large unentangled loop situated at the corresponding cell of the confining tube (e.g. loop BC at cell 2 in Fig. 5a). The dynamics of the chain in an array of fixed obstacles is mapped onto the dynamics of reptons along a one-dimensional (1-d) lattice by two simple axioms:

1) Never vacate a site in the middle of the cluster and always preserve the order of reptons: $x_{i+1} - x_i$ is either 0 or 1. Here x_i is the coordinate of the i-th repton along the one-dimensional lattice.

2) Hoping probabilities of reptons along the 1-d lattice are related to the coordination number z of the cells, formed by the array of obstacles. (A typical coordination number $z=6$). Internal jumps, allowed by axiom 1, occur with probability $1/z$. Jumps onto an empty site, allowed by axiom 1, occur with probability $(z-1)/z$.

The stress relaxation function is determined by the set of sites on the 1-d lattice (cells of the corresponding array of obstacles) that have not been vacated between time 0 (the moment of the step strain) and time t. The number of these "always occupied" sites is the difference between the furthest propagation of the right end of the cluster to the left $x_R(t) = min(x_N(t))$ and the furthest propagation of the left end of the cluster to the right $x_L(t) = max(x_1(t))$. The stress relaxation function is the fraction of unvacated sites

$$\mu(t) \approx < x_R(t) - x_L(t)>/N \tag{41}$$

averaged over the different time histories and over only positive values of $x_R(t) - x_L(t)$.

Zero-shear-rate viscosity is determined by the integration (Eq. 39) of the stress relaxation function (Eq. 41) $\eta = \int\mu(t)\, dt$. The numerically calculated dependence of viscosity on the number of reptons (for coordination number $z = 6$) is $\eta \sim N^{3.4}$ for the entire range of repton numbers N studied (almost two decades).

A cluster of N reptons with coordinates $\{x_i\}$ can be represented by the coordinate of its center of mass and a set of $N-1$ nearest neighbor bonds $\{\sigma_i\}$, where $\sigma_i = x_{i+1} - x_i$. From axiom 1 we see that bonds σ_i can either be 0 or 1. Thus any state of a cluster of reptons can be represented by a set of $N-1$ "0-s" and "1-s", or spins up $\uparrow$ and spins down $\downarrow$. Repton dynamics in this "spin" representation takes the form of Kawasaki dynamics. The probability of the interchange of the nearest neighbors $\sigma_i \leftrightarrow \sigma_{i+1}$ is $1/z$. The end variables (σ_1 and σ_{N-1} are set equal to 0 (spin down $\downarrow$) with probability 1/z and to 1 (spin up $\uparrow$) with probability $(z-1)/z$ (see the second axiom).

4.4 EVANS-EDWARDS MODEL

Kawasaki dynamics of spins (random exchange of neighboring spins) can be generalized to an arbitrary array of vectors $\{\mathbf{a}_i\}$ (e.g. on a cubic lattice). The first such set of models was developed by Stockmayer and coworkers [48 - 50] to represent the motion of unentangled polymers. A single time step in these models corresponds to a random exchange of neighboring vectors $[\mathbf{a}_i, \mathbf{a}_{i+1}] \rightarrow [\mathbf{a}_{i+1}, \mathbf{a}_i]$ (see Fig. 6a). If $\mathbf{a}_i = \mathbf{a}_{i+1}$, such exchange does not lead to a new configuration (e.g. $[\mathbf{a}_4, \mathbf{a}_5]$ in Fig. 6a). If $\mathbf{a}_i$ is

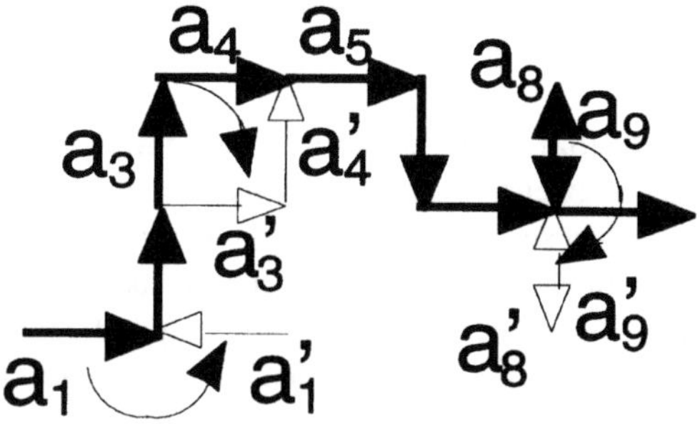

Figure 6a. Allowed moves of theVerdier-Stockmayer model.

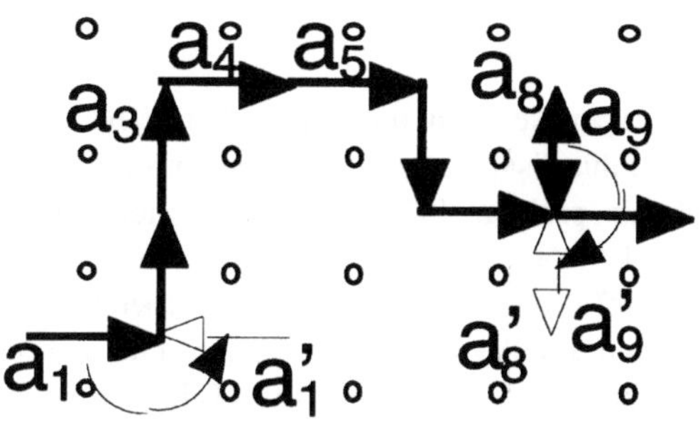

Figure 6b. Allowed moves of the Evans-Edwards model.

perpendicular to a_{i+1}, this move leads to a bond flip (with monomer i moving to the opposite corner of the square) (e.g. $[a_3, a_4]$ in Fig. 6a). If $a_i = -a_{i+1}$ the exchange corresponds to a bond flip along the same line (e.g. $[a_8, a_9]$ in Fig. 6a). The end vectors a_1 and a_N are usually allowed to choose any orientation at random (see Fig. 6a).

These local uncorrelated moves correspond to "local friction" Rouse model. We will use this Verdier-Stockmayer model [48 - 50] in Section 5 to represent the transverse motion of the confining tube due to constraint release.

The Verdier-Stockmayer model of polymer dynamics can be generalized to represent the motion of a chain in an array of fixed obstacles [51, 52]. Let us imagine placing obstacles on the lattice dual to the one on which the chain is defined (open circles in Fig. 6b). The exchange of the neighboring perpendicular vectors across the obstacles is forbidden (e.g. $[a_3, a_4] \rightarrow [a'_3, a'_4]$ in Fig. 6a is not allowed in Fig. 6b). All other Verdier-Stockmayer moves (for $a_i = -a_{i+1}$ and for end vectors a_1 and a_N) are allowed (see Fig. 6b). This Doi-Evans-Edwards model [51, 52], which is a modification of Orwell-

Verdier-Stockmayer model [48 - 50], is a direct representation of the chain dynamics in an array of fixed obstacles.

The early results [52] of this model confirmed the predictions of "pure" reptation. But more recent and more extensive simulations [43 - 45] give $\tau \sim \eta \sim N^{3.4}$, where the viscosity is defined in a way similar to that in the repton model (see Section 4.3). Thus the numerical solutions of the reptation model that includes all modes of a chain in an array of obstacles (both "pure" reptation and tube length fluctuations modes) give the molecular weight dependence of both relaxation time and viscosity in agreement with experiments.

But the same computer simulations [43 -45] predict the self-diffusion coefficient

$$D_{3D} \sim N^{2.5} \tag{42}$$

in disagreement with experiments in melts (Eq. 29a). The theoretical challenge is to understand the source of this disagreement. At the end of next section we will hint to one possible solution of this problem [53]. One should note that many experiments in entangled solutions [54 - 56] are consistent with this stronger molecular weight dependence of the self-diffusion coefficient (Eq. 42). Therefore entangled solutions and melts may have different molecular weight dependence of the self-diffusion coefficient, but similar molecular weight dependence of viscosity. This fact presents another theoretical challenge. The third related challenge is to understand how a chain in polymer melts can diffuse many times its size without relaxing stress [57]. Indeed, the mean square distance diffused during relaxation time, is

$$D\,\tau \approx b^2 N^{1.4} > R \approx b^2 N \tag{43}$$

This fact cannot be explained by the single chain models. At the end of the next section we briefly discuss the memory effects in polymer melts and introduce a collective reptation model [53].

4.5 REPTATION WITH TOO MANY ENDS AND WITHOUT ANY ENDS

In section 4 above we have discussed the dynamics of a single chain in an array of fixed obstacles on the basis of tube models. All the discussions above were limited to linear chains. The topology of the chain (star vs. linear vs. ring) significantly modifies polymer dynamics.

Snake-like reptation is impossible for stars (see the contribution of T. McLeish's to the present volume for the dynamics of stress relaxation in branched polymers) [2, 58 - 60]. Stars move by arm retraction, which are large (exponentially unlikely) fluctuations of the tube lengths of their arms. The relaxation time of a star in an array of fixed obstacles is exponentially long in the number of entanglements of their arms [2, 58 -60]

$$\tau_{star} \approx \tau_0 \exp(\nu N/N_e) \qquad (44)$$

independent of the number of arms f of the star. The viscosity of the melt of stars is also independent of the number of arms f (for $f>3$, but not too large to cause arm stretching). But the self-diffusion coefficient of a star does depend on the number of arms. The dependence of the self-diffusion coefficient of a star $D_{star}(f)$ on the number f of arms was originally thought to be exponential $\tau_{star} \approx \tau_0 \exp[\nu(f-2) N/N_e]$ [2, 61, 62]. But it was soon realized that it is much weaker than exponential [63]

$$D_{star}(f) \approx \{[1 + (z-1)(f-1)/2]\, z!\, f!/(z+f-1)!\}\, a^2/\tau_{star} \qquad (45)$$

where z is the coordination number of the cell in the array of obstacles ($z \approx 6$), a is the tube diameter and τ_{star} is the relaxation time of an arm of the star (Eq. 44). This expression is different from the ones, proposed earlier [2, 61, 62], because the arm retractions of $(f-2)$ arms, needed to make a step, do not have to occur simultaneously (see [63] for details). Experimental measurements of the self-diffusion coefficient of the multi-arm stars [64] confirm this result (Eq. 45 with $z=6$).

Another very interesting problem is the dynamics of a ring polymer in an array of fixed obstacles. The ring is not permanently trapped by the obstacles, but is able to diffuse. The ring does not have free ends and, therefore, "classical" snake-like reptation [62] is not expected for it. The ring in an array of fixed obstacles takes a hyperbranched animal-like configuration [63, 66 - 68] with the size

$$R_{ring} \approx b\, N^{1/4} \qquad (46)$$

if one ignores the excluded volume effects. There were several studies of the dynamics of a ring in an array of fixed obstacles [63, 65, 66] that predicted relaxation time of the ring $\tau \sim N^3$ similar to the linear chain (Eq. 28a). More recent analysis [68] showed that the dynamics of rings is significantly different from linear chains with relaxation time

$$\tau_{ring} \sim N^{2.5} \qquad (47)$$

and self-diffusion coefficient

$$D_{ring} \sim N^{-2} \qquad (48)$$

Computer simulations of a ring in an array of fixed obstacles confirm these new predictions [68].

5. Many-Chain Effects

5.1 CONSTRAINT RELEASE

In the previous section we have discussed the motion of the chain along its confining tube. The topological constraints, defining the tube, were assumed to be fixed. These topological constraints are formed by neighboring chains, which also move along their respective tubes. As one of these chains (say chain A) moves away, the topological constraint it used to impose on a given chain (say B) disappears (Fig. 7). A new set of configurations is now available for chain B. Another chain (say chain C) moves in and imposes a new topological constraint on chain B. The total number of topological constraints imposed on a given chain by its neighbors is, on average, unchanged.

Therefore, as some neighbors move away and remove their constraints on a given chain, others move in and place new constraints on it.

This exchange of neighbors and their constraints imposed on a given chain, corresponding to the change of the tube a given polymer is confined to, is called constraint release.

This process was first introduced by de Gennes [58] and later analyzed by many researchers[61, 69-73]. The simplest model of the constraint release process is a Verdier-Stockmayer model [48-50] of the primitive path. When a neighboring chain A moves

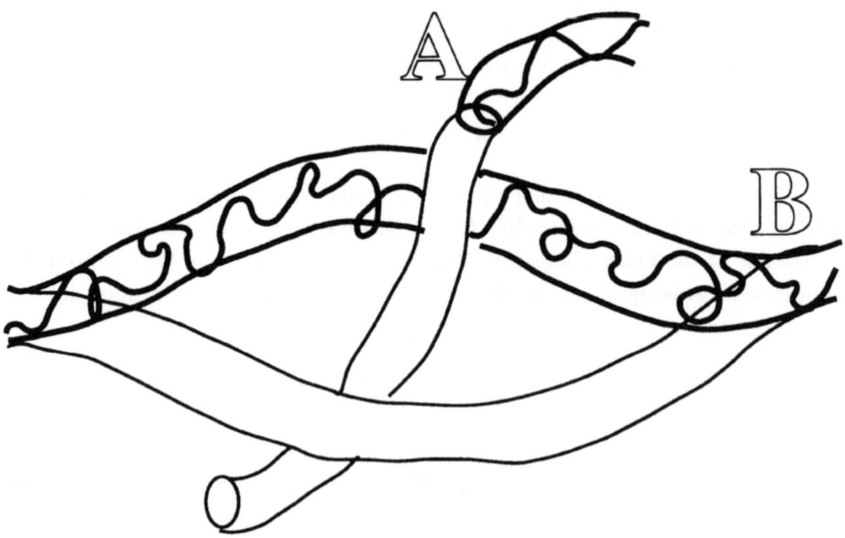

Figure 7. Constraint release process. As chain A reptates away, it releases its topological constraint on chain B.

away, chain B can explore an additional volume of the order of an entanglement mesh size a^3. If a new chain C moves in, it can locally confine chain B to this new volume, deforming the tube of B. This deformation can be modeled by a local jump of the Verdier-Stockmayer model [48-50]. The rate of these local jumps of the primitive path are reciprocally proportional to the lifetime of the constraints [61].

5.2 REPTATION VS. CONSTRAINT RELEASE

Each chain in an entangled solution or melt, simultaneously participates in two types of motion:

I. Single chain motion along its confining tube - reptation and tube length fluctuations.

II. Transverse motion of its tube (primitive path) due to the release of constraints, caused by the motion of neighboring chains.

The diffusion coefficient of a *P*-mer in a solution or melt of *N*-mers can be written as a sum of contributions from each of these two types of motion

$$D \approx R^2/\tau_{rep} + R^2/\tau_{tube} \tag{49}$$

where the reptation time of the *P*-mer (see Eq. 28) is

$$\tau_{rep} \approx \tau_1 P^3/N_e \tag{50}$$

and the constraint release time can be estimated as a Rouse time of P/N_e sections with relaxation time τ for each

$$\tau_{tube} \approx \tau(P/N_e)^2 \tag{51}$$

The relaxation time τ of an entanglement strand is the lifetime of its constraints. If the constraints are formed by *N*-mers, their lifetime is the reptation time of *N*-mers (see Eq. 28)

$$\tau \approx \tau_1 N^3/N_e \tag{52}$$

Combining equations (51 and 52) we find the constraint release time

$$\tau_{tube} \approx \tau_1 \, N^3 P^2/N_e^3 \tag{53}$$

The faster of the two types of motions wins and controls the diffusion of the *P*-mer [70].

For very long *P*-mers

$$P > N^3/N_e^2 \tag{54}$$

constraint release time (Eq. 53) is shorter than their reptation time (Eq. 50). The diffusion of these long chains occurs by Rouse-like constraint release motion of their tubes. For shorter P-mers

$$P < N^3/N_e^2 \tag{55}$$

the reptation time (Eq. 50) is shorter and dominates the diffusion of these chains. This is the case in monodisperse solutions and melts (for $P = N$).

Experiments on tracer diffusion of deuterated polystyrene into the melt of hydrogenated polystyrene were performed by Green and Kramer [74] using forward recoil spectroscopy. These experiments confirmed that when the molecular weight of matrix chains is comparable or higher than that of tracer chains, the diffusion coefficient of these tracer chains is independent of the molecular weight of the matrix. It only depends on the molecular weight of diffusing chains as

$$D \sim P^{-2} \tag{56}$$

Note that this result agrees with the predictions of "pure" reptation model (Eq. 29a) [37] (without tube length fluctuations) and does not agree with simulations of polymer in an array of fixed obstacles (Eq. 42) [43 - 45].

The cross-over to the constraint-release-dominated regime at shorter matrix molecular weights was observed to occur at

$$P \sim N^{2.5} \tag{57}$$

This suggests that constraint release process is faster than predicted by Eqs. 53-55 and has weaker matrix molecular weight dependence ($\sim N^{2.5}$ instead of $\sim N^3$).

5.3 COHERENT CONSTRAINT RELEASE

It was suggested by Klein [62] that this discrepancy between the simple constraint release theory (Eqs. 53-55) and experiments (Eq. 57) may be due to the coherence in the release of constraints. A typical neighboring chain (A) can impose several constraints on a given chain (B) if it entangles with it in many places. When this neighboring chain A reptates away, it releases all of these constraints on B, making the lifetime of each of them shorter [62].

Let E be an average number of entanglements between two entangled neighboring chains. The effective constraint lifetime is

$$\tau_{eff} \approx \tau/E \tag{58}$$

There are many different conjectures about the value of E [62, 75-79]. The most recent one was proposed by Brochard, Gay and de Gennes. [79] Their argument is summarized below.

The average number of entanglements per N-mer, N/N_e, grows linearly with its degree of polymerization N. The number of neighboring N-mers a given one overlaps with in a melt is $R^3/(b^3N) \sim N^{1/2}$. When a neighboring N-mer moves away it releases, on average, E entanglements, where

$$E \approx N^{1/2}/N_e \tag{59}$$

Note that this statement is valid for $N > N_e^2$. For shorter chains ($N < N_e^2$) this model [79] predicts no coherent constraint release ($E = 1$). Experimentally, a more effective constraint release (Eq. 57) was observed for shorter chains [74].

Another conjecture [77, 78] is that the number of entanglements E between a pair of neighboring chains is proportional to the number of overlaps between their tubes

$$E \approx (N/N_e)^{1/2} \tag{60}$$

The predictions of this conjecture are compared with experimental data [74] in Fig. 8.

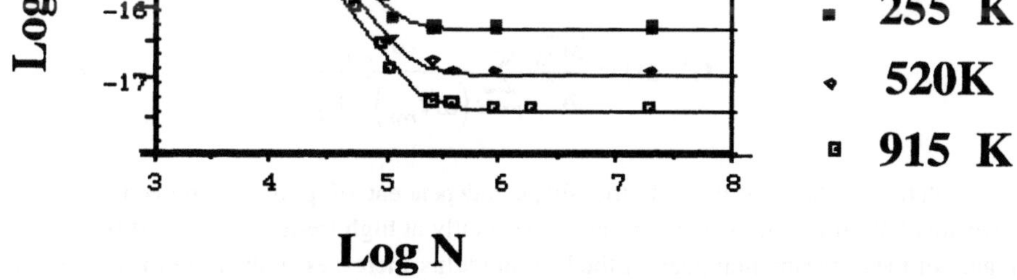

Figure 8. Self-diffusion coefficient of Z-mers into the melt of N-mers. Experimental data [74] for five different molecular weights of Z-mers: *55000, 110000, 255000, 520000* and *915000*. Lines are the predictions of the coherent constraint release model [77].

5.4 STRESS RELAXATION

A more detailed test of the molecular models of polymer dynamics is the comparison of their predictions with the stress relaxation experiments. The stress relaxation modulus $G(t)$ is the ratio of stress $\sigma(t)$ after a step strain to the magnitude of this strain γ [2, 10]

$$G(t) = \sigma(t)/\gamma \tag{61}$$

A very sensitive test of the models is the loss modulus $G''(\omega)$ - the out-of-phase response to the oscillatory strain at frequency ω. Loss modulus is a sine transform of the stress relaxation function

$$G''(\omega) = \int G(t)\, sin(\omega t)\, dt \tag{62}$$

In monodisperse polymer melts $G''(\omega)$ has a characteristic peak that can be compared with predictions of different models.

5.4.1 *Single Chain in a Tube Models.*

The stress relaxation function G(t) for the "pure" reptation model was calculated by Doi and Edwards [2, 80] by solving the first passage problem for the diffusion of a chain in a tube (see T. McLeish, this volume)

$$G(t) = G_0 \frac{8}{\pi^2} \sum_{p:odd} \frac{1}{p^2} \exp\left(-\frac{p^2 t}{\tau_{rep}} \right) \tag{63}$$

Note that the main contribution comes from the first mode $p = 1$ and the function is almost single exponential. The corresponding loss modulus

$$G''(\omega) = \frac{8G_0}{\pi^2} \sum_{p:odd} \frac{\omega\tau_{rep}}{\left(\omega\tau_{rep}\right)^2 + p^4} \tag{64}$$

has a relatively sharp peak with the shape independent of polymer molecular weight. Experimentally the peak is much broader, especially at high frequencies (short times). The breadth of the experimental peak of the loss modulus increases with decreasing molecular weight. This comparison suggests that "pure" reptation model underestimates stress relaxation at higher frequencies. This high frequency relaxation is due to the tube length fluctuation modes [39, 40], described in section 4. Doi's approximation of the stress relaxation function contains two parts:

(i) Tube length fluctuations at times shorter than Rouse time;

(ii) Reptation at longer times.

More accurate calculations have been recently performed [46, 47], but they all lead to qualitatively similar results.

The results of all models that include tube length fluctuation modes are in much better agreement with the experimentally measured loss modulus $G''(\omega)$ of monodispersed melts, than the prediction of "pure" reptation model (Eq.64). Tube length fluctuations corrections lead to a loss peak which widens with decreasing molecular weight.

5.4.2 *Effect of Constraint Release on Stress Relaxation.*

In Section 5.2 we have argued that constraint release has a limited effect on self-diffusion coefficient: it is important only for the diffusion of very long chains in the matrix of much shorter ones and can be neglected in monodisperse solutions and melts.

It was demonstrated by Graessley [61] that the effect of constraint release on the stress relaxation is much more important than on the self-diffusion and cannot be neglected even for monodisperse systems. As we mentioned above, constraint release can be described by the Rouse model of the primitive path sections. Let us estimate the stress relaxation function of this Rouse model [1, 2, 6].

The p-th mode of the N-mer involves a coherent motion of N/p monomers. The Rouse relaxation time of each such section (there are p of them per chain) is

$$\tau_p \approx \tau\,(N/p)^2 \qquad (65)$$

At time τ_p there is one unrelaxed mode per section (its end-to-end vector) or p unrelaxed modes per N-mer. The value of the stress relaxation function per chain at time τ_p is proportional to thermal energy kT times p/N

$$G(\tau_p)/chain \sim kT\, p/N \qquad (66)$$

From equation 65 we find that

$$p/N \approx (\tau_p/\tau)^{-1/2} \qquad (67)$$

Therefore the stress relaxation function for the Rouse model decays as the square root of time

$$G(t) \sim (t/\tau)^{-1/2} \qquad (68)$$

Thus a finite fraction of stress relaxes at time scales of the order of constraint lifetime τ (reptation time of surrounding chains) in the Rouse model of constraint release for the monodisperse solutions or melts. This is also the time scale at which the stress relaxes by reptation. Both processes simultaneously contribute to the relaxation of stress.

Therefore constraint release has to be taken into account for stress relaxation even in monodisperse systems. The contribution of constraint release to stress relaxation in polydisperse solutions and melts is even more important as will be discussed below.

5.4.3 *Models of Constraint Release.*

Single chain models, such as Doi-Edwards model of "pure" reptation (Eq. 63) or Doi's model of tube length fluctuation, assume linear contribution to the stress relaxation function from each component of the polydisperse system

$$G(t) = \Sigma_N \phi_N \, \mu_N(t) \tag{69}$$

where ϕ_N is the volume fraction of N-mers and $\mu_N(t)$ is the single chain stress relaxation function of N-mers. For example, for binary blends these models predict

$$G(t) = \phi_L \, \mu_L(t) + \phi_S \, \mu_S(t) \tag{70}$$

But it was observed in many experiments that the amount of stress, relaxed at the time scale of shorter chain reptation time τ_S, is much higher than the volume fraction of short chains [71 - 73].

This phenomena can be explained by the constraint release models. The stress relaxed at time scales τ_S was supported not only by short chains, but also by parts of long chains. This partial relaxation of stress, supported by long chains, occurs by releasing constraints between long and short chains. Release of constraints imposed on long chains by short ones makes a significant contribution to the stress relaxation at reptation times of short chains τ_S.

If topological constraints can be represented by pairwise entanglements between chains, there are three types of them in a binary blend: between two long chains (L-L); between two short ones (S-S) and between a short and a long chain (S-L). If one represents the dynamics of each chain along its tube by "pure" reptation, there are two time scales in the problem - reptation times of long (τ_L) and short (τ_S) polymers. If the constraint on a given chain is caused by a long neighbor, its lifetime is τ_L, while if the neighbor is short, the lifetime of the constraint is τ_S. The constraint release process can be represented by a Rouse model with two mobilities of the effective beads:
(i) slow, corresponding to entanglements with long chains;
(ii) fast, corresponding to entanglements with short chains.
These two types of beads can be assumed to be randomly distributed along the Rouse chain with relative concentrations, corresponding to probabilities of entanglements with a chain of each type [72]. The simplest assumption is that these relative concentrations are proportional to the volume fractions of each types of chains (for binary entanglements).

The combined stress relaxation function for both reptation and constraint release of a binary blend is

$$G(t) = \phi_L \, \mu_L(t) \, R_L(t) + \phi_S \, \mu_S(t) \, R_S(t) \qquad (71)$$

where $R_L(t)$ and $R_S(t)$ are the stress relaxation functions due to Rouse-like constraint release of tubes of long and short chains (with two bead mobilities). This model is in reasonable agreement with experiments [72]. Adding constraint release modes to "pure" reptation improves the dependence of the heights of the loss modulus peaks on the volume fractions ϕ_L and ϕ_S of components. But this model lacks the higher frequency modes because it does not include the tube length fluctuations.

These tube length fluctuation modes (see sections 4.2 and 5.4.1) of the neighboring chains affect the constraint release modes of a given one. If entanglements between chains are assumed to be binary, there should be a *duality* between constraint release events and "chain in a tube" relaxation events [73]. A release of an entanglement by reptation or tube length fluctuation of one chain in its tube leads to a release of the constraint of the second chain. If this duality is accepted, the distribution of constraint release rates $P(\varepsilon)$ can be determined from the stress relaxation function of the "chain-in-a-tube" (type I) motion.

The constraint release process in this self-consistent model is described by a Rouse model with random bead mobilities. The distribution of these mobilities is given by $P(\varepsilon)$. The contribution of the constraint release process to the stress relaxation, as defined by the random Rouse model can be calculated [73] by a method similar to the one introduced in a different context by Dean [81].

The predictions of this self-consistent model are in very good agreement with experiments on binary blends [73] (see Fig. 9).

5.5 CONSTRAINT RELEASE VS. TUBE DILATION

The constraint release models represent the effect of surrounding polymers on the motion of a given one by allowing transverse motion of the tube, controlled by the rate of release of entanglements. But the width of the confining tube stays, on average, constant.

An alternative picture was proposed by Marrucci [82] and Viovy [83]. As constraints are released, the confining tube of a given chain is assumed to widen. For example, after a step strain the relaxed sections of the chains are assumed to be unable to confine the polymer any more, effectively becoming solvent-like.

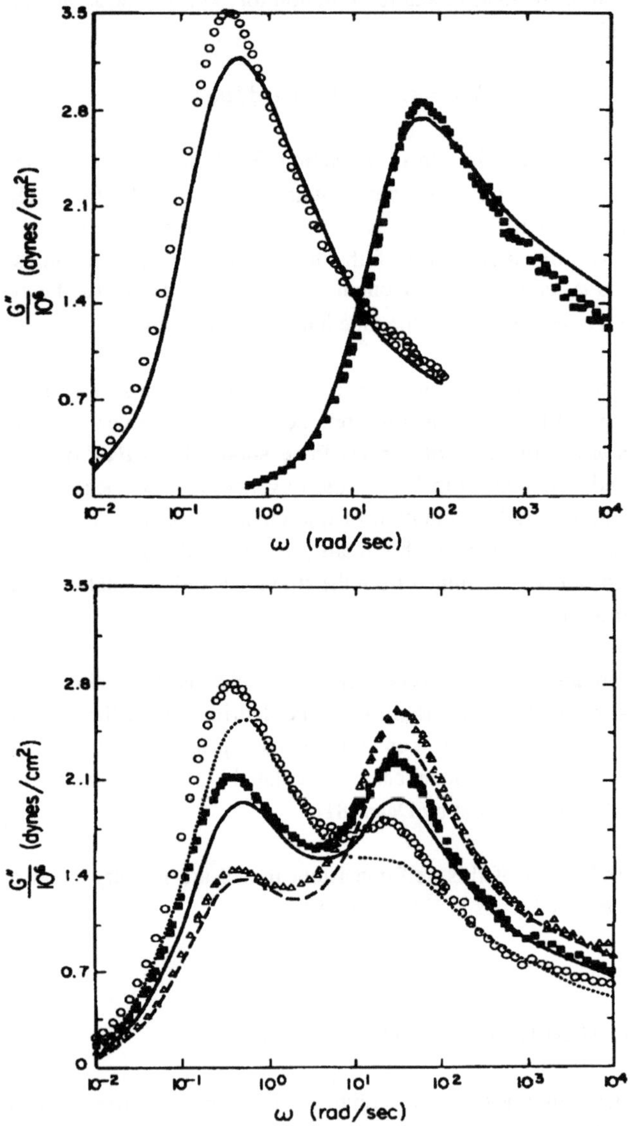

Figure 9. (a) Simultaneous fit of loss modulus data of two mononodisperse polybutadiene samples (filled squares are for $M = 70\ 900$ and open circles are for $355\ 000$ by self-consistent model with two fitting parameter ($G_0 = 1.1\ 10^7$ dyn/cm^2 and $\tau_L = 6$s).

(b) Comparison of blend predictions of the self-consistent model (no adjustable parameters) with the data for three compositions (open circles $\phi_L = 0.882$, filled squares $\phi_L = 0.768$, open triangles $\phi_L = 0.638$).

Recall the dependence of the degree of polymerization between entanglements N_e in Θ-solvents on polymer concentration c (Section 3.2)

$$N_e \sim c^{1-\alpha} \tag{72}$$

where α is the exponent, different for various conjectures about the nature of entanglements. Polymer concentration c is replaced in the tube dilation models by the fraction of unrelaxed chains $\mu(t)$. The degree of polymerization between entanglements increases with time as

$$N_e(t) = N_e(0)\,\mu(t)^{1-\alpha} \tag{73}$$

The stress relaxation function in these models is [82, 83]

$$G(t) = G_0(t)\,\mu(t) \approx [kT/N_e(t)]\,\mu(t) \approx G_0(0)\,\mu(t)^{\alpha} \tag{74}$$

where modulus $G_0(t)$ (Eq. 13) changes with time inversely proportional to the degree of polymerization between entanglements $N_e(t)$. Here $\mu(t)$ is the single chain in a tube (type I) relaxation function [83].

Note that for $\alpha = 2$ one recovers the celebrated double reptation model of des Cloizeaux [47, 84, 85]. Thus double reptation is a special case (for $\alpha = 2$) of the tube dilation models. In section 3.2 we have described several conjectures about entanglements, leading to various values of exponent α (from $\alpha = 2$ [20] to $\alpha = 3$ [19]). Experiments seem to point towards an intermediate value $\alpha = 2.3$ [21, 22]. Thus $\alpha = 2$ of double reptation may not be the best choice from the class of tube dilation models.

There is a conceptual difference between the tube dilation models (such as double reptation) [82 - 85] and the constraint release models [58, 59, 69-73]. This difference can be demonstrated by the following example. Consider a long probe P-mer entangled in a melt of shorter N-mers. Tube dilation assumes that as soon as short chains relax, stress in the long P-mer drops to zero. In particular, double reptation imposes an exact symmetry between type I and type II processes. As one chain reptates away, stress at a common entanglement (stress point) is relaxed *completely*. In constraint release models this stress relaxes only partially due to connectivity of the P-mer. The relaxation time of P-mer is $(P/N_e)^2\,\tau_{rep}(N)$ in the constraint release models, where $\tau_{rep}(N)$ is the reptation time of matrix N-mers. In double reptation model an isolated P-mer relaxes at the same time scales $\tau_{rep}(N)$ as the matrix N-mers.

Which of the two approaches is the correct one? The author of the present review favors the constraint release one [86]. But this issue is far from being completely resolved. In particular, tube dilation models are currently used for branched polymers (see T. McLeish's contribution to this volume). To demonstrate the limits of validity of the tube dilation models remains an important theoretical challenge.

There are also interesting regimes, predicted by the constraint release models, such as reptation of the tube of a P-mer in the super-tube formed by its entanglements with other long P-mers in a matrix of shorter N-mers [86].

Another important and very interesting theoretical challenge is to generalize these many-chain models to the blends of dissimilar polymers.

5.6 COLLECTIVE REPTATION

In Section 4.4 we have outlined some problems with interpretation of self-diffusion and viscosity results in entangled solutions and melts. In particular, it was not clear how a chain can diffuse many times its size without relaxing stress (Eq. 43) [57].

It was pointed out by Deutsch [87,88] that the reptation model encounters potential problems in entangled solutions and melts. When a polymer reptates, it effectively removes mass from one end of its tube and places it at the other end. This depletes the entanglement network at one end of the tube and inflates it at the other end. These deformations create stress in the entanglement network that, due to topological constraints, cannot relax by local chain rearrangements. This stress can only relax if this excess and deficit of mass is balanced by that at the neighboring ends of other tubes. But the ends of neighboring tubes (and chains) are separated by large distances

$$d \sim N^{1/3} \tag{75}$$

in high molecular weight solutions and melts. This local deformation of entanglement network is the source of the potential barrier for the reptation process [87-90].

As chain A reptates away, it leaves behind a trace in the form of elastic deformation of entanglement network. Other chains, moving nearby, are attracted to this trace and tend to follow it, partially recovering the configuration (and stress) of chain A [53]. Thus the memory of chain A is left in the form of elastic deformation of the entanglement net even after the chain diffused away. Therefore part of the stress, originally supported by chain A, is left in its trace and is partially picked up by the chains that follow its trace. This is the essence of the proposed explanation [53] of how the chain can diffuse many times its size without relaxing stress. Detailed analysis shows that the ratio of relaxation time τ_{relax} and diffusion time τ_{dif}

$$\tau_{relax}/\tau_{dif} \sim N^{1/3} \tag{76}$$

in reasonable agreement with experiments in melts. This effect is expected to be weaker in polymer solutions, suggesting a possible reason for the difference in observed molecular weight dependence in self-diffusion coefficients (Eqs. 29a and 42).

We have sketched above the possible explanations for the existing puzzles in polymer dynamics, but much more work needs to be done, before all questions are answered and all problems are solved. This presents an ultimate theoretical (and experimental) challenge in the field of polymer physics.

6. Summary of Challenges

1. Entanglements. One of the main unsolved problems in polymer physics is how to define entanglements in many-chain systems. Many conjectures about the nature of entanlements have been proposed [18-22, 29-35]. Some are in reasonable agreement with experiments in polymer solutions [21,22]. Others seem to describe entanglements in melts well [19, 31-35]. But there is still no comprehensive picure of entanglements applicable in all regimes. A solution of this important problem will lead to advances in many areas of polmer science including rheology and rubber elasticity.

2. Dynamics of Linear Solutions and melts. Despite significant progress in this area of polymer physics, there is still a number of fundamental open questions.
a) What is the source of the difference between the molecular weight dependence of the diffusion coefficient in solutions $D \sim N^{-2.5}$ [54-56] and melts $D \sim N^{-2}$ [38].
b) There are several predictions for the asymptotic molecular weight dependence of viscosity of linear chains

(i) $\eta \sim M^{3.4}$ (extrapolation of experiments) (ii) $\eta \sim M^3$ (reptation model)

(iii) $\eta \sim exp(\alpha M^{1/3})$ [87,88] (iv) $\eta \sim exp(\alpha M^{2/3})$ [89,90]

Which one (if any) is correct?

3. *Many chain effects*

(a) Is there an effect of the coherent constraint release on polymer dynamics?

(b) What is the range of validity of the tube dilation model?

4. *Rings.* What is the dynamics of the entangled solution and melt of rings?

7. References

1. de Gennes, P.G. (1979) *Scaling Concepts in Polymer Physics*, Cornell University Press, Ithaca.
2. Doi, M. and Edwards, S.F. (1986) *The Theory of Polymer Dynamics*, Clarendon Press, Oxford.
3. Pearson, D.S. (1987) *Rubb. Chem. Tech.* **60**, 439.
4. Leger, L., Hervet, H., Marmonier, M.F. and Viovy, J.L. (1989) in *New Trends in Physics and Physical Chemistry of Polymers*, Lee, L.-H. editor, Plenum, New York.
5. Lodge, T.P., Rotstein, N.A. and Prager, S. (1990) *Adv. Chem. Phys.* **LXXIX**, 1.
6. Rouse, P.E. (1953) *J. Chem. Phys.* **21**, 1272.
7. B. H. Zimm, B.H. (1956) *J. Chem. Phys.* **24**, 269.

8. Colby, R.H. (1996) *J. Phys. (France),* in press.
9. Joanny, J.F., Grant, P., Turkevich, L.A. and Pincus, P. (1981) *J. Phys. (France)* **42**, 1045.
10. Ferry, J.D. (1980) *Viscoelastic Properties of Polymers*, John Wiley & Sons, Inc., New York.
11. Wigel, F.W. (1986) *Introduction to Path-Integral Methods in Physics and Polymer Science*, World Scientific, Singapore.
12. Nechaev, S., (1990) *Int. J. Mod. Phys.*, **4**, 1809.
13. Vologotskii, A.V., Lukashin, A.V., Frank-Kamenetskii, M.D. and Anshelevich, V.V. (1974) *Sov. Phys.-JETP*, **39**, 1059. Vologotskii, A.V., Lukashin, A.V. and Frank-Kamenetskii, M.D. (1975) *Sov. Phys.-JETP*, **40**, 932.
14. Gauss, C.F. (1877) *Koenig Ges. Wiss.* Goettingen **5**, 602.
15. Edwards, S.F. (1967) *Proc. Phys. Soc*, **91**, 513.
16. des Cloizeaux, J. (1981) *J. Phys. (France) Lett.* **41**, L-433.
17. Tanaka, F. (1982) *Prog. Theor. Phys.* **68**, 148, 164.
18. Graessley, W.W. and Edwards, S.F. (1981) *Polymer*, **22**, 1329.
19. Kavassalis, T.A. and Noolandi, J. (1989) *Macromolecules*, **22**, 2709.
20. Brochard, F. and de Gennes, P.G. (1977) *Macromolecules*, **10**, 1157.
21. Colby, R. H. and Rubinstein, M. (1990) *Macromolecules*, **23**, 2753.
22. Colby, R. H., Rubinstein, M. and Viovy, J. L. (1992) *Macromolecules*, **25**, 996.
23. Flory, P.F. (1953) *Principles of Polymer Chemistry,* Cornell University Press, Ithaca.
24. Adam, M. and Delsanti, M. (1984) *J. Phys. (France),* **45**, 1513.
25. Takahashi, Y., Noda, I. and Nagasawa, M. (1985) *Macromolecules*, **18**, 2220.
26. Plazek, D.J., Seoul, C. and Bero, C.A. (1991) *J. Non-Cryst. Solids*, 131, 570.
27. Colby, R.H., Fetters, L.J., Funk, W.G. and Graessley, W.W. (1991) *Macromolecules*, **24**, 3873.
28. Daoud, M., Cotton, J.P., Farnoux, B., Jannink, G, Sarma, G., Benoit, H., Duplessix, R., Picot, C. and de Dennes, P.G. (1975) *Macromolecules*, **8**, 804.
29. Edwards, S.F. (1988) *Proc. Royal Soc. London.*, **A419**, 221.
30. Iwata, K. and Edwards, S.F. (1989) *J. Chem. Phys.*, **90**, 4567.
31. Ronca, G. (1983) *J. Chem. Phys.*, **79**, 1031.
32. Lin, Y.H. (1987) *Macromolecules*, **20**, 3080.
33. Kavassalis, T.A. and Noolandi, J. (1987) *Phys. Rev. Lett.*, **59**, 2674.
34. Fetters, L.J., Lohse, D.J., Richter, D., Witten, T.A. and Zirkel, A (1994) *Macromolecules* **27**, 4639.
35. Fetters, L.J., Lohse, D.J. and Colby, R.H. (1996) in *Physical Properties of Polymers Handbook,* Mark, J.E. editor, AIP Press.
36. Edwards, S.F. (1967) *Proc. Phys. Soc. London*, **92**, 9.
37. de Gennes, P.G. (1971) *J. Chem. Phys.*, **55**, 572.
38. Tirrell, M. (1984) *Rubber Chem. Technol.*, **57**, 523.
39. Doi, M. (1981) *J. Polym. Sci., Polym. Lett. Ed.*, **19**, 265.
40. Doi, M. (1983) *J. Polym. Sci., Polym. Phys. Ed.*, **21**, 667.
41. Rubinstein, M. (1987) *Phys. Rev. Lett.*, **59**, 1946.
42. Rubinstein, M. (1989) in *New Trends in Physics and Physical Chemistry of Polymers*, Lee, L.-H. editor, Plenum, New York.
43. Chang, I. (1990) Ph.D. thesis, University of **Rochester**, NY.
44. Deutsch, J.M. and Madden, T.L. (1989) *J. Chem. Phys.*, **91**, 3252.
45. Reiter, J. (1991) *J. Chem. Phys.*, **94**, 3222.
46. Ketzmerick, R. and Ottinger, H.C. *Continuum Mec. Thernodyn.*, **1**, 113.
47. des Cloizeaux, J. (1990) *Macromolecules*, **23**, 4678.
48. Verdier, P.H. and Stockmayer ,W.H. (1962) *J. Chem. Phys.*, **36**, 227.

49. Orwoll, R.A. and Stockmayer, W.H. (1969) *Adv. Chem. Phys.*, **15**, 305.
50. Verdier, P.H. (1966) *J. Chem. Pys.*, **45**, 2118, 2122; (1970) *ibid.* **52**, 5512.
51. Doi, M. (1973) *Polymer J.*, **5**, 288.
52. Evans, K.E. and Edwards, S.F. (1981) *J. Chem. Soc., Faraday Trans. 2*, **77**, 1891, 1913.
53. Rubinstein, M. and Obukhov, S.P. (1993) *Phys. Rev. Lett.*, **71**, 1856.
54. Nemoto, N., Kojima, T., Inohue, T., Kishine, M., Hirayama, T. and Kurata,M. (1989) *Macromolecules*, **22**, 3793.
55. Nemoto, N., Kishine, M., Inohue, T. and Osaki, K. (1990) *Macromolecules*, **23**, 22.
56. Nemoto, N., Landry, M.R., Noh, I., Kitano, T., Wesson, J.A. and Yu, H. (1985) *Macromolecules*, **18**, 308.
57. Pearson, D.S., Ver Strate, G, von Meerwall, E. and Shilling, F.C. (1987) *Macromolecules*, **20**, 1133.
58. de Gennes, P.G. (1975) *J. Phys. (Paris)*, **36**, 1199.
59. Doi, M. and Kuzuu, N.Y. (1980) *J. Polym. Sci., Polym. Lett. Ed.*, **18**, 775.
60. Pearson, D.S. and Helfand, E. (1984) *Macromolecules*, **17**, 888.
61. Graessley, W.W. (1982) *Adv. Polym. Sci.*, **47**, 67.
62. Klein, J. (1986) *Macromolecules*, ***19***, 105.
63. Rubinstein, M. (1986) *Phys. Rev. Lett.*, **57**, 3023.
64. Shull, K.R., Kramer, E.S. and Fetters, L.J. (1990) *Nature*, **345**, 790.
65. Cates, M.E. and Deutsch, J.M. (1986) *J. Phys. (Paris)*, **47**, 2121.
66. Nechaev, S.K., Semenov, A.N. and Koleva, M.K. (1987) *Physica (Amsterdam)*, **140A**, 506.
67. Khokhlov, A.R. and Nechaev, S.K. (1985) *Phys. Lett.*, **112A**, 156.
68. Obukhov, S.P., Rubinstein, M. and Duke, T. (1994) *Phys. Rev. Lett.*, **73**, 1263.
69. Klein, J. (1978) *Macromolecules*, **11**, 872.
70. Daoud, M. and de Gennes, P.G. (1979) *J. Polym. Sci., Polym. Phys. Ed.*, **17**, 1981.
71. Graessley, W.W. and Struglinski, M.J. (1986) *Macromolecules*, **19**, 1754.
72. Rubinstein, M., Helfand, E. and Pearson, D.S. (1987) . *Macromolecules*, **20**, 822.
73. Rubinstein, M. and Colby, R.H. (1988) . *J. Chem. Phys.*, **89**, 5291.
74. Green, P.F. and Kramer, E.J. (1986) *Macromolecules*, **19**, 1108.
Green, P.F., Mills, P.J., Palmstrom, C.J., Mayer, J.W. and Kramer, E.J. (1984);. *Phys. Rev. Lett*, **53**, 2145.
75. Montfort, J.R., Marin, G. and Monge, P. (1986) *Macromolecules*, **19**, 1979.
76. Brochard, F. and de Gennes, P.G. (1992) *Langmuir*, **8**, 3033.
77. Brochard-Wyart, F., Ajdari, A., Leibler, L., Rubinstein, M. and Viovy, J.L. (1994) *Macromolecules*, **27**, 803.
78. Ajdari, A., Brochard-Wyart, F., de Gennes, P.G., Leibler, L., Viovy, J.L. and Rubinstein, M. (1994) *Phyica A*, **204**, 17.
79. Brochard-Wyart, F., Gay, C. and de Gennes, P.G. (1996) *Macromolecules*, **29**, 377.
80. Doi, M. and Edwards, S.F. (1978) *J. Chem. Soc. Faraday Trans. 2*, **74**, 1789, 1802, 1818.
81. Dean, P. (1964), *Proc. Phys. Soc.*, **84**, 727.
82. Marrucci, G. (1985) *J. Polym. Sci., Polym. Phys. Ed.*, **23**, 159.
83. Viovy, J.L. (1985) *J. Phys. (Les Ulis, Fr.)*, **46**, 847.
84. des Cloizeaux, J. (1988), *Europhys. Lett*, **5**, 437.
85. des Cloizeaux, J. (1990) *Macromolecules*, **23**, 3992.
86. Viovy, J.L., Rubinstein, M. and Colby, R.H. (1991) *Macromolecules*, **24**, 3587.
87. Deutsch, J.M. (1985) *Phys. Rev. Lett.*, **54**, 56.
88. Deutsch, J.M. (1987 *J. Phys (Paris)*, **48**, 141.
89. Semenov, A.N. (1989) *Zh. Eksp. Theor. Fiz.*, **96**, 194 [*Sov. Phys. JETP*, **69**, 107.
90. Semenov, A.N. (1991) *Physica (Amsterdam)*, **171A**, 517.

SLOW MODES IN DYNAMIC LIGHT SCATTERING FROM SEMI-DILUTE SOLUTIONS OF POLYMERIC AND POLYMER-LIKE SYSTEMS

S.J. CANDAU
Laboratoire d'Ultrasons et de Dynamique des Fluides Complexes
U.R.A. au CNRS n°851, Université Louis Pasteur
4, rue Blaise Pascal, 67070 Strasbourg Cedex, France

1. Introduction

An approach to the dynamical properties of semi-dilute solutions of polymeric systems rests on the analysis by Dynamic Light Scattering (DLS) of the long time behaviour of the correlation function of the concentration fluctuations. It was shown theoretically that when the coupling of the concentration fluctuations to the viscoelasticity is present, the time correlation function of the concentration fluctuations is multi-exponential [1-5] (see article by A.N. Semenov, this volume). If one excepts the regime of very small scattering wave-vectors, generally unaccessible experimentally, where the autocorrelation function is almost single-exponential, two noticeable stages of relaxation of concentration fluctuations are predicted. The first stage is due to cooperative deformation of the entanglement network and is characterized by a single relaxation time. The second stage is associated with the relaxation of the longitudinal stress tensor occurring through the reptation process (see article by M. Rubinstein, this volume) and is characterized by a spectrum of relaxation times. The relative amplitude of the slow reptational mode is determined by the ratio of the elastic modulus M_g of the entanglement network to the bulk osmotic modulus K.

Conflicting results and/or interpretations have been reported for DLS experiments performed on different polymeric or polymer-like systems. In most cases, the autocorrelation function of scattered light is found to be bimodal as expected from the theoretical models but the variations of the amplitudes and characteristic times associated with the slow process, as a function of scattering wavevector and polymer concentration are generally in disagreement with the theoretical predictions.

In this paper, after a brief theoretical reminder, we review the main DLS results obtained in different systems : polymers in good and theta solvents, solutions of giant micelles, polyelectrolytes.

T. McLeish (ed.), Theoretical Challenges in the Dynamics of Complex Fluids, 53-62.

This review allows one to put in evidence the main unsolved issues in this field.

2. Theoretical Reminder

A detailed treatment of the dynamics of concentration fluctuations is given in the chapter presented by A. Semenov in this book. Here we simply recall the conclusions of the theoretical models needed to discuss the DLS results.

DLS experiments provide a determination of the normalized dynamical correlation function $g^{(1)}(q,t)$ of polymer concentration fluctuations in semidilute solutions of unbreakable polymers, defined as

$$g^{(1)}(q,t) = \frac{< \delta c^*(q,0)\delta c(q,t) >}{< |\delta c(q,0)|^2 >} \qquad (1)$$

where $\delta c(q,t)$ and $\delta c(q,0)$ represent a fluctuation of polymer concentration of wave vector q at time t and zero, respectively, and q is given by :

$$q = (4\pi n/\lambda_0)\sin(\theta/2) \qquad (2)$$

where n is the refractive index of the solution, λ_0 the wavelength in the vacuum and θ the scattering angle.

A model for the dynamics of semi-dilute polymer solutions in theta solvent has been derived by Brochard and de Gennes [1][2].

It was shown that, whereas in the long-wavelength, low frequency limit, $g^{(1)}(q,t)$ was a single exponential decaying as $\exp(-D_c q^2 t)$, above a small q threshold beyond which τ_c is smaller than the reptational time, two relaxational modes were predicted : a fast gel mode with the characteristic time

$$\tau_c = \left[\frac{K}{M_g+K}\right]\left[\frac{k_B T q^2}{6\pi\eta_0\xi_H}\right]^{-1} \qquad (3)$$

where ξ_H is the hydrodynamic length and η_0 is the solvent viscosity, and a slow hydrodynamic mode arising from the coupling between stress and concentration fluctuations with the time :

$$\tau_s \simeq \tau_{max}\frac{M_g+K}{K} \qquad (4)$$

where τ_{max} is the longest time of conformational relaxation (which is just the reptation time in the case of linear polymers) and is q independent. The relative amplitude of the slow mode is also q independent.

Later on, several other theoretical studies were published on this subject [3-5]. In particular Semenov derived a full treatment where the relaxation of concentration fluctuations takes into account both deformation of the entanglement network and the direct effect of the reptation motion of polymer chains [3]. In references (1,2), only the first effect was considered, leading to the prediction given by equation (4).

In Semenov's approach, three stages of relaxation of concentration fluctuations are predicted. The first stage is due to cooperative deformation of entanglement network and is characterized by a time $\tau_c^{-1} \propto q^2 N^0$ where N is the number of links per polymer chain. The second stage is a Rouse-type relaxation of the macromolecular tension along the tube with a relaxation time τ_R. The last stage, associated with the reptation process, is characterized by a spectrum of relaxation times between τ_{slow} and τ_{max} : τ_{max} is the terminal time of the stress relaxation function and is proportional to N^3, and τ_{slow} depends on q and N on a complicated way depending on the relative value of the size of the macromolecule with respect to the inverse of the scattering vector. The variation of the three characteristic relaxation times τ_c, τ_R and τ_{slow} as a function of q^{-1}, as calculated by Semenov is reported in Figure 1. It must be noted that the q^{-1} domain investigated in most experiments extends on both sides of the value $q^{-1} = R$ (R, end-to-end distance), in a range where, τ_{slow} is roughly q-independent and varies as :

$$\tau_{slow} \sim \tau_0 N^2 q^0 \tag{5}$$

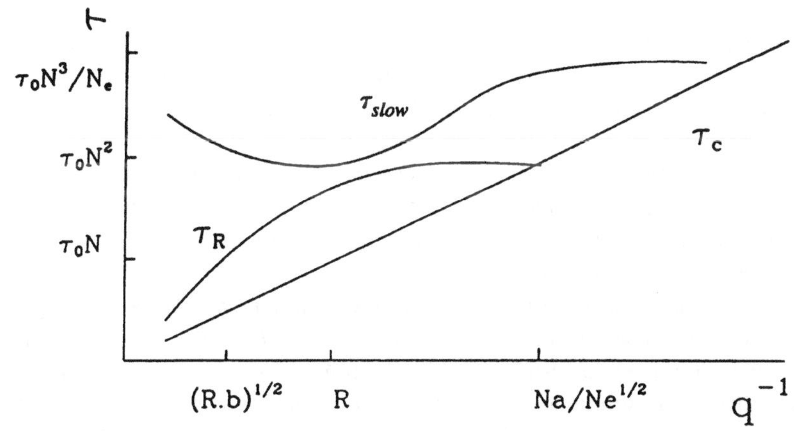

Figure 1. - Schematic (double logarithmic) plot of cooperative (τ_c), Rouse (τ_R) and slow (τ_{slow}) relaxation times *versus* q^{-1} ; b is the tube diameter, and a, the length of the statistical unit (from Ref.[3]).

where τ_0 is the characteristic microscopic time. In fact, τ_{slow} as given by Eq.(5) corresponds to a crude average of a distribution between τ_{slow} and τ_{max}. It is only in the limit of inverse scattering vectors larger than the tube length i.e. $q^{-1} \geq Na\, N_e^{-1/2}$ (a, size of a polymer link, N_e mean number of links per entanglement) that the behaviour given by Eq.(4) is recovered. Such a limit is generally not accessible experimentally.

Semenov has also calculated the amplitudes of the different relaxational contributions. As a rule, the amplitude of the first cooperative stage A_c is the largest, and the amplitude A_R of the second "Rouse" stage is the smallest.

The relative amplitude of the slow mode A_{slow} is found to be approximately equal to (5/6) (M_g/K) for $qR \ll 1$ and (8/15)(M_g/K) for $qR \ll 1$.

3. Experimental Results

In the following we discuss successively the results obtained in different polymeric or polymer-like systems.

3.1. SEMI-DILUTE SOLUTIONS OF POLYMERS IN GOOD SOLVENT

A series of DLS experiments performed in the eighties has raised a controversy concerning mainly the shape of the autocorrelation function of scattered light. While some authors fit the experimental autocorrelation function to a single exponential [6-8], others claim that two modes, a "fast" and a "slow" mode, are present [9-11]. In the latter case, it is found that, as the concentration increases, the correlation function is increasingly dominated by the slow relaxation process when $qR \ll 1$. This slow mode was tentatively attributed to the self-diffusion of the chains, as the concentration dependence of the relaxation time was consistent with the reptation model [9-10]. However, the slow mode is between one and two orders of magnitude slower than self-diffusion as measured by pulsed field gradient NMR experiments [11].

Although the origin of the above behaviour is not yet elucidated, now the experimentalists seem to agree that for solutions in very good solvents and at polymer volume fractions less than 0.1, the autocorrelation function is fitted by a single exponential within the experimental accuracy.

Such a conclusion is in agreement with the theoretical prediction concerning the amplitude of the slow mode, that is controlled by the ratio M_g/K. The ratio E/K, where E is the shear plateau modulus that is of the order of M_g, has been recently determined in different semi-dilute solutions in good solvent by combining measurements of the osmotic pressure and of the elastic plateau modulus [12]. This ratio was found to vary between 5 10^{-3} and 2.5 10^{-2} depending on the system.

These values remain constant in a large range of polymer concentration

($10^{-2} - 0.12$ g cm^{-3}) both moduli scaling with the polymer concentration as $c^{\sim 2.3}$. Contributions with such amplitudes cannot be detected within the experimental accuracy in the autocorrelation function of scattered light.

3.2. SEMI-DILUTE SOLUTIONS OF POLYMERS IN THETA SOLVENT

In theta solvent one expects a behaviour markedly different, as the elastic modulus M_g and the osmotic modulus K scale differently with polymer concentration. The elastic modulus M_g is proportional to the density of entanglements and follows a behaviour rather close to that observed in good solvent, that is a scaling to the polymer concentration with an exponent ~ 2. In fact, it was found experimentally that the plateau modulus was about the same in a theta and in a good solvent [12]. As for K, it becomes much smaller than in a good solvent because the binary thermodynamic interactions vanish, and it scales like C^3 [12-14]. In the range of concentrations investigated, the ratio M_G/K is large enough that a slow mode could be detected.

Many experiments were performed on semi-dilute solutions in theta solvent. All the authors agree that in the gel-like regime ($\tau \ll \tau_{max}$), the autocorrelation function can be approximated by a bimodal shape. They also agree on the q^{-2} dependence of the fast relaxation time τ_{fast}. However, discordant results were reported concerning the slow mode and more specifically on the q dependence of the corresponding characteristic time τ_{slow}. This is not surprising considering the complicated shape of the autocorrelation function predicted by Semenov : obviously the relaxation time τ_{slow} determined experimentally will depend on the data analysis.

The results reported by Adam and Delsanti [15] on the q, c and N dependences of the slow mode are consistent with the theory of Brochard and de Gennes. Also the experimental values of τ_{slow} are very close to τ_{max}. According to Semenov, such regime is attained only in the limit where q^{-1} is larger than the tube length, which is not the case experimentally probed. In the q regime explored, the characteristic time τ_{slow} is much smaller than τ_{max} (cf. Fig. 1). However τ_{slow} is a crude average corresponding approximately to a decrease of the correlation function from 1 to 1/e. The longest relaxation time is of the order of τ_{max}.

In the ref.[15] the authors used a rather unorthodox method that allowed them to pick up the longest relaxation time. Instead of using the classical cumulants or Laplace transform methods, they scanned the correlation function with a variable correlation time window, forcing a single exponential fit, with a free base-line, and taking as τ_{fast} and τ_{slow} the two extreme relaxation times obtained in this scan.

At the moment, there has not been any attempt to fit directly the Semenov prediction to an experimental autocorrelation function.

3.3. SEMI-DILUTE SOLUTIONS OF GIANT MICELLES

The structural and dynamical properties of worm-like micellar systems are described in other chapters of this book (see F. Lequeux and S.J. Candau, and M. Cates). Numerous experimental studies have stressed the analogy between polymers and worm-like micelles. In particular, the osmotic compressibility and the plateau modulus scale as power laws of the surfactant concentration with exponents close to those predicted for semi-dilute polymer solutions in good solvent [16].

There are not at the moment theoretical models available to describe the dynamics of concentration fluctuations in worm-like systems.

One could however expect that the general description given for classical polymers still applies with however a caveat related to the fact that the lifetime of the tube is generally much less than the reptation time because of the breaking-recombination process. In fact, the different regimes represented in Fig. 1 might be shifted along the horizontal axis. In particular, one could expect that the regime where $\tau_{slow} = \tau_{max}$ be attained when q^{-1} is larger than the length λ along which the chain reptates between two breakages (cf. Cates chapter). This condition that would replace that specifying q^{-1} larger than the tube length is now fulfilled in the q-range explored experimentally.

Recent DLS experiments have reported bimodal autocorrelation function of scattered light [17,18] (Fig. 2).

The two relaxation processes are widely separated in time so that a classical analysis of each mode using cumulants method or inverse Laplace transform can be performed. The fast relaxation time τ_{fast} was found to vary as q^{-2} and could be assigned as the diffusive gel mode. The slow relaxation time τ_{slow} is q independent and is very close to the mechanical relaxation time which is obtained from the dynamic shear modulus measurements (cf. Cates chapter).

These results are in good agreement with Semenov's predictions in the regime where the relaxation of concentration fluctuation is a consequence of elastic deformations of the entanglement networks.

Furthermore, the relative amplitude of the slow mode was found in ref.[17] of the order of 0.15, and independent of q and of polymer concentration within the experimental accuracy. This is what expected since, both M_g and K whose ratio gives the amplitude of the slow mode scale with the same exponent ($\sim$ 2.3) with polymer concentration. The ratio M_g/K is found larger than for classical polymers in good solvent, due likely to a shorter length between entanglements resulting in a larger plateau modulus.

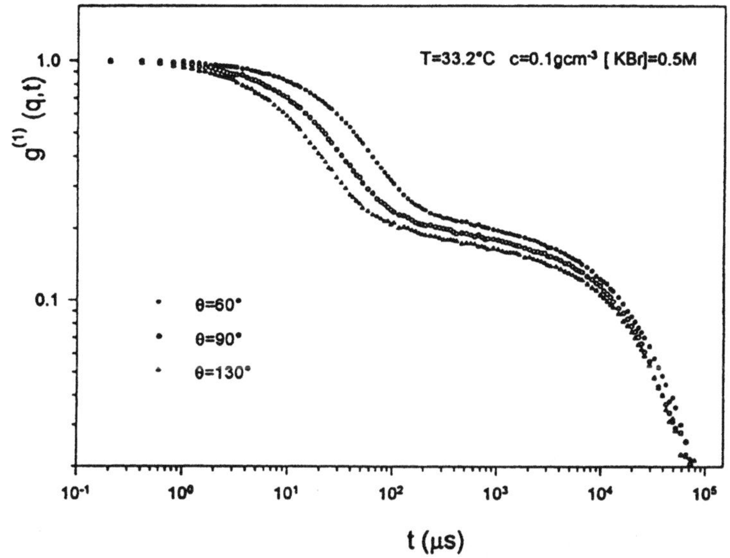

Figure 2. – Time autocorrelation function of the scattered field for different scattering angle for aqueous solution of CTAB in presence of 0.5 M KBr. The fast relaxation time varies as q^{-2} whereas the slow relaxation time is q independent (from Ref.[17]).

3.4. SEMI-DILUTE SALT-FREE AQUEOUS SOLUTIONS OF POLYELECTROLYTES

The combination of both polymer and electrolyte properties results in high complexity of the structural and dynamic properties of these systems. The description of the structural properties of polymers in good solvent in terms of a single characteristic length breaks down and one must consider a series of different concentration regimes whose description is given in a recent paper of Dobrynin et al. [19]. In particular there exists a large concentration domain where the polyelectrolyte chains are overlapped but unentangled. In such a regime one wouldn't expect to observe in DLS experiments any reptational mode. Furthermore, the osmotic modulus that scales like c in polyelectrolyte solutions is very large so that one would expect a concentration fluctuation to relax predominantly through the fast cooperative mode. In any case, the dynamical properties of polyelectrolyte solutions are complex and the dynamics of concentration fluctuations have not yet been described theoretically, if one excepts a study concerning the case of weakly charged polyelectrolytes in poor solvent [20].

Static and Dynamic Light Scattering experiments have revealed a complex picture of the properties of polyelectrolytes in solution that cannot be fully explained within the framework of the approaches derived for neutral polymers.

A review of the ensemble of experimental studies allows to draw out the following general features [21] :

- The autocorrelation function of scattered light exhibits two diffusive modes.

- The relaxation time of the fast decay remains almost concentration-independent (after an initial rise) whereas the relaxation time of the slow process decreases with increasing concentration. At very high dilution the two relaxation times tend to coincide so that only one diffusive process is observed (Fig. 3) [21-23].

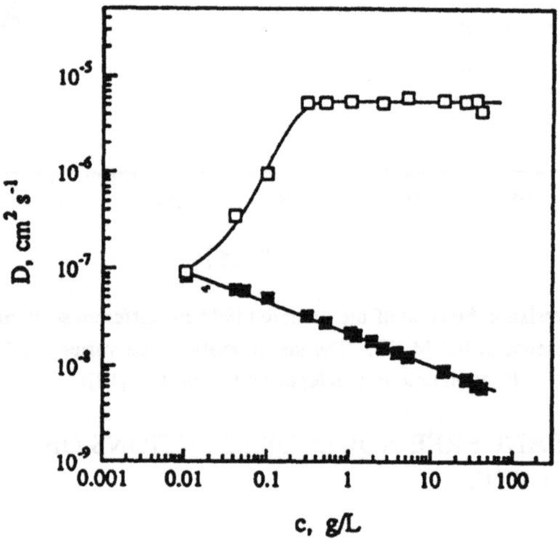

Figure 3. – Dependence of the fast diffusion coefficient $D_{fast} = (\tau_{fast}\, q^2)^{-1}$ (□) and slow diffusion coefficient $D_{slow} = (\tau_{slow}\, q^2)^{-1}$ (■) on concentration for solutions of a sodium polystyrene sulfonate with (M_w = 100 000) from Ref.[22]

- The relative amplitude of the slow relaxation process decreases with decreasing concentration and eventually disappears completely

- The static structure factor exhibits a very sharp upturn in the low q range

- The more hydrophobic the backbone of the polyelectrolyte, the larger the relative amplitude of the slow mode. The slow mode is very weak and even undetectable for solutions of a polyelectrolyte with hydrophilic backbone like polyacrylic acid.

The above observations suggest that the fast mode is a cooperative diffusive mode as in semi-dilute solutions of neutral polymers, with its amplitude controlled by the osmotic compressibility of the counterions. The slow mode is envisioned as the translational diffusion of clusters of chains whose dimensions can be obtained from the angle dependence of either static or dynamic light scattering. The domains are found in the semi-dilute regime, i.e. when the polyions chains physically overlap. Recent experiments using fluorescence spectroscopy have revealed the presence of hydrophobic domains in semi-dilute solutions of sodium polystyrene sulfonate [24].

Upon addition of salt, the onset of splitting of the two modes is shifted toward higher polyelectrolyte concentrations so that in a rather large range of polymer concentration one observes a single diffusive process. The appearance of a slow process upon decreasing the salt content of a semi-dilute polyelectrolyte solution has been well described in the literature and referred to as an "ordinary-extraordinary" transition [21].

If the interchain clusters description is valid, then the addition of salt would destroy these clusters.

The behaviour of polyelectrolyte solutions in DLS experiments seems to be now well described experimentally. But the understanding of this behaviour requires to answer many questions concerning more specifically the following issues : the mechanism of association of chains in salt-free systems, the nature of the attractive interactions involved, the effect of the electrostatic screening by the salt, the influence of the hydrophobicity of the backbone.

4. Conclusions

The results reported in this paper show that there are still many challenges for theoreticians and experimentalists to fully understand the dynamics of concentration fluctuations in polymeric or polymer like systems. In particular the theoretical models to describe the behavior of worm-like micelles (neutral or charged) and that of polyelectrolytes are lacking.

From the experimental point of view, the main results obtained on neutral polymers have been analyzed in a framework, rather close to the reality, assuming two main processes linked with the deformation of the entanglement network, but a full study, albeit very difficult, exploring all the regimes predicted by A. Semenov remains to be done.

References

1. Brochard, F. and de Gennes, P.G. (1977) Dynamical scaling for polymers in theta solvents, *Macromolecules* **10**, 1157-1161.
2. Brochard, F. (1983) Gel-like modes of polymer solutions in theta solvents, *J. de Phys. (France)* **44**, 39-43.
3. Semenov, A.N. (1990) Dynamical correlation function of polymer density fluctuations in concentrated solutions, *Physica A* **166**, 263-287.

4. Doi, M. and Onuki, A. (1992) Dynamic coupling between stress and composition in polymer solution and blends, *J. Phys. II (France)* **2**, 1631-1656.
5. Wang, C.H. (1992) Dynamic light scattering and viscoelasticity of a binary polymer solution, *Macromolecules* **25**, 1524-1529.
6. Munch, J.P., Lemaréchal, P., Candau, S.J. and Herz, J. (1977) Light scattering spectroscopy of polydimethylsiloxane-toluene gels, *J. Phys. (France)* **38**, 1499-1509.
7. Wiltzius, P., Haller, H.R., Cannell, D.S. and Schaeffer, D.W. (1984) Dynamics of long-wavelength concentration fluctuations in solution of linear polymers, *Phys. Rev. Lett.* **53**, 834-837.
8. Wiltzius, P. and Cannell, D.S. (1986) Wavevector dependence of the initial decay rate of fluctuations in polymer solutions, *Phys. Rev. Lett.* **56**, 61-64.
9. Schaeffer, D.W., Han, C.C. (1985) Quasielastic light scattering from dilute and semi-dilute polymer solutions, in R. Pecora (ed.), *Dynamic Light Scattering*, Plenum, pp. 181-243, and references therein.
10. Joanny, J.F., Candau, S.J. (1989) Hydrodynamic properties, in C. Booth, C. Price (eds), *Comprehensive Polymer Science*, Pergamon, Oxford, pp. 199-236, and references therein.
11. Brown, W. (1984) The concentration dependence of diffusion in semi-dilute polymer solutions, *Polymer* **25**, 680-689.
12. Raspaud, E., Lairez, D. and Adam, M. (1995) On the number of blobs per entanglement in semi-dilute and good solvent solutions : melt influence, *Macromolecules* **28**, 927-944.
13. Noda, I., Kato, N., Kitano, T. and Nagasawa, M. (1981) Thermodynamic properties of moderately concentrated solutions of linear polymers, *Macromolecules* **14**, 668-676.
14. Adam, M., Fetters, L.J., Graessley, W.W. and Witten, T.A. (1991) Concentration dependence of static and dynamic properties for polymeric stars in a good solvent, *Macromolecules* **24**, 2434-2440.
15. Adam, M. and Delsanti, M. (1985) Dynamical behavior of semi-dilute polymer solutions in a θ-solvent : quasi-elastic light scattering experiments, *Macromolecules* **18**, 1760-1770.
16. Cates, M.E. and Candau, S.J. (1990) Statics and dynamics of worm-like surfactant micelles, *J. Phys. Condens. Matter* **2**, 6869-6892 and references therein.
17. Buhler, E., Munch, J.P. and Candau, S.J. (1995) Dynamical properties of worm-like micelles : a light scattering study, *J. Phys. II (France)* **5**, 765-787.
18. Nemoto, N., Kuwahara, M., Yao, M. and Osaki, K. (1995) Dynamic light scattering of CTAB/NaSal threadlike micelles in a semi-dilute regime. 3. Dynamical coupling between concentration fluctuation and stress, *Langmuir* **11**, 30-36.
19. Dobrynin, A., Colby, R.H. and Rubinstein, M. (1995) Scaling theory of polyelectrolyte solutions, *Macromolecules* **28**, 1859-1871.
20. Ajdari, A., Leibler, L., Joanny, J.F. (1991) Cooperative diffusion in weakly charged polyelectrolyte solutions, *J. Chem. Phys.* **95**, 4580-4583.
21. Schmitz, K.S. (1994) An overview of polyelectrolytes : six topics in search of an author, in K.S. Schmitz (ed.), *Macroion Characterization. From Dilute Solutions to Complex Fluids*, ACS Symposium Series 548, Washington D.C., pp. 1-22.
22. Amis, E.S., Valachovic, D.E., Sedlák, M. (1994) Structure and dynamics of linear flexible polyelectrolytes in salt-free solution as seen by light scattering, in K.S. Schmitz (ed.), *Macroion Characterization. From Dilute Solutions to Complex Fluids*, ACS Symposium Series 548, Washington D.C., pp. 322-336.
23. Förster, S., Schmidt, M., Antonietti, M. (1990) Static and dynamic light scattering by aqueous polyelectrolyte solutions : effect of molecular weight, charge density and added salt, *Polymer* **31**, 781-792.
24. Essafi, W., (1996) Structure des polyélectrolytes fortement chargés, *PhD Thesis*, Paris VI.

I. RELAXATION OF CONCENTRATION FLUCTUATIONS IN ENTANGLED POLYMER MIXTURES

A.N.SEMENOV
Department of Applied Mathematics, University of Leeds
Leeds LS2 9JT, U.K.

Abstract. The dynamics of composition fluctuations in entangled compatible polymer blends are considered theoretically using the reptation model. Both symmetric and asymmetric (long + short chains) mixtures are considered. It is shown that the relaxation dynamics is generally non-exponential and is characterized by 4 sets of modes: two reptation modes (fast and slow) corresponding to diffusion of mainly short chains and mainly long chains respectively, and two Rouse modes. It is also shown that two theoretical concepts of fast and slow diffusion are both valid in a limited region of wave-lengths, where the theories adequately describe the corresponding relaxation processes.

1. Introduction

Interdiffusion of polymers is a problem of considerable interest for both basic knowledge and applications. In particular, interdiffusion of entangled polymers attracted much attention recently [1, 2, 3, 4]. Most of the theories interpret the polymer mixing process in terms of the mutual diffusion constant, D_m. This interpretation implies that the process is (1) diffusive; (2) is characterized by a single relaxation time.

The aim of this paper is to describe the dynamics of concentration (composition) fluctuations in polymer blends without using the above assumptions. In fact we show that the dynamics might be non-diffusive and also is characterized by many relaxation modes even if the characteristic scale is much larger than the polymer coil size, R. The problem of fast/slow diffusion in asymmetric mixtures [2, 3, 4] is also discussed; it is shown that both concepts are correct in some sense.

T McLeish (ed), Theoretical Challenges in the Dynamics of Complex Fluids, 63-76.

Generally, the dynamics in polymer mixtures is not much different from that of polymer solutions [5, 6, 7]; so, the problem is considered here along the lines of ref. [6]. Although in most of the regimes a rigorous quantitative mean-field treatment is possible (see ref. [6]), below we present a crude scaling picture of relaxation processes in the blends. This is done in order to simplify the argumentation.

2. Symmetric compatible mixtures

Let us consider an incompressible blend of two polymers $N_A = N_B = N$ in the regime of entanglements, $N \gg N_e$. The mean volume fractions of polymers are $\phi_A = \phi_0$ and $\phi_B = 1 - \phi_0$. The local polymer concentration is fluctuating, $\phi_A(\mathbf{r}) = \phi_0 + \delta\phi(\mathbf{r})$, giving rise to scattering of these fluctuations which is often characterized by the dynamic structure factor (scattering function). This may be directly measured by techniques such as neutron spin-echo, or photon-correlation light scattering:

$$S(q,t) = \langle \delta\phi_q(t)\delta\phi_{-q}(0)\rangle \tag{1}$$

where $\delta\phi_q$ is the corresponding Fourier component

$$\delta\phi_q = \int \delta\phi(\mathbf{r})e^{-i\mathbf{qr}}d^3r \tag{2}$$

A fluctuation $\delta\phi_q(t=0)$ will relax in time on the average according to the law $\delta\phi_q(t) = \delta\phi_q(0)f(q,t)$, where the function f is related to the structure factor, eq. 1,

$$f(q,t) = \frac{S(q,t)}{S(q,0)}$$

In the following we aim to estimate the form of $f(q,t)$ at different $\lambda \sim 1/q$.

2.1. STATIC STRUCTURE FACTOR

Let us consider first an equilibrium situation. The free energy density for a slightly inhomogeneous state $\phi_A(\mathbf{r}) \equiv \phi(\mathbf{r}) = \phi_0 + \delta\phi(\mathbf{r})$ in the Flory-Huggins approximation [8, 9] is

$$F = \frac{\phi}{N_A}\ln\phi + \frac{1-\phi}{N_B}\ln(1-\phi) + \chi\phi(1-\phi) + \frac{b^2}{24}\frac{(\nabla\phi)^2}{\phi(1-\phi)} \simeq$$

$$F_0 + const\ \delta\phi + \frac{1}{2}\Gamma(q)\delta\phi^2 \tag{3}$$

where the volume per link, v, and the thermal energy k_BT are set to 1, b is the monomer size, χ is the Flory parameter characterizing interactions between A and B links, and

$$\Gamma(q) = \frac{1}{\phi(1-\phi)N} + x + \frac{b^2q^2}{12\phi(1-\phi)} \tag{4}$$

Here $x \equiv 2|\chi|$, and we assume that $\chi < 0$, i.e. the polymers are compatible. The wave-vector q in eq. 4 stands for $i\nabla \equiv i\frac{\partial}{\partial \mathbf{r}}$. Below we assume that $xN\phi(1-\phi) \gg 1$, so that the first term in eq. 4 can be neglected.

Eq. 3 together with Boltzmann statistics implies that [10] $S(q,0) = \frac{1}{\Gamma(q)} \simeq \frac{S(0,0)}{1+q^2\xi^2}$, where $S(0,0) \simeq \frac{1}{x}$ and $\xi \simeq \frac{b}{2\sqrt{3}}\frac{1}{\sqrt{x\phi(1-\phi)}}$ is the static correlation length. In the region $q\xi \ll 1$ (corresponding to scales larger than ξ) the static structure factor is nearly constant: $S(q,0) \simeq \frac{1}{x}$.

A concentration fluctuation $\delta\phi$ is relaxing because it implies a non-equilibrium state: the free energy $\delta F = \frac{1}{2}\Gamma(\delta\phi)^2$ associated with the fluctuation is positive. The driving force for the relaxation is thus proportional to $\Gamma(q)$: the larger Γ the faster the relaxation. If we set formally $x = 0$, i.e. consider the ideal system with no interaction between A and B links, then in the low-q limit $\Gamma_0 = \frac{1}{\phi(1-\phi)N}$. For a finite x the driving force is stronger by a factor $X \equiv 1 + \phi(1-\phi)xN$, $X \gg 1$, which we will call the thermodynamic enhancement (or acceleration) factor. This factor is roughly equal to the ratio of the thermodynamic force to the entropic force. Note that in the case $X \gg 1$ the fluctuation energy δF is dominated by the interaction contribution $\delta F \simeq \delta F_{int} = \frac{1}{2}x(\delta\phi)^2$. Physically the relaxation in this case is driven by the *molecular field*

$$U_{mol} = \frac{\partial(\delta F_{int})}{\partial(\delta\phi)} = x\delta\phi \tag{5}$$

The dynamics of concentration fluctuations is considered below in the region of long enough wave-lengths: $\lambda = 1/q$ is assumed to be larger than both the static correlation length ξ and the tube diameter, $a \sim bN_e^{1/2}$ (for definitions see articles of McLeish and Rubinstein, this volume).

2.2. RELAXATION OF LONG WAVES, $1/Q \gg R$

Let us consider first the case of wave-length, $\lambda = 1/q$, longer than the coil size $R = N^{1/2}b$. Here a concentration fluctuation relaxes by a mutual diffusion of polymer chains: the relaxation factor $f(q,t) = e^{-D_m q^2 t}$. For an ideal system ($x = 0$) the mutual diffusion constant is equal to the polymer self-diffusion constant (the constant is the same for both polymers as we are considering a symmetric mixture): $D_{m0} = D_{self}$. The self-diffusion constant of a reptating chain is [9, 11]

$$D_{self} \simeq D_0 \frac{N_e}{N^2}$$

where D_0 is a monomer diffusion coefficient. With the thermodynamic acceleration the mutual diffusion constant becomes

$$D_m = D_{m0}X \simeq D_0 \frac{N_e}{N} x\phi(1-\phi)$$

The corresponding relaxation time is [9]

$$\tau = \frac{1}{D_m q^2} \sim \frac{N}{N_e} \frac{\lambda^2}{x\phi(1-\phi)} \tag{6}$$

where the monomer relaxation time $\tau_0 = b^2/D_0$ is set to unity, and λ is measured in monomer sizes (b).

2.3. SHORT WAVE-LENGTHS: THE PSEUDO-ROUSE MODE

Let us turn to the case of short wave-lengths, $\lambda = 1/q < R$. In order to understand qualitatively the relaxation in this regime, let us consider a region $\mathcal{R}$ with characteristic size $\sim \lambda$, where $\delta\phi_0$ is, say, positive. A decrease of $\delta\phi$ implies a flow of monomers outside the region. A reptation motion of a chain with both ends far from the region can not provide the flow: the number of monomers of this chain inside the region is constant, the inward and outward "currents" of monomers are equal and compensate each other.

However the situation is different if one end of the chain occurs in $\mathcal{R}$. The chain is then pushed out of this region where the monomer potential is higher than on the average by $\delta U_{mol} = x\delta\phi$. As the chain is confined in a tube, it will *reptate* in the outward direction. The driving force for this motion is the difference between the molecular potentials probed by two chain ends, δU_{mol}. The curvilinear velocity w of the chain is proportional to δU_{mol} and inversely proportional to the friction ζ along the tube: $w \sim \delta U_{mol}/\zeta$. Let us temporarily set $N_e = 1$ in order to simplify the argument.[1] Then ζ is of order of the tube length, which is $\sim N$. Thus we get $w \sim x\delta\phi/N$. Note that w is equal to the number of links escaping from $\mathcal{R}$ per unit time per chain. Since the number of chain ends per unit volume is $\sim 1/N$, the concentration relaxation rate is

$$\frac{\partial(\delta\phi)}{\partial t} \sim -\frac{1}{N} w \sim -\frac{x}{N^2}\delta\phi$$

This implies the relaxation time $\tau_{pR} \sim N^2/x$, which we will call pseudo-Rouse time as it is proportional to N^2 (like the longest Rouse time), but is physically due to the reptation.

[1]Here and below we also assume for simplicity that composition is symmetric, $\langle\phi_A\rangle = \phi = 1/2$; however the ϕ-dependence will be included in the final equations.

In order to get back the N_e and ϕ−dependencies we assume a smooth crossover between τ_{pR} and the diffusion time, eq. 6, at the boundary between the regimes, $\lambda \sim R$. Thus we get [12, 13]

$$\tau_{pR} \sim \frac{N^2}{N_e x \phi (1-\phi)} \tag{7}$$

2.4. SHORT WAVE-LENGTHS: RANDOM REPTATION

The only type of chain motion considered in the above derivation is the biased reptation induced by the molecular field. As is shown below a random curvilinear diffusion (random reptation) of chains along their tubes might be important for even shorter wave-lengths. The displacement of the chain *along the tube* during a time τ is

$$\Delta s \sim N\sqrt{\tau/\tau_{dis}} \tag{8}$$

where $\tau_{dis} \sim N^3/N_e$ is the longest reptation (disentanglement) time, and s is a reduced coordinate along the tube which measures a curvilinear "distance" as the mean number of links between any two points on the primitive path.

The spatial displacement of a monomer (in particular – a chain end) is $\Delta r \sim b\sqrt{\Delta s} \sim R\left(\frac{\tau}{\tau_{dis}}\right)^{1/4}$. With $\tau = \tau_{pR}$, eq. 7, we get $\Delta r \sim b\left(\frac{N}{x\phi(1-\phi)}\right)^{1/4} \equiv \xi_2$. Note that the new *dynamic* scale, ξ_2, is a geometrical average between the static correlation length, ξ, and the coil size, R:

$$\xi_2 \sim \sqrt{\xi R} \tag{9}$$

The random displacement is not important if $\lambda > \xi_2$. However if the opposite inequality holds, then the chain ends moves through many λ during the relaxation time. Therefore a chain end would escape out of the region $\mathcal{R}$ before a relaxation driven by the molecular field is accomplished thus decreasing the rate of the relaxation process. Hence we expect that the relaxation time becomes longer in the regime $\lambda < \xi_2$. In order to estimate the relaxation time in this regime we note that during a time $\tau > \tau_{pR}$ the tail part Δs (defined in eq. 8) randomly tries different conformations in a spatial region of size Δr which is larger than λ. Therefore this tail part must come to a virtual equilibrium with the molecular field. Hence the spatial distribution of monomers belonging to these tails, ϕ_{tail}, must become inhomogeneous: less tail monomers in the regions where U_{mol} is positive and vice versa

$$\phi_{tail} = \langle \phi_{tail} \rangle + \delta\phi_{tail}; \quad \delta\phi_{tail} = -\langle \phi_{tail} \rangle \kappa(q) U_{mol}$$

where $\kappa(q)$ is the static susceptibility of ideal chains to the external field, $\kappa(q) \sim 1/q^2$ in the region of short wave-lengths [9] (i.e. if λ is shorter than the chain size, or, in our case, than the tail size Δr). Taking into account that $\langle \phi_{tail} \rangle \sim \frac{\Delta_s}{N}$, and using eqs. 5, 8, we obtain

$$\delta\phi_{tail} \sim -\frac{1}{q^2}\sqrt{\frac{\tau}{\tau_{dis}}}x\delta\phi$$

Obviously the relaxation occurs (i.e. a redistribution of the tail parts "heals" the initial inhomogeneity $\delta\phi_0$) if $\delta\phi_{tail} \sim \delta\phi$. Thus we get the relaxation time $\tau \sim \tau_{dis}\left(\frac{q^2}{x}\right)^2$. Adding a ϕ-dependent prefactor $\frac{1}{\phi^2(1-\phi)^2}$ in order to get a smooth crossover between the pseudo-Rouse (eq. 7) and the random reptation regimes at $\lambda \sim \xi_2$, we finally obtain

$$\tau_{rr} \sim \frac{N^3}{N_e}\frac{1}{\lambda^4 x^2}\frac{1}{\phi^2(1-\phi)^2} \tag{10}$$

Note that the relaxation time is increasing as the wave-length λ is *decreased:* this is an anomalous dependence. Also note that the relaxation in this regime is not exponential: the relaxation factor $f(q,t)$ is not equal to $\exp(-t/\tau_{rr})$, but is a more complicated multiple mode relaxation function, with the characteristic time scale τ_{rr}. Eq. 10 is valid in the region $\xi < \lambda < \xi_2$; for $\lambda \sim \xi$ the relaxation time is of order of the disentanglement time $\tau_{dis} \sim \frac{N^3}{N_e}$.

2.5. ROUSE MODE

Another possibility to "heal" a short-scale inhomogeneity $\delta\phi_0$ involves internal motion of chains inside their tubes – the curvilinear Rouse modes.[2]

The idea is that an initial fluctuation $\delta\phi_0$ can be relaxed via a 1-dimensional deformation of chains inside their tubes: in the regions where $\delta\phi$ is positive the chains become extended along the tubes (thus becoming thinner), and vice versa. Obviously this is a very fast way to adapt the fluctuation as it involves a short-scale curvilinear Rouse motion of the chain fragments. The driving force for this motion is again the molecular field, $U_{mol} = x\delta\phi$, which pushes the chain fragments along the tubes in the proper direction. During this process the thermodynamic energy, $\delta F_{int} = \frac{x}{2}(\delta\phi)^2$ partly transforms into elastic energy of the chains deformation. Let us estimate this elastic energy. An increase of the 1-dimensional

[2]The author is grateful to Prof. Marrucci, who pointed out this possibility. The Rouse relaxation has been considered for the case of polymer melts [14] and solutions [6]. For the case of polymer mixtures the description of the Rouse mode is the same; yet it turns out that this mode is more important for mixtures than for solutions.

chain density implies that the number of links between two entanglements, n, increases: $n = N_e + \Delta n$. With $\Delta n \sim N_e$ the corresponding elastic energy is ~ 1 (in $k_B T$ units) per entanglement. Therefore the energy per monomer is $\sim \frac{1}{2N_e}\left(\frac{\Delta n}{N_e}\right)^2$. Considering now contributions of all A-chains, and taking into account that the relative changes of 3-dimensional and 1-dimensional densities must coincide: $\frac{\Delta\phi}{\phi} = \frac{\Delta n}{N_e}$ we get the elastic energy of A-chains per unit volume: $F_{elA} \sim \frac{\phi}{2N_e}\left(\frac{\Delta\phi}{\phi}\right)^2$. Here $\Delta\phi$ is the concentration changed implied by the deformation, so that the total fluctuation after the deformation is $\delta\phi = \delta\phi_0 + \Delta\phi$. The contribution of B-chains is $F_{elB} \sim \frac{1-\phi}{2N}\left(\frac{\Delta\phi}{1-\phi}\right)^2$. Minimizing the total free energy $\delta F = \delta F_{int} + F_{elA} + F_{elB}$ we get the fluctuation amplitude after the Rouse relaxation process

$$\delta\phi = \delta\phi_0 + \Delta\phi = \frac{\delta\phi_0}{1+\alpha}$$

where

$$\alpha \sim x N_e \phi(1-\phi) \tag{11}$$

Therefore the fraction $I_R = \frac{\alpha}{1+\alpha}$ of the initial fluctuation is relaxed during the Rouse stage. The remaining fluctuation part, $\frac{1}{1+\alpha}\delta\phi_0$, is relaxing via the reptation processes considered in the previous sections, $I_{rept} = \frac{1}{1+\alpha}$.

It is easy to check that after the Rouse process the fluctuation free energy decreases by a factor $1+\alpha$: $\delta F = \frac{\delta F_0}{1+\alpha} = \frac{x^*}{2}(\delta\phi_0)^2$, where

$$x^* = \frac{x}{1+\alpha} \tag{12}$$

is the renormalized interaction parameter which governs the rate of the reptation relaxation stage.

Let us estimate the relaxation time τ_R of Rouse stage. If $\alpha \sim 1$, then this time is simply the Rouse time of a chain fragment (g) of spatial size $\lambda = 1/q$: $\tau_R \sim \tau_R(q) \sim \tau_0 g^2 \sim \frac{\tau_0}{q^4 b^4}$. It is important that the Rouse process is not a single-mode (exponential) relaxation; in particular for short times $\Delta\phi \sim -\delta\phi_0 (t/\tau_R(q))^{1/2}$ (see ref. [14]). If the parameter α is large, then at the beginning the rate of the relaxation process must be higher by the factor α since the process is driven by the molecular field and therefore the rate must be proportional to the thermodynamic parameter $\alpha \propto x$. Thus $\Delta\phi \sim \alpha\delta\phi_0(t/\tau_R(q))^{1/2}$ at the early stage ($t \ll \tau_R$). The characteristic relaxation time is defined by the condition $\delta\phi \sim \delta\phi_0/2$. Hence we get

$$\tau_R \sim \tau_R(q)/\alpha^2 \sim \frac{\lambda^4}{(x N_e \phi(1-\phi))^2} \tag{13}$$

To be definite let us assume that the parameter α is large, $xN_e\phi(1-\phi) \gg 1$. In this case the renormalization of the interaction parameter, $x \to x^* \sim \frac{1}{N_e\phi(1-\phi)}$, eq. 12, is important. Imposing this renormalization for the pseudo-Rouse time we get

$$\tau_{pR}^* \sim N^2 \tag{14}$$

(by chance this time is now exactly the longest Rouse time independently of N_e). The dynamic correlation length is now

$$\xi_2^* \sim b(NN_e)^{1/4}$$

i.e. ξ_2^* is roughly the geometrical average between the coil size $R \sim bN^{1/2}$ and the tube diameter $a \sim bN_e^{1/2}$. The random reptation time is

$$\tau_{rr}^* \sim \frac{N^3 N_e}{\lambda^4} \tag{15}$$

A smooth crossover between the short-wave, eq. 14, and the long-wave, eq. 6, predictions occurs at $\lambda \sim R\sqrt{\alpha}$. Note that for $\lambda \sim R\sqrt{\alpha}$ the Rouse mode, the pseudo-Rouse mode and the diffusion mode (see eqs. 13, 14, 6) are all characterized by the same time $\tau \sim N^2$. The Rouse mode dominates in the whole region $\lambda < R\sqrt{\alpha}$. The amplitude of the reptation (pseudo-Rouse/random reptation) modes, I_{rept}, is $\sim 1/\alpha$ in the region $a < \lambda < R$. In the region $R < \lambda < R\sqrt{\alpha}$ the amplitude of the pseudo-Rouse mode is increasing with λ so that it becomes of order 1 for $\lambda \sim R\sqrt{\alpha}$:

$$I_{rept} \sim \frac{1}{\alpha}\frac{\lambda^2}{N}$$

In the region $\lambda > R\sqrt{\alpha}$ the Rouse mode still exists, its relaxation time is $\tau_R \sim N^2$, and its amplitude is decreasing with λ, $I_R \sim \frac{\alpha N}{\lambda^2}$. (The $1/\lambda^2$ behavior of the amplitude of the Rouse mode in the long wave-length limit is universal, it was considered in ref. [15]).

Thus we considered relaxation processes in a symmetric polymer blend corresponding to scales larger than the tube diameter, $a \sim bN_e^{1/2}$. Two sets of modes due to mainly Rouse-like and reptation-like dynamics are identified. The characteristic relaxation time of the (faster) Rouse mode is

$$\tau_R \sim \begin{cases} \frac{\lambda^4}{\alpha^2} & , \quad a < \lambda < R\sqrt{\alpha} \\ N^2 & , \quad \lambda > R\sqrt{\alpha} \end{cases} \tag{16}$$

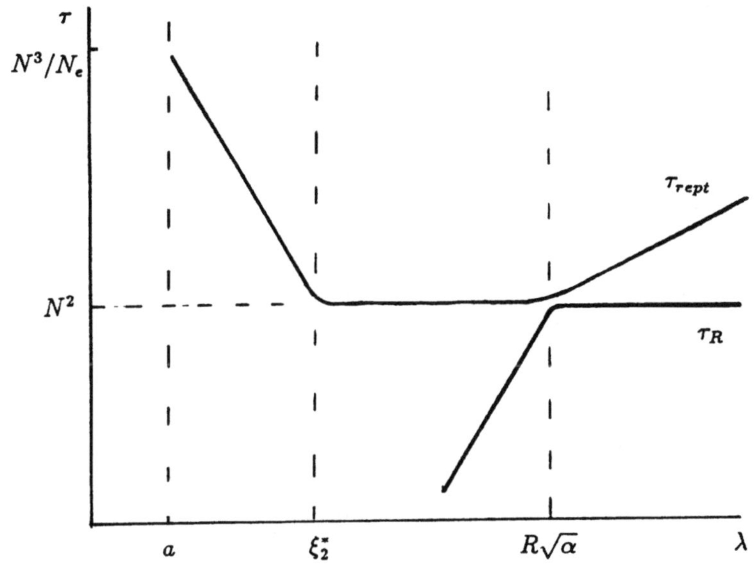

Figure 1. Schematic log-log plot of relaxation times vs. wave-length λ.

The relaxation time of the (slower) reptation mode scales as

$$\tau_{rept} \sim \begin{cases} \frac{N^3 N_e}{\lambda^4} & , \quad a < \lambda < \xi_2^* \sim \sqrt{aR} \\ N^2 & , \quad \xi_2^* < \lambda < R\sqrt{\alpha} \\ \frac{N\lambda^2}{\alpha} & , \quad \lambda > R\sqrt{\alpha} \end{cases} \tag{17}$$

3. Asymmetric mixtures

Let us consider an asymmetric polymer blend: $N_A \gg N_B > N_e$. The question is: what is the nature of concentration fluctuation dynamics in the region of long wave-lengths, $\lambda > R_A = N_A^{1/2} b$. A simple (inter) diffusion dynamics was assumed in a number of theoretical works [2, 3, 4], where the mutual diffusion constant D_m was calculated. The next question concerns D_m: whether it is controlled by diffusion of shorter chains (fast diffusion), or of longer chains (slow diffusion). The fast diffusion idea was supported in refs. [3, 4], it implies that shorter (B) molecules could rapidly diffuse into the long chain (A) matrix thus providing a mixing (a relaxation of an initial fluctuation). The slow diffusion concept developed in refs. [2] is based on the notion that in order for B molecules to reptate into some region, A molecules must reptate out of the region to give space to the newcomers as the system is virtually incompressible. Thus B chains must wait for A chains to move, and hence the diffusion constant is controlled by the slow component.

Below we show that both concepts are actually correct. However the relaxation is not a simple (single-mode) diffusion process. In a wide range of wave-lengths the relaxation is characterized by at least two modes: fast and slow (there is also an intermediate Rouse mode). The slow mode is connected with the reptation motion of long A-chains. The fast mode is due to diffusion of the fast (B) component + continuous deformation of the entanglement network formed by A chains.

3.1. THE FAST MODE

During the fast process the B molecules can be considered as an effective *solvent* for long A chains. The process is thus similar to a cooperative diffusion of entangled polymers in a solution [9]: the system of entangled A chains is moving as a whole without changing its topological state. However for a finite λ the motion implies a *deformation* of the entanglement network. Therefore the fast cooperative stage can not provide a complete relaxation. Let us calculate the amplitude of the fast mode – the fraction of the initial concentration fluctuation that could be relaxed during the fast stage. The situation is governed by the balance between the thermodynamic (interaction) energy $\delta F_{int} = \frac{1}{2}x(\delta\phi)^2$ and the elastic energy of the network deformation $F_{el} = \frac{1}{2}G\epsilon^2$, where $G = \frac{\phi}{N_e(\phi)}$ is the longitudinal elastic modulus of the entanglement network, and ϵ is a longitudinal deformation (strain), which implies a concentration change, $\Delta\phi = \epsilon\phi$. Here $N_e(\phi)$ depends on ϕ as the entanglement network is created by A-chains only; an empirical rule [16] $N_e(\phi) = N_e/\phi$ is used below. Taking into account that $\delta\phi = \delta\phi_0 + \Delta\phi$, we write the total free energy (per unit volume):

$$\delta F = \delta F_{int} + F_{el} = \frac{1}{2}\left(x(\delta\phi_0 + \Delta\phi)^2 + \frac{1}{N_e}(\Delta\phi)^2\right)$$

Minimization of δF leads to $\Delta\phi = \delta\phi_0\frac{\alpha'}{1+\alpha'}$, $\delta\phi = \delta\phi_0\frac{1}{1+\alpha'}$, where $\alpha' = xN_e$. Therefore the amplitude of the fast mode is $I_{fast} = \frac{\alpha'}{1+\alpha'}$ and the amplitude of all slow modes is $I_{slow} = \frac{1}{1+\alpha'}$. We will assume that $\alpha' \gg 1$ in order to be consistent with the previous sections. After the fast relaxation process the free energy of the residual fluctuation is $\delta F = \frac{\delta F_0}{1+\alpha'} = \frac{1}{2}x'(\delta\phi_0)^2$, where the parameter $x' = \frac{x}{1+\alpha'} \simeq \frac{1}{N_e}$ is much smaller than the initial interaction parameter x. Therefore most of the initial fluctuation energy is dissipated during the fast stage, and hence this stage is thermodynamically driven: its rate is proportional to x. The energy δF that has to be dissipated during the slow stage is proportional to $x' \simeq \frac{1}{N_e}$, so that this stage is driven by the elastic energy stored in the deformed network.

Let us calculate the fast relaxation time, which is actually determined solely by the motion of B component. The fast process can be described in exactly the same way as the relaxation in a symmetric mixture with $N = N_B$ described in the previous sections (see eqs. 16, 17). The only difference is that A component is now passive which slows down the relaxation by the factor $1/\phi$. [3]

In particular, in the regime $\lambda > R_B\sqrt{\alpha}$, the fast relaxation time is the diffusion time of B-chains (the Rouse-B mode is negligible in this regime):

$$\tau_{fast} = \frac{N_B \lambda^2}{x N_e \phi^2 (1-\phi)} \tag{18}$$

Eq. 18 is in agreement with the corresponding result of the fast diffusion theory [4].

3.2. THE SLOW MODES

The slow relaxation process is due to a Rouse/reptation motion of A chains in the "solution" of B chains. The dynamics of A-chains is similar to that in symmetric mixtures with $N = N_A$. The difference now is that the other (B) component is very mobile, so that the whole driving force $f_A = f = -\nabla U_{mol}$ is acting on the A chains, whereas in a symmetric mixture this force is split between the components: $f_A = (1-\phi)f$, $f_B = -\phi f$. Thus f_A is now larger by a factor $1/(1-\phi)$, which is equivalent to an additional renormalization of the interaction parameter, $x' \to x'' = x'/(1-\phi) \sim \frac{1}{N_e(1-\phi)}$. Therefore the *effective* parameter α (see eq. 11) is now $\alpha = x'' N_e \phi(1-\phi) = \phi$. Further we will assume that ϕ is not small in order to avoid complications connected with a difference between N_e and $N_e(\phi)$. Then the relaxation times of two slow modes – Rouse and reptation – are defined by eqs. 16, 17 with $N = N_A$, $R = R_A = N_A^{1/2} b$ and $\alpha \sim 1$:

$$\tau_{R-A} \sim \begin{cases} \lambda^4 & , \quad a < \lambda < R_A \\ N_A^2 & , \quad \lambda > R_A \end{cases} \tag{19}$$

$$\tau_{slow} \sim \begin{cases} N_A^3 N_e/\lambda^4 & , \quad a < \lambda < \xi_2^* \sim \sqrt{a R_A} \\ N_A^2 & , \quad \xi_2^* < \lambda < R_A \\ N_A \lambda^2 & , \quad \lambda > R_A \end{cases} \tag{20}$$

[3] In the symmetric case the A component is moving with respect to the common matrix with velocity $v_A = -(1-\phi)v$, and B component - with $v_B = \phi v$; the relative velocity is thus v. In the case of passive A component the relative velocity is equal to $v_B = \phi v$, that is the relaxation rate is slower by $1/\phi$.

The amplitudes of both modes are comparable, $I_{R-A} \sim I_{slow} \sim \frac{1}{\alpha'} = \frac{1}{xN_e}$, in the region $\lambda < R_A$; for longer waves the reptation slow mode dominates: $I_{R-A} \sim \frac{1}{xN_e}\frac{R_A^2}{\lambda^2}$, $I_{slow} \sim \frac{1}{xN_e}$.

It is important to remember that the slow relaxation processes are driven by the elastic energy stored in the entanglement network. Therefore the slow relaxation time can not become longer than the disentanglement time of A-chains, $\tau_{dis-A} \sim N_A^3/N_e$. Hence the slow relaxation time is actually governed by a diffusion of A-chains ($\tau_{slow} \sim N_A\lambda^2$) only in a limited region $R_A < \lambda < \lambda_1$, where λ_1 is defined by the condition $\tau_{slow} \sim \tau_{dis-A}$:

$$\lambda_1 \sim \frac{N_A}{\sqrt{N_e}}$$

Note that λ_1 is of order of the real tube length. In the region $\lambda > \lambda_1$, the slow relaxation is mainly due to the disentanglement process: $\tau_{slow} \sim N_A^3/N_e$. Since τ_{fast}, eq. 18, is proportional to λ^2, it will formally become larger than τ_{slow} at $\lambda \gtrsim \lambda_2 \sim \frac{N_A^{3/2}}{N_B^{1/2}} x^{1/2}(1-\phi)^{1/2}$.

In the regime $\lambda > \lambda_2$ we should exchange the definitions of the fast and slow modes. So the slow relaxation time in this regime is given by eq. 18, and the fast relaxation time is $\tau_{fast} \sim \tau_{dis-A}$. The amplitude of the fast mode is small and is further decreasing with increasing λ: $I_{fast} \sim \frac{1}{xN_e}\left(\frac{\lambda_2}{\lambda}\right)^4$.

The dynamic scales λ_1 and λ_2 were predicted for the case of a polymer solution in ref. [6]. The dynamic scale λ_2 and the two reptation modes (fast and slow) in a theta-solution were also considered before in ref. [17].

4. Conclusions

Dynamics of concentration fluctuations in entangled polymer blends is a complicated process which is characterized by a number of relaxation modes.

For a symmetric blend two sets of modes originated from the scales larger than the tube diameter, a, are predicted. These are the Rouse modes and the reptation modes. The characteristic time of the Rouse relaxation scales as $\min(\lambda^4 b^4, N^2)$. The main qualitative predictions for the case $xN_e \sim 1$, $\phi \sim 1/2$ are as follows: The amplitude of the Rouse relaxation dominates in the high q region, $1/q = \lambda < R = N^{1/2}b$. The reptation mode is due to an inter-diffusion of two components. The mutual diffusion constant is $D_m \sim 1/N$ in the long-wave region, $\lambda > R$.

The reptation relaxation time $\tau_{rept} \sim N^2$ is independent of the wave-vector q in the intermediate regime $\sqrt{Ra} < \lambda < R$; we call the corresponding process a pseudo-Rouse relaxation as it is due a biased reptation of polymer chains. In the region $a < \lambda < \sqrt{Ra}$ an anomalous wave-vector dependence of the reptation time is predicted: $\tau_{rept} \propto N^3/\lambda^4$.

The dynamics in asymmetric blends ($N_B \ll N_A$) turns out to be even more complicated. In general case 4 modes (sets of modes) are predicted: two Rouse modes - B and A, and two reptation modes - fast and slow. Roughly speaking the Rouse relaxation modes are important if the wave-length is shorter than the corresponding coil size (R_B or R_A). In the region of long wave-lengths, $\lambda > \max(R_B, R_A)$, we found that the relaxation dynamics is *not* always a simple diffusion. A single mode diffusion relaxation is revealed only in the regime of extremely long wave-lengths, $\lambda > \lambda_2 \propto N_A^{3/2}/N_B^{1/2}$, where the mutual diffusion constant is governed by the fast component: $D_m \sim 1/N_B$. We thus support the fast diffusion concept in this regime.

Another interesting result concerns an intermediate regime $R_A < \lambda < \lambda_1 \propto N_A$. Two diffusion relaxation processes – fast and slow – are predicted in this regime. The ratio of the amplitudes (intensities) of the processes (fast/slow) is proportional to $\alpha \sim xN_e$. The relaxation rates are determined by two mutual diffusion constant, $D_{m-fast} \sim 1/N_B$ in agreement with fast diffusion theories [3, 4], and $D_{m-slow} \sim 1/N_A$ in agreement with slow diffusion theories [2]. The two theoretical concepts are thus bridged together in this regime.

Challenges

An experimental verification of the main qualitative trends discussed above would be a challenge as it would provide a test of both the particular theory of polymer interdiffusion and the general reptation model.

Acknowledgements

I am grateful to G.Marrucci and M.Rubinstein for useful discussions. This work was supported in part by a NATO ASI conference grant and also by NATO's Scientific Affairs Division in the framework of the Science for Stability Programme.

References

1. Klein, J. (1990) The interdiffusion of polymers, Science 250, 640-646.
   Composto, R.J., Kramer, E.J. and White, D.M. (1987) Fast macromolecules control mutual diffusion in polymer blends, Nature 328, 234-236.
2. Brochard, F. , Jouffroy, J., and Levinson, P. (1983) Polymer-polymer diffusion in melts, Macromolecules 16, 1638.
   Brochard, F. and De Gennes, P.G. (1986) Polymer-polymer interdiffusion, Europhys. Lett. 1, 221.

3. Kramer, E.J., Green, P.F. and Palmström, C.J. (1984) Interdiffusion and marker movements in concentrated polymer-polymer diffusion couples, Polymer 25, 473.
Green, P.F., Palmström, C.J., Mayer, J.W. and Kramer, E.J. (1985) Marker displacement measurements of polymer-polymer interdiffusion, Macromolecules 18, 501.
Composto, R.J., Mayer, J.W., Kramer, E.J. and White, D.M. (1986) Fast mutual diffusion in polymer blends, Phys. Rev. Lett. 59, 1312.
Sillescu, H. (1984) Relation of interdiffusion and self-diffusion in polymer mixtures, Makromol. Chem. Rapid Commun. 5, 519-523.
4. Brochard-Wyart, F. (1987) Kinetics of polymer-polymer interdiffusion, in M. Nagasawa (ed.), Studies in Polymer Science, v.2, Elsevier Science Publishers, Amsterdam, pp. 249-256.
5. Semenov, A.N. (1986) Relaxation of long-wave density fluctuations in a concentrated polymer solution, Sov.Phys.JETP 63, 717-720.
6. Semenov, A.N. (1990) Dynamical correlation function of polymer density fluctuations in concentrated solutions, Physica A 166, 263-287.
7. Semenov, A.N. (1991) Concentration fluctuation effects for dynamics of polymer solutions, Makromol. Chem. Macromol. Symp. 44, 285-294.
8. Flory, P.J. (1953) Principles of Polymer Chemistry, Cornell Univ. Press, Ithaca.
9. De Gennes, P.G. (1979) Scaling Concepts in Polymer Physics, Cornell Univ. Press, Ithaca.
10. Edwards, S.F. (1966) Proc. Phys. Soc. London 88, 265.
11. Rubinstein, M. Lecture I in this volume.
12. Pincus, P. (1981) J. Chem. Phys. 75, 1996.
13. Binder, K. (1983) J. Chem. Phys. 79, 6327.
14. De Gennes, P.G. (1981) Coherent scattering by one reptating chain, J. Phys. Fr. 42, 735-740.
15. Anastasiadis, S.H. and Semenov, A.N. (1997) Macromolecules.
16. Graessley, W.W. (1974) Adv. Polym. Sci. 16, 1.
17. Brochard, F. (1983) J. Phys. Fr. 44, 39.

II. THE EFFECT OF POLYMER DENSITY FLUCTUATIONS ON DYNAMICS OF HIGH-MOLECULAR-WEIGHT POLYMERS

A.N.SEMENOV

Department of Applied Mathematics, University of Leeds
Leeds LS2 9JT, U.K.

Abstract. The dynamics of entangled polymer chains in concentrated solutions and melts are discussed in terms of coupling between reptation motions and excluded volume interactions. It is shown that this coupling gives rise to an exponential increase of the disentanglement time and viscosity in the region of very high molecular weights. It is also shown that this effect should be even stronger in thin melted polymer layers. In particular, an exponential dependence of viscosity vs. the layer thickness is predicted.

1. Introduction

The reptation theory [1] has been definitely successful in accounting for many aspects of polymer dynamics (see lectures [2]). However there is a long standing controversy between the reptation predictions for the (longest conformation) relaxation time $\tau = \tau_{dis}$ and zero-shear viscosity η and the experimental data: the theory predicts that $\eta \propto \tau_{dis} \propto N^3$ in the entanglement regime, $N > N_e$, whereas the observed dependence is $\eta \propto N^z$, where $z \simeq 3.4 \pm 0.2$ (here N is the number of links per chain). [3]

In recent years a number of theoretical models has been developed in order to explain the observed molecular weight dependence of the viscosity [4, 5, 6]. The main idea of the theories [4, 5] which are consistently based on the reptation model was to account for the internal modes of polymer dynamics "inside the tube" – the so called tube-length fluctuation effect. It was recognized that this effect does give rise to a stronger increase of η with N, but only in a limited molecular-weight range. In fact the tube-length

T McLeish (ed), Theoretical Challenges in the Dynamics of Complex Fluids, 77-86

fluctuation effects provides a correction to the viscosity [4]:

$$\eta \propto N^3 \left\{ 1 - \mathcal{O}\left[\left(\frac{N_e}{N} \right)^{1/2} \right] \right\}^3$$

which implies a higher than 3 *apparent* power, $\frac{\partial \ln \eta}{\partial \ln N} > 3$. However in the asymptotic limit $N \to \infty$ the predicted behavior is still N^3.

One may put a formal question: what is the $\lim_{N\to\infty} \frac{\eta(N)}{N^3}$ – is it finite or infinite? Experimental data do not provide an unambiguous answer, although the observed trend $\frac{\eta(N)}{N^3} \propto N^{0.4}$ supports the infinite limit. If we assume that this answer is true (that is that the asymptotic dependence, $\eta(N)$, is stronger than N^3), then the next question might be: which effect is responsible for this behavior. It is natural to expect that it is *interactions* that are important here. One can distinguish two main types of interactions in polymer systems: (1) "topological" (which ensure that chain fragments do not intersect each other); (2) excluded volume. In order to understand whether the excluded volume interactions are important or not, let us switch off these interactions, that is consider an ideal but entangled polymer system (no excluded volume but chain fragments can not penetrate through each other). The reptation theory accounts for topological interactions in this system via the tube concept, which is an approximation. Is it possible that a complete set of topological constraints might give rise to a stronger $\eta(N)$ dependence than N^3? The rigorous answer to this question is no [7]: with no excluded volume interactions $\eta(N) \leq \mathcal{O}(N^3)$. The idea of the proof is the following: first we attribute anisotropic diffusivities ($D_{0\perp}$, $D_{0\parallel}$) to each polymer segment setting the transversal diffusion constant to be lower that the longitudinal one, $D_{0\perp} < D_{0\parallel}$. In the limit $D_{0\perp} \to 0$ the only possible motion of a polymer chain is a reptation along its own contour. Therefore the classical prediction of the reptation theory must be strictly valid in this case: $\eta(N) \propto \tau(N) \propto N^3$. On the other hand it is possible to show using the dynamic variational principle [8] that an increase of $D_{0\perp}$ back to a finite value must result in a decrease of the longest conformation relaxation time, and thus a decrease of viscosity.

Therefore if the limit $\eta(N)/N^3$ is infinite, this *must* be due to excluded volume interactions.

2. The effect of a frozen molecular field

Let us consider a concentrated system (solution or melt) of long flexible polymer chains in an inhomogeneous state $\phi(\mathbf{r}) = \phi_0 + \delta\phi(\mathbf{r})$, where $\phi_0 \sim 1/2$ is the mean monomer concentration. The excess free energy density of the excluded-volume interactions can be written as (in $k_B T$ units):

$F_{int} = \frac{x}{2}(\delta\phi)^2$, where x is the excluded-volume parameter (here and below we assume that the monomer volume v is set to 1). The corresponding molecular field which affects the motion of each polymer chain is as in the preceding chapter

$$U_{mol} = \frac{\partial F_{int}}{\partial \phi} = x\delta\phi$$

It is known that in concentrated polymer systems (in particular in polymer melts) fluctuations of both polymer density and molecular field are small and are not important for equilibrium properties [9] (in particular the chain statistics remains nearly Gaussian). Below we show that nevertheless the molecular field fluctuations are very important for dynamics.

Let us first consider the effect of a *frozen* molecular field. The density fluctuations near two points are virtually independent if the distance between the points is larger than the static correlation length ξ (see [10]). Therefore a snap-shot of the molecular field can be considered as a spatial white noise on scales larger than ξ. Let us average the molecular potential U_{mol} over a large volume V:

$$\bar{U}_{mol} = \frac{1}{V}\int_V U_{mol}(\mathbf{r})d^3r$$

Obviously the mean-square fluctuation of the average value must be proportional $1/V$:

$$\left\langle \bar{U}^2_{mol} \right\rangle = \frac{1}{V^2}\int_V d^3r \int_V d^3r' \left\langle U_{mol}(r)U_{mol}(r')\right\rangle = \frac{const}{V}$$

Knowing the static correlation function of the system [9, 11, 10] it is possible to show that const in the last equation is equal to the interaction parameter x.

Let us now set V to the volume of a polymer coil, $V \sim R^3 \simeq N^{3/2}b^3$, where b is the monomer size, and R is the coil size. The mean potential energy of a chain, which tries different conformations inside $V \sim R^3$, is $W \simeq N\bar{U}_{mol}$, so that the mean-square fluctuation is $\langle(\delta W)^2\rangle \simeq N^2\langle(\delta\bar{U}_{mol})^2\rangle = N^2\frac{x}{V} \sim \frac{x}{b^3}N^{1/2}$. Let us divide the space into cells of size R. The typical fluctuation of the mean potential energy from one cell to another is large:

$$\delta W \sim x^{1/2}N^{1/4} \tag{1}$$

(here we assume that $b \sim 1$ in units of $v^{1/3}$). Therefore a polymer chain probes deep potential wells and high potential barriers as it moves from cell to cell. In particular the chain will be trapped for a long time in a cell with lower potential surrounded by the cells with higher mean potential.

The time for a chain to out of a typical well (which also determines the conformation relaxation time) is thus exponentially large:

$$\tau \sim \tau_{dis}^{(0)} \exp(\delta W) \propto \exp\left(const\ x^{1/2} N^{1/4}\right)$$

where $\tau_{dis}^{(0)}$ is the relaxation time of an ideal system (with no excluded volume interactions), $\tau_{dis}^{(0)} \sim N^3/N_e$. Therefore the (frozen) molecular field exponentially slows down the dynamics if $Nx^2 \gg 1$.

In the opposite limit $Nx^2 < 1$ the effect of the field is small:

$$\tau \simeq \tau_{dis}^{(0)} [1+\Delta], \quad \Delta = const \left\langle \delta W^2 \right\rangle = const\ x N^{1/2} \qquad (2)$$

3. Annealed field: perturbations

Let us now turn to a real situation: the molecular field $U_{mol} = x\delta\phi$ is time-dependent. Hence a potential well (discussed in the previous section) confining a chain can disappear after some relaxation time, so that the effect of the molecular field must be weaker in the annealed case. Let us consider the effect of the annealed molecular field first assuming that the field is weak and using a perturbation approach.

Consider a polymer chain and the corresponding tube at some moment, $t = 0$. During a time $t < \tau_{dis}$ a tail part of the chain creeps out of the original tube and probes different conformations. It is this tail part that is influenced by the molecular field inhomogeneities: the rest of the chain is still confined inside the original tube, its environment did not change. Since reptation is a curvilinear random diffusion process, the displacement along the tube is proportional to $\sqrt{t}$. Therefore the number of links in the tail part is $g(t) \sim N\sqrt{\frac{t}{\tau_{dis}}}$ (according to the definition of the disentanglement time g must be $\sim N$ for $t \sim \tau_{dis}$). The corresponding spatial scale (the size of the tail) is $\lambda \sim b\sqrt{g(t)} \sim R\left(\frac{t}{\tau_{dis}}\right)^{1/4}$. Alternatively we define the time of reptation on distance λ, $t(\lambda) \sim \tau_{dis}\left(\frac{\lambda}{R}\right)^4$. The molecular field does affect the tail dynamics on the time-scale t only if the relaxation time $\tau(\lambda)$ of the molecular field fluctuations with spatial scale $\sim \lambda$ is longer than t.

As U_{mol} is proportional to $\delta\phi$, its relaxation time coincide with that of the corresponding concentration fluctuation. The dynamics of concentration fluctuations in a polymer solution is nearly the same as for asymmetric mixtures, component A being the polymer, $N_A = N$, and B being the solvent, $N_B = 1$. As discussed in the previous chapter, three relaxation modes are predicted for asymmetric mixtures for spatial scales λ larger than both the tube diameter $a \sim bN_e^{1/2}$ and the size of shorter molecules, R_B: fast,

Rouse-A, and slow (the Rouse-B mode does not exist as $N_B = 1$). It is easy to check that the fast relaxation time is always much shorter than the time of reptation, $t(\lambda)$. The Rouse-A mode is also not important as its amplitude is of the same order as the slow-mode amplitude, but it relaxes faster. Therefore we should focus on the slow relaxation stage.[1]

The relaxation time $\tau(\lambda)$ of the slow mode is decreasing with λ in the range $a < \lambda < R = R_A$:

$$\tau(\lambda) \sim \begin{cases} N^3 N_e/\lambda^4 & , \quad a < \lambda < \xi_2^* \simeq \sqrt{aR} \\ N^2 & , \quad \xi_2^* < \lambda < R \end{cases}$$

On the other hand the reptation time, $t(\lambda)$, is increasing with λ. The two times become equal for $\lambda \sim \xi_2^* \simeq \sqrt{aR}$; in this region both times are of order of N^2. Therefore the dynamic scale

$$\lambda^* = \xi_2^* \tag{3}$$

defines the critical size of the tail part, the critical number of links being $g^* \sim (\xi_2^*/b)^2 \sim \sqrt{NN_e}$. The molecular field fluctuations on scales larger than λ^* are not important as the fluctuations are relaxing faster than the chain reptates. On the other hand a tail part is reptating in a nearly fixed environment (virtually frozen molecular field) on scales $\lambda < \lambda^*$, $g < g^*$. The effective curvilinear diffusion constant is thus slowed down, and the longest reptation (disentanglement) time is increased. The relative correction Δ to these quantities is due to the effect of the field on scales $\lambda \lesssim \lambda^*$. Therefore Δ must be defined by eq. 2 with g^* instead of N. In addition we have to take into account that the effective *amplitude* of the molecular field fluctuations for the slow relaxation stage is reduced because of the preceding fast relaxation. This reduction can be taken into account by the renormalization $x \to x' \sim \frac{1}{N_e}$ discussed in the previous chapter. Thus

$$\Delta \sim x' (g^*)^{1/2} \sim \left(\frac{N}{N_e^3}\right)^{1/4}$$

and the corrected disentanglement time is

$$\tau_{dis} \sim \frac{N^3}{N_e}\left[1 + const\ \left(\frac{N}{N_e^3}\right)^{1/4}\right] \tag{4}$$

The result, eq. 4, has been obtained before [7] using a quantitative rigorous perturbation approach (rather than scaling arguments). The const in this equation turns out to be dependent on the monomer concentration ϕ_0 and

[1] The properties of this process does not depend on N_B, see [10]. Therefore it does not matter if $N_B = 1$ or $N_B = N_e$.

the polymer rigidity parameter v/b^3: $const = 1.15\frac{v}{b^3}\frac{1}{\phi_0}$ (in general case $N_e = N_e(\phi_0)$ in eq. 4).

Thus the relative fluctuation correction Δ is increasing with N. It becomes large for $N > N_e^3$; here the perturbation approach fails and a qualitatively different (stronger than N^3) behavior of τ_{dis} vs. N is expected. This regime is considered below.

4. Relaxation time for $N > N_e^3$

Here we expect that τ_{dis} is exponentially large (this assumption is verified at the end). Let us consider the process of reptation of a chain tail out of the original tube into the new environment. The tail size $\lambda \sim bg^{1/2}$ is assumed to be much smaller than R, so that the corresponding time scale t is smaller than τ_{dis} and therefore the network structure is nearly fixed during the process.

Let us calculate the effective barrier arising from interactions between the tail part (consisting of $g \sim \frac{\lambda^2}{b^2}$ links) and other chains. The main idea is that tail monomers have to be inserted into a region $V \sim \lambda^3$ and thus they must swell the entanglement network in this region. The corresponding additional elastic energy is $\delta F_{el} \sim \frac{1}{2}\int_V G\epsilon^2 d^3r$, where $G \sim \frac{1}{N_e}$ is the relevant elastic modulus, and ϵ is the relative deformation. We can substitute ϵ by an average value $\bar{\epsilon}$ in order to get the lowest boundary estimate of δF_{el}. In turn $\bar{\epsilon}$ can be estimated as $\bar{\epsilon} \sim \frac{gv}{V} \sim \frac{g}{g^{3/2}}$, where gv is the intrinsic volume of the tail (we assume that $v \sim b^3$). Thus we get $\delta F_{el}(\lambda) \sim GV\bar{\epsilon}^2 \sim \frac{g^{1/2}}{N_e} \sim \frac{\lambda}{N_e b}$ so that the potential barrier δF_{el} is increasing with λ. The barrier can not be relaxed until an end of another chain occurs in the region V as the structure of the entanglement network in this region is fixed otherwise. On the other hand if the region V happens to include another chain end, the corresponding chain can move (reptate) out of the deformed region thus relaxing the elastic energy there. Therefore the critical size λ^* is of order of the mean distance between the chain ends, $\lambda^* \sim bN^{1/3}$ (so that $g^* \sim N^{2/3}$), and the corresponding critical barrier is

$$W^* = \delta F_{el}(\lambda^*) \sim \frac{N^{1/3}}{N_e}$$

The relaxation time (actually both the time of reptation on distance λ^* and the global disentanglement time) must be increased by the Boltzmann factor $\exp(W^*)$ due to the barrier:

$$\tau_{dis} \sim \tau_{dis}^{(0)} \exp\left(const\,\frac{N^{1/3}}{N_e}\right)$$

Thus the following molecular weight dependence of both the viscosity and the terminal relaxation time is predicted [7]

$$\eta \propto \tau_{dis} \propto N^3 \begin{cases} 1 + const \left(\frac{N}{N_e^3}\right)^{1/4} & , \quad N_e < N < N_e^3 \\ \exp\left(const \left(\frac{N}{N_e^3}\right)^{1/3}\right) & , \quad N > N_e^3 \end{cases} \tag{5}$$

Hence the classical reptation behavior ($\eta \propto \tau \propto N^3$) is valid in a limited region $N_e < N < N^*$, where $N^* \sim N_e^3$. In the region $N > N^*$ strong deviations from N^3 law are expected. However it is dangerous to consider these results as an explanation of the apparent $N^{3.4}$ viscosity dependence as the critical chain length N^* is typically very large – it apparently corresponds to the largest molecular weights accessible in experiments.

An exponential increase of the relaxation time in the high molecular weight regime was also predicted in ref. [12] using the same idea as discussed above which was however implemented in a different way. In particular the barrier W was argued to be proportional to the number of links g in the tail part, rather than to $\lambda \propto g^{1/2}$. In other words it was suggested [12] that reptation of a chain tail from its original tube to a new environment implies a finite extra potential energy, Δu, *per link*. A critical barrier $W^* \propto g^* \propto N^{2/3}$ was predicted as a result [12]:

$$\tau_{dis} \propto \exp\left(const\ N^{2/3}/N_e^{11/6}\right) \tag{6}$$

Eq. 6 disagrees with the output of the perturbation theory, eq. 4, which has been rigorously derived [7]. Therefore at least the N_e power (11/6) in eq. 6 does not seem to be correct.

I also comment that the assumption of a finite potential barrier per link, Δu, is dangerous due to the following reason. Let us adopt the picture of ref. [12]. Let u_0 be the effective potential energy per link in an equilibrium tube. The ergodicity principle ensures that u_0 must be also equal to the time-averaged potential of a link. A reptation of a chain can be considered as a sequence of "critical" steps: during each step a critical tail part g^* moves into an unfavorable environment with $u = u_0 + \Delta u$ and create a new tube part. A finite fraction of the links inside the new tube part must experience a higher than equilibrium potential $u = u_0 + \Delta u$. The difference $u - u_0$ can be relaxed only if a finite fraction of the neighboring chains, "creating" the tube, would relax. The corresponding relaxation time is of order of τ_{dis}, the global disentanglement time, which is much longer than the time of one step (as $g^* \ll N$). Therefore the potential per link in the new parts of the tube created at each step must be systematically higher than u_0. Hence the time-average potential of a link must be larger than u_0 by a finite fraction of Δu, in disagreement with the ergodicity principle.

5. Generalizations

5.1. D-DIMENSIONS

Eq. 5 can be easily generalized for a d-dimensional space. The result is

$$\tau_{dis} \sim \tau_{dis}^{(0)} \begin{cases} 1 + const \, \frac{N^{1-d/4}}{N_e^{d/4}} & , \quad N_e < N < N^* \\ \exp\left(const \, \frac{N^{4/d-1}}{N_e}\right) & , \quad N > N^* \end{cases} \tag{7}$$

where $N^* \sim N_e^{\frac{d}{4-d}}$. For $d = 3$ these equations reduce to eq. 5. Note that $N^* \to \infty$ for $d = 4$, which means that the molecular field fluctuations are not important in this case. On the other hand $N^* \sim N_e$ for $d = 2$: here

$$\tau_{dis} \propto \exp\left(const \, \frac{N}{N_e}\right) \tag{8}$$

in the whole entanglement regime, $N > N_e$. Thus the fluctuations are extremely important for two-dimensional systems.

5.2. POLYMER LAYERS

Let us consider polymer dynamics in a melted layer. This system provides a bridge between 2-dimensional and 3-dimensional systems depending on the layer thickness H. Let us assume that polymer chains are shorter than $N_{3d}^* \sim N_e^3$. In this case the critical 3d (bulk) scale is (see eq. 3) $\lambda^*(N) \sim b(NN_e)^{1/4}$. The upper boundary for the critical scale in this region is $H^* \equiv \lambda^*(N_{3d}^*) \sim bN_e$. Let us assume that $H < H^*$. In this case three basic entangled regimes can be identified:

(1) If the chains are short so that $\lambda^*(N) < H$, then the situation is nearly 3-dimensional: the boundaries of the layer do not affect much the reptation process. The disentanglement relaxation time in this regime is given by eq. 4: the relative correction $\Delta \sim \left(\frac{N}{N_e^3}\right)^{1/4}$.

(2) For longer chains ($N > N_1$) the layer confinement is important and thus the molecular field fluctuations affect the reptation dynamics in a 2-dimensional way: in this regime $\Delta \propto N^{1/2}$ (see the upper line in eq. 7).

(3) Finally for even longer chains ($N > N_2$, where N_2 is defined by the condition $\Delta(N_2) \sim 1$) the relaxation time starts to increase exponentially (see eq. 8), $\tau \propto \exp(CN)$, where C depends on H and N_e.

Imposing a smooth crossover between these three regimes we get the following "scaling" results:

$$\tau_{dis}/\tau_{dis}^{(0)} \sim \begin{cases} 1 + const_1 \left(\frac{N}{N_e}\right)^{1/4} & , \quad N_e < N < N_1 \\ 1 + const_2 \frac{b}{H} \left(\frac{N}{N_e}\right)^{1/2} & , \quad N_1 < N < N_2 \\ \exp\left[const_3 \left(\frac{b}{H}\right)^2 \frac{N}{N_e}\right] & , \quad N > N_2 \end{cases} \tag{9}$$

where the crossover chain lengths are

$$N_1 \sim \frac{1}{N_e}\left(\frac{H}{b}\right)^4, \quad N_2 \sim N_e \left(\frac{H}{b}\right)^2$$

Thus we predict that the relaxation time and the viscosity of a polymer layer are exponentially increasing with molecular weight if $N > N_2$; the dependence of η and τ on the layer thickness H is also exponentially strong in this regime. It is important to note that N_2 is not much larger than N_e if the layer is thin enough ($H \gtrsim b$). Therefore the region $N > N_2$ seems to be experimentally accessible.

6. Conclusions

(1) Fluctuations of the molecular field are extremely important for dynamics of very high molecular-weight polymers.

(2) The effect of fluctuations in bulk concentrated solutions and melts on the terminal stress-relaxation time is governed by the parameter $\Delta \sim \left(\frac{N}{N_e^3}\right)^{1/4}$.

(3) The classical reptation model is valid in a limited molecular weight region

$$N_e < N < N_e^3$$

(4) The disentanglement time and zero-shear viscosity exponentially increase with N in the region $N > N_e^3$.

(5) The effect of molecular field is even stronger in thin polymer layers: here an exponential increase onsets at $N > N_2 \sim \left(\frac{H}{b}\right)^2 N_e$.

Challenges

An experimental verification of the strong slowing down effect of excluded volume interactions would be certainly a challenge. Computer simulations should provide another possibility to test the results mentioned above. In particular it seems an attractive idea to simulate the reptation dynamics in thin layers where the predicted effect of the interactions is exponentially strong.

Finally a deeper insight into the mechanism of coupling between reptation and molecular field (preferably on a more quantitative basis) would certainly be very desirable.

Acknowledgements

I acknowledge useful discussions with M. Rubinstein. This work was supported in part by a NATO ASI conference grant and also by NATO's Scientific Affairs Division in the framework of the Science for Stability Programme.

References

1. Doi, M.D. and Edwards, S.F. (1986) The Theory of Polymer Dynamics, Oxford Univ. Press, Oxford, U.K.
2. Rubinstein, M. Lectures in this volume.
3. Graessley, W.W. (1974) Adv. Polym. Sci. 16, 1.
4. Doi, M. (1983) J. Polym. Sci. B 21, 667.
5. Des Cloizeaux, J. (1894) J. Phys. Fr. 45, L17.
   Lin, Y.H. (1984) Macromolecules 17, 2846.
   Needs, R.J. (1984) Macromolecules 17, 437.
   Rubinstein, M. (1987) Phys. Rev. Lett. 59, 1946.
6. Wendel, H. and Noolandi, J. (1982) Macromolecules 15, 1318.
   Rendell, R.W., Ngai, K.L. and McKenna, G.B. (1987) Macromolecules 20, 2250.
   Scher, H. and Schlesinger, M.F. (1986) J. Chem. Phys. 84, 5922.
   Weiss, G.H., Bendler, J.T. and Schlesinger, M.F. (1988) Macromolecules 21, 521.
   Fixman, M. (1988) J. Chem. Phys., 89, 3892.
7. Semenov, A.N. (1989) Fluctuation effects in the dynamics of concentrated polymer solutions, Sov.Phys.JETP 69, 107-118.
   Semenov, A.N. (1991) Concentration fluctuations and reptation dynamics in polymer solutions and melts, Physica A 171, 517-553.
8. Lomakin, A.V. and Semenov, A.N. (1988) Dynamics of Gaussian chain with topological constraints. Sov.Phys.JETP 67, 1812-1819.
9. Edwards, S.F. (1966) Proc. Phys. Soc. London 88, 265.
10. Semenov, A.N. Lecture I in this volume.
11. Semenov, A.N. (1990) Dynamical correlation function of polymer density fluctuations in concentrated solutions, Physica A 166, 263-287.
12. Deutsch, J.M. (1985) Phys. Rev. Lett. 54, 56.
    Deutsch, J.M. (1987) J. Phys. Fr. 48, 141.

RHEOLOGY OF ENTANGLED POLYMERS: TOPOLOGICAL AND NON-LINEAR EFFECTS

Tom McLEISH
IRC in Polymer Science and Technology
Department of Physics
University of Leeds
Leeds, LS2 9JT, UK

1. Introduction

The contribution by M. Rubinstein (this volume) has introduced the fruitful and compelling approach to the dynamics of entangled polymers provided by the tube model. The theoretical framework is economic in that the number of free parameters required to make predictions is very limited: as well as the Kuhn step length b, one more static parameter is needed in the tube diameter, a (or equivalently the plateau modulus G_0) and one dynamic parameter - the monomeric friction coefficient ζ (or equivalently the Rouse time of an entanglement length). Once these parameters are determined for a polymer of fixed chemistry, quantitative predictions for the linear rheology, as determined by the stress relaxation modulus $G(t)$ or its Fourier transform $G^*(\omega)$ are in principle calculable. Because the model is a molecular one, albeit coarse-grained on the level of Gaussian sub-chains, it also provides predictions for other, more direct probes of the molecular dynamics such as the dynamic structure factor $S(\mathbf{k},t)$ (see below and articles of J. Candau and A.N. Semenov, this volume). This very important advantage of molecular theories has yet to be fully exploited in the context of polymers, mainly because the associated timescales are so long. We will see below how single-chain structure-factors may be calculated within the theory alongside the rheological response.

A second appealing feature of tube model theories is that they provide a natural hierarchy of effects which one can incorporate or ignore at will in a calculation, depending on the accuracy desired. We have already seen how, in the case of linear polymers, bare reptation in a fixed tube provides a first-order calculation; more accurate levels of the theory may incorporate the co-operative effects of constraint release and further refinements such as path-length fluctuation via the Rouse modes of the chains.

Thirdly the implicit claim is that entangled polymer dynamics are completely dominated at long times by the *topological* interactions of the chains. The conclusion is that rheological behaviour should show a high degree of universality. For example, two monodisperse polymer melts of different chemistries but with the same number of entanglements per chain (M/M_e equal for both) should exhibit stress relaxation functions $G(t)$ which may be superimposed by simple scaling in modulus and time. This is a stronger requirement than simply demanding that the molecular weight scaling of the viscosity is

T. McLeish (ed.), Theoretical Challenges in the Dynamics of Complex Fluids, 87-113.

universal for all linear polymers. It is indeed true, and an experimental foundation-stone of the entanglement concept, that all monodisperse polymer melts exhibit a viscosity η which scales as M for low molecular weight and $M^{3.4}$ for high molecular weight. However, the viscosity it just the integral of the stress relaxation function:

$$\eta = \int_0^\infty G(t)dt \tag{1}$$

which contains much more information than η alone, and ought also to be universal. Figure 1 shows published data on three polymers which may be manufactured with very narrow polydispersity by anionic polymerisation: polystyrene (PS), polybutadiene (PB) and polyisoprene (PI) [1]. The three have similar degrees of entanglement. The dissipative modulus as a function of frequency $G''(\omega)$ is the function plotted, because is tends to show more structure than either $G(t)$ or $G'(\omega)$.

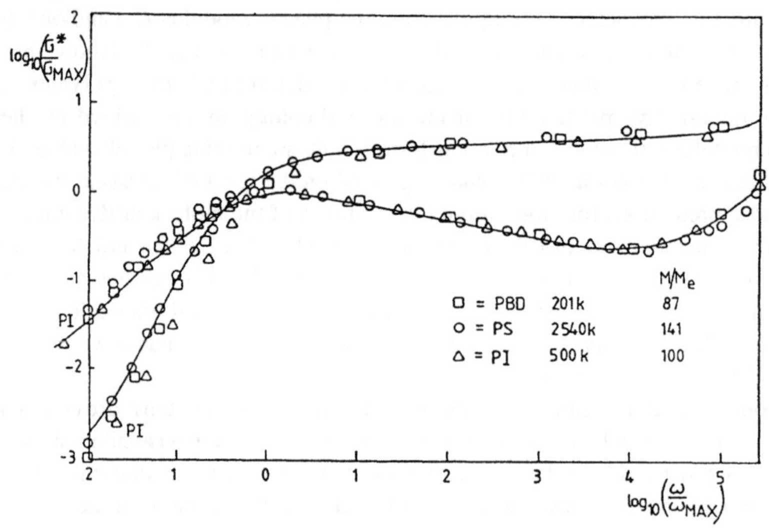

Figure 1. $G''(\omega)$ for monodisperse linear polymers of PI, PB and PS. The curves have been shifted so that the plateau moduli and terminal times coincide.

Both the shape of the peak around the dominant relaxation time τ_{max}^{-1} and the frequency range before the minimum in the curve are very similar, once allowance has been made for differences in plateau modulus and monomeric friction. This is strong support for theories based on universal aspects of polymer structure. In particular, a purely *topological* theory of dynamics leads naturally to the conjecture that changes in the *molecular topology* itself will radically alter the motions of entangled molecules. The simplest change one can imagine is to introduce a single branch-point into the linear molecule, creating a "star polymer". Melts of monodisperse star polymers with controlled numbers of arms radiating from a central branch point have been available for 30 years [2], and do indeed have very striking rheological behaviour. We shall see how these and more complex molecular architectures may be treated within the tube model.

Fourthly it is possible to extend the model to make predictions of response in highly non-linear deformations and flows [3]. This is naturally of great interest in applications, since most of polymer-processing involves extremely large and rapid deformations, but is also proving of value as a strong experimental test of theoretical assumptions and of polymer structures such as branching. We will look briefly at some of the challenges these regimes of flow present.

2. Monodisperse Linear Polymers

The fundamental example of the tube model's application is the simplest case of linear chains of identical molecular weight. It will provide the starting point for more complex applications.

2.1 EXPRESSION FOR THE STRESS

The first ingredient in any theory for the rheology of a complex fluid is the expression for the stress in terms of the microscopic structure variables. In the case of flexible polymers the total stress arises to a good approximation from the entropy of the chain paths. At equilibrium the polymer paths are random walks - of maximal entropy. We derive an expression for the stress-tensor here from the *principle of virtual work* (for analogous derivation for the stress in a complex fluid containing interfaces see the chapter of M. Doi). A deformation ε_{ij} induces preferred orientation of the steps of the walks, which are therefore no longer random - the entropy has decreased and the free energy density f increased. So

$$\sigma_{ij} = \frac{\partial f}{\partial \varepsilon_{ij}} \tag{2}$$

which is valid for timescales longer than those contributing to the sampling of microstates in f but shorter than the slow degrees of freedom. In a polymer melt, the free energy density can be expressed in terms of subchains of (arbitrary) degree of polymerisation N and end-to-end distance R. If these subchains have a density c_N in the melt, we have from the properties of Gaussian chains [3]

$$f = 3kTc_N \frac{\langle R^2 \rangle}{2Nb^2} \tag{3}$$

where the angular brackets denote an average over configurations of subchains. The tube model provides a *specific choice* of the scale of subchain (R and N) which couples to the bulk imposed strain: assuming that the tubes themselves deform affinely with the bulk, then chains confined within them are constrained to deform at lengthscales larger than the tube diameter a but not at smaller scales. These are the "slow degrees of freedom" in this case. So choosing $Nb^2 = R^2 = a^2$ and carrying out the differentiation in equation (2) leads to the simple form:

$$\sigma_{ij} = \frac{3kT}{a} c_{N_e} \, a\langle u_i u_j \rangle \tag{4}$$

where $\boldsymbol{u}$ is the unit vector tangent to the tube axis containing a segment of chain length N_e. The cancelling terms in a have been left in to make explicit terms in tension and strand length (see section 3.1.1 below). As in rubber elasticity, the stress arises from the anisotropic orientation of the elastically-active segments, and is dominated by the entropy associated with this orientation. In contrast to a cross-linked rubber, it is the deformed tube (deforming affinely with the bulk by assumption) which confers its orientation on any chain which occupies it, rather than permanent cross-links.

2.2 STRESS RELAXATION

The curvilinear diffusion along the contour of the tube gives an immediate theory for stress relaxation, once the special physics of the free ends of the chains are accounted for. The usual assumption is that the chain ends can explore the melt in any direction (but see the contribution in this volume by A.N. Semenov for possible exceptions). They create new tube segments for the chain which follows them, but this new tube is chosen isotropically, so as time proceeds after a small step-strain there are two populations of occupied tube segments: a decaying fraction of anisotropically oriented segments and a growing fraction of isotropic segments (figure 2).

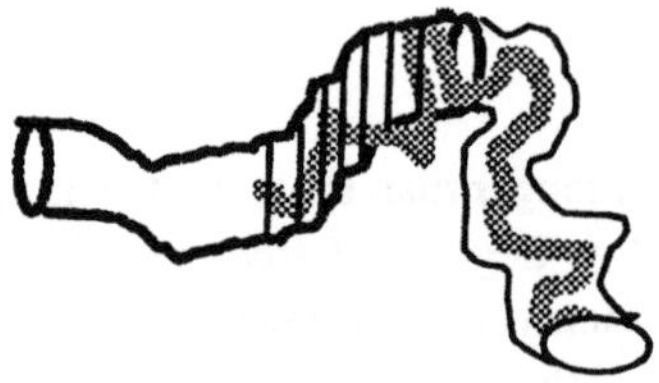

Figure 2. By curvilinear diffusion a chain evacuates its original tube and creates new tube segments from an isotropic distribution. Remaining portions of the original tube are shown shaded.

If we assume that the anisotropic part of the stress is now just proportional to the fraction of chain segments still constrained by the original (and deformed) tube segments, then we just need to calculate the fraction of original tube surviving at time t to calculate the stress relaxation function $G(t)$. The easiest way so do this is to recognise an equivalence between a picture in which the chain diffuses curvilinearly along the tube and one in which the *tube segments* diffuse curvilinearly along the *chains*. In this frame in which the chain is fixed, a tube segment from the population at $t=0$ ceases to exist when it reaches a chain end. So if $p(s,t)$ is the probability that a tube segment sited a curvilinear distance s from the end of a chain at time t is one of the original population, then p will satisfy a linear 1-dimensional diffusion equation - the physics is identical to that of diffusion in one dimension if we map the population of original tube segments onto diffusing particles. The disappearance of original tube segments at the ends of the chain correspond to absorbing boundary conditions, so finally,

$$\frac{\partial}{\partial t} p(s,t) = D_C \frac{\partial^2}{\partial s^2} p(s,t) \quad ; \qquad p(0,t) = p(L,t) = 0 \tag{5}$$

Here D_c is the curvilinear centre-of-mass diffusion constant of the chain, and is given in terms of the monomeric friction constant ζ by $D_c=kT/N\zeta$. L is the length of the primitive path, or tube length of the chain, and is just Nb^2/a. Finally we need the initial condition on $p(s,t)$, which is just that at $t=0$ the survival probability is unity everywhere on the domain $(0,L)$. A standard method of solving a partial differential equation set such as this is to expand in eigenfunctions of the spatial operator $\partial^2/\partial s^2$ and its boundary conditions. These are the sine modes $\phi_p(s,t)=\sqrt{(2/L)}\sin(p\pi s/L)\exp(-p^2\pi^2 D_c t/L^2)$,
in terms of which the full solution is given by a linear expansion

$$p(s,t) = \sum_{p=1}^{\infty} a_p \phi_p(s,t) \tag{6}$$

The coefficients a_p are in this case just the Fourier sine series for the "top-hat function" which is unity between 0 and 1 and zero elsewhere (the initial condition on p). A standard calculation of coefficients gives

$$a_p = \begin{cases} \sqrt{\dfrac{8L}{p^2\pi^2}} & p \text{ odd} \\ 0 & p \text{ even} \end{cases} \tag{7}$$

The relaxation modulus $G(t)$ is just the plateau modulus multiplied by the fraction of remaining tube, which in turn is just the integral of the survival probabilities $p(s,t)$ over the tube co-ordinate s:

$$G(t) = G_0 \int_0^L p(s,t)ds \tag{8}$$

Substitution of (5), (6) and (7) into (8) gives the well-known "Doi-Edwards" relaxation spectrum:

$$G(t) = G_0 \mu(t) \qquad ; \quad \mu(t) = \sum_{p\,\mathrm{odd}} \frac{8}{p^2 \pi^2} e^{-p^2 t / \tau_{rep}} \tag{9}$$

The sum over weighted relaxation times is heavily dominated by the longest time (the reptation time) $\tau_{rep} = L^2/\pi^2 D_c$. Because of this the frequency-dependent dissipative modulus, $G'(\omega)$ is expected to show a sharp maximum (see chapter by M. Rubinstein). The higher modes do modify the prediction from that of a single-mode "Maxwell" model, which is predicted for the case of "living" polymers (see chapter by M.E. Cates), but only to the extent of reducing the form of $G''(\omega)$ to the right of the maximum from $\sim\omega^{-1}$ to $\sim\omega^{-1/2}$. In fact, experiments on monodisperse linear polymers show a still broader maximum, with $G''(\omega) \sim \omega^{-1/4}$ to the right of the peak, as shown in figure 1. This is the starting point for the additional effects of path-length fluctuations and co-operative constraint release discussed by Rubinstein.

2.3 SINGLE-CHAIN STRUCTURE FACTOR

There is a vast and continually growing collection of data on the rheological response of entangled polymeric fluids. This is due both to the relative ease with which these data are obtained and to the sensitivity of the data to structural features such as polydispersity (see chapter by M. Rubinstein) and branching (see below). However, the molecular theory for polymer dynamics makes predictions of other quantities which can provide much more specific checks on the behaviour at the molecular level than the macroscopic rheological response. A common example is the scattering structure factor (see chapters by J. Candau and A.N. Semenov). In the case of polymers, the ability to replace hydrogen atoms selectively with deuterium allows wide application of the technique of neutron scattering [4]. If uncorrelated single chains are deuterated in a melt of chemically identical but hydrogenous chains, scattering can measure the *single chain structure factor* $S(\mathbf{k}, t)$:

$$S(\mathbf{k},t) = \frac{N}{L^2} \int_0^L ds \int_0^L ds' \left\langle \exp[i\mathbf{k}.(\mathbf{R}(s,t) - \mathbf{R}(s',0))] \right\rangle \tag{10}$$

Here $\mathbf{R}(s,t)$ is the chain trajectory as a function of arc co-ordinate and time, and $\mathbf{k}$ is the scattering vector of the experiment, related to the neutron wavelength and scattering angle [4]. When $t=0$ this expression gives the "static structure factor", measured typically by small-angle neutron scattering (SANS). At finite time the delay necessary between effectively interfering scattering events may be introduced by the technique of neutron spin-echo (NSE) [5]. Full calculation of $S(\mathbf{k},t)$ is given in reference [3] for the cases of both unentangled (Rouse) and entangled (reptation) dynamics. In particular for the latter a similar one-dimensional diffusion equation is solved as for the stress relaxation, but with differing boundary conditions. Here we will discuss the limiting forms of the expression only, in distinct regimes of time and scattering vector (or inverse length-scale).

2.3.1 *Unentangled motion* $t<\tau_e$, $kR_g>>1$

For times less than the Rouse time of an entanglement segment, τ_e, the chain behaves as if it were free since no section has moved far enough to be strongly affected by the tube constraint. The characteristic decay-rate of the scattering function at wavevector k is dominated by the Rouse-time of chain segments whose size is the order of k^{-1}, $\tau_\kappa \sim k^{-4}$. A detailed calculation gives for $t>>\tau_k$ [3]

$$S(\mathbf{k},t) \cong S(\mathbf{k},0)\exp\left[-\frac{2}{\sqrt{\pi}}\left(t/\tau_k\right)^{1/2}\right] \quad ; \quad \tau_k = \frac{12\zeta_0}{k_B T k^4 b^2} \tag{11}$$

The form of the time-dependence can be understood from the "anomalous" diffusion of a piece of Rouse chain, which displaces in time so that $<r^2(t)>\sim t^{1/2}$ so that the $\exp(-k^2Dt)$ scattering from Fickian diffusers is replaced by $\exp(-k^2Dt^{1/2})$. The initial structure factor $S(\mathbf{k}, 0)$ is just the static scattering function from an ideal Gaussian random walk. Known as the Debye function, at lengthscales well within the coil radius it is just the Fourier transform of the coulomb-like density profile of monomers on the same chain as a given monomer, and asymptotes to $12/\mathbf{k}^2b^2$.

2.3.1. *Entangled motion* $t>>\tau_e$, $kR_g>>1$

For times longer than τ_e, the free Rouse motion of segments is restricted to the tube contour and eventually, for $t>\tau_R$, the structure factor is dominated by reptation. In the high wavenumber limit relative to the coil size (though still on lengthscales larger than the tube diameter), a simple argument again leads to the correct asymptotic result. If $\mathbf{R}(s,t)$ is still in a piece of original tube then on average it contributes with the structure factor as $S(\mathbf{k},0)$. This is true with a probability of $\mu(t)$. Otherwise it has left the original tube and cannot interfere constructively with the piece at $\mathbf{R}(s',0)$, contributing zero to the structure factor. Thus

$$S(\mathbf{k},t) \cong \frac{12}{\mathbf{k}^2 b^2}\mu(t) \tag{12}$$

The general form of these predictions for the dynamic scattering function is that after an initial decay rate dependent on $\mathbf{k}$ and characteristic of free chains, a much slower decay takes over, independent of $\mathbf{k}$ but depending on the molecular weight of the labelled chain. At present the correlation times available from NSE techniques are limited to a few hundred nanoseconds. This is not long enough to probe the long-time dependence of the entangled regime (12), but has conclusively demonstrated the cross-over from free to hindered timescales [5].

3. Monodisperse Star-Branched Polymers

As conjectured above in the introduction to this chapter, the fundamental role of topology in this approach to entangled polymer dynamics would indicate that changes to the topology of the molecules themselves would radically affect the dynamical response of the melts. In fact rheological data on monodisperse star-branched polymers, in which a number of anionically-polymerised "arms" are coupled by a multifunctional core molecule, pre-dated the first application of tube theory in the presence of branching [6]. Just the addition of one branch point per molecule has a remarkable effect, as may be seen by comparing the dissipative moduli of comparable linear and star polymer melts in figure 3.

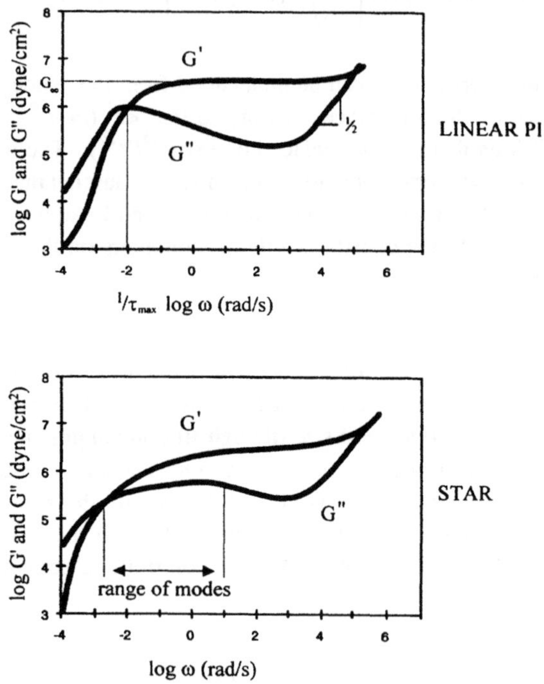

Figure 3. The dissipative modulus $G''(\omega)$ for linear (top) and three arm star branched (bottom) PI from reference [2]. Note the broad range of relaxation times indicated by the width of the peak in the star-polymer.

Three experimental observations are particularly striking: (1) the range of relaxation times in star polymers is much broader than for monodisperse linear polymers. Rather than representing a dominant single relaxation time, the mechanical spectra require a range of comparably-weighted modes; (2) the range of timescales for this spectrum varies exponentially with the molecular weight of the arms, M_a so that both the terminal time and viscosity of the star polymer melt shows a dependence of roughly exp($-\nu\, M_a/M_e$) with $\nu\sim O(1)$; (3) the terminal time and viscosity are dependent *only* on M_a and *not* on the number of arms when the polymers are well-entangled.

3.1 TUBE MODEL FOR STARS IN A FIXED NETWORK

The tube model gives a direct indication of why one might expect the strange observations on star melts described above to be true. Because the presence of the branch points themselves in a high molecular weight star-polymer melt is at an extremely dilute level, the physics of local entanglements is expected to be identical to the linear case: each segment of polymer chain behaves as if it were in a tube of diameter a. However, in this case reptation is suppressed because the diffusion of the centre of mass of a molecule along the tube of one of the arms would necessitate dragging the two other arms into the same tube. This is entropically unfavourable - $3kT$ of free energy (via each of the three degrees of orientational freedom of a subchain of length M_e) is lost for each tube segment dragged a distance a into the tube of another chain segment. Thus the branch point is effectively pinned up to very rare fluctuations of the configuration. Instead we must rely on the *fluctuations* in the length of primitive path of the star arms (which must be present due to the Rouse "breathing" modes along the tube contours) to renew the configurations of the entangled arms as in figure 4. These will lose stress contributions from each arm independently, and it is not surprising that the probability of the largest fluctuations which completely release a star arm into a new configuration are exponentially rare in the arm length. We now proceed to treat this insight a little more quantitatively.

Figure 4. Proposed mechanism of entangled dynamics of a star polymer in a melt. Retractions as shown partially renew the tube, beginning with rapid retractions near the free end and much more rarely renewing deeper parts of the molecule.

3.1.1 *Brownian Chain Tension in a Melt and the Tube Potential*

An alternative way of viewing the entropy loss on constraining entangled polymers is to see it as equivalent to a *tension* of $3kT/a$ along every entangled chain. So the free energy change above becomes just this tension multiplied by the distance translated curvilinearly along the tube, a. This is a very useful insight - it provides an alternative way of understanding the stress formula (3): the tension $3kT/a$ is carried by each entanglement segment of length a. The components of the total stress are calculated by counting strands transferring tension across the corresponding planes [3].

The chain tension arises in a physical way: at timescales short enough for the tube constraints to be effectively permanent, each chain end is subject to random Brownian motion at the scale of an entanglement strand such that it may make a random choice of exploration of possible paths into the surrounding melt. One of these choices corresponds to retracing the chain back along its tube (thus shortening the primitive path), but far more choices correspond to extending the primitive path. The net effect is the chain tension, sustained by the free ends.

In fact without the inclusion of the chain-end tension, the equilibrium path-length of the chain is not maintained. We can write a potential $U(s)$ for the length of the primitive path by including both the (quadratic) curvilinear rubber-elastic term and the (linear) end-tension term as follows:

$$\begin{aligned} U(s) &= \frac{3kT}{2Nb^2}s^2 - \frac{3kT}{a}s \\ &= \frac{3kT}{2Nb^2}(s-L)^2 + const. \end{aligned} \tag{13}$$

where $L = Nb^2/a$ is the equilibrium primitive path length of the chain. This quadratic potential will determine the fluctuation dynamics of an arm of an entangled star polymer.

3.1.2 *Approximate Theory for Stress-Relaxation in Star Polymers*

The observations above can be rapidly turned into a semi-quantitative theory for star-polymer stress-relaxation [7] which is amenable to more quantitative refinement [8]. The key observation is that the diffusion equation for stress release via the one free end passing a deformed tube segment is now modified by the potential of (13). Apart from small displacements of the end, the diffusion to any position s along the arm will now be *activated* and so exponentially suppressed. Each position along the arm, s, will possess its own characteristic stress relaxation time $\tau(s)$ given approximately by

$$\tau(s) = \tau_0 \exp\left(\frac{U(s)}{kT}\right) \tag{14}$$

Here τ_0 is an inverse "attempt frequency" for explorations of the potential by the free end, and may be expected to scale as the Rouse time for the star arm (in fact this is not quite true - the actual scaling is as $M_a^{3/2}$ [8,10]). The relaxation modulus can be calculated exactly as for linear polymers, using the expression (8), but this time with the simple (Poisson process) expression for $p(s,t)$ from the activated diffusion picture:

$$p(s,t) = e^{-t/\tau(s)} \tag{15}$$

The form of $p(s,t)$ is well-approximated for highly entangled arms by a step function in s : at any time intermediate between the relaxation time if the first entangled segments near the end of the arm and the core-segments of the star, segments exterior to the segment $s(t)$ (given by the inverse of the function $\tau(s)$ with $\tau=t$) are almost certain to have relaxed, while segments nearer to the core are almost certainly unrelaxed. In this theory the diffusion constant of the star molecule and the viscosity are both determined by the longest of the relaxation times of (14), so depend exponentially on the arm molecular weight M_a via $\tau_{max}=\tau(0)\sim\tau_0\exp(-U(0)/kT)\sim\tau_0\exp(\nu' M_a/M_e)$.

This theory was able to account for both the molecular-weight scaling of the dynamic quantities D_0, η, and τ_{max} as well as for the shape of the relaxation spectrum (see figure 3) apart from one important feature - the constant ν' in the leading exponential behaviour multiplying the dimensionless arm molecular weight needed to be adjusted. This can be seen as follows. The prediction of the tube model for the plateau modulus from the stress formula (3) is (see section 5.1.1 below):

$$G_0 = \frac{4}{15} c_N f_{eq} L \tag{16}$$

where f_{eq} is the equilibrium tension $3kT/a$. From the definition of the entanglement molecular weight enshrined in

$$G_0 = \frac{\rho RT}{M_e} \tag{17}$$

[1] in terms of the density of entanglement strands ρ/M_e, we can deduce a quantitative relationship between a and b as $a^2=(4/5)N_e b^2$ and find from (13) that $U(0)=15/8N_a/N_e$. So the tube model gives a prediction for the constant ν' of 15/8. This was in marked contrast to a value of about 0.6 required to fit experimental data [8] and taken at face value gives enormous discrepancies in predictions for, for example, the viscosity, overpredicting values by many orders of magnitude for well-entangled stars. The resolution of this difficulty comes from recognising the importance of constraint-release (see chapter by M. Rubinstein, this volume) especially in the case of branched polymers.

3.2 THEORY OF STAR POLYMER MELTS

3.2.1 *Approximate Theory for Constraint Release in Star Polymer Melts*

The much more significant contribution of constraint release to the dynamics of entangled star polymers in comparison to linear polymers arises from the very broad distribution of relaxation timescales discussed above. Fortunately, the same breadth of timescales provides a simple way of calculating the effect [9]. The idea is that at the time that a given tube segment s in the population is relaxing, all chain segments occupying segments of tube s' such that $s'>s$ (nearer a chain end) have renewed their configurations typically *many* times. So chain segments at s and nearer the star cores do not entangle with these fast segments at the timescale $\tau(s)$ and beyond. Alternatively the tube is widened due to this effective dilution of the entanglement network - fast-relaxing segments act as solvent for the slower relaxing ones. Such an idea applied to constraint-release in linear polymers is problematical [10,11] because of the dominance of the single relaxation time τ_{rep}, but becomes applicable in the case of stars, and branched polymers generally.

The new information necessary is the dependence of the effective entanglement molecular weight on the concentration, ϕ of unrelaxed (so entangling) segments. This is known from experiments on dilution of polymer melts by theta-solvents to be approximately $M_e(\phi)=M_{e0}/\phi$, which corresponds, via (17), to the approximately quadratic concentration dependence of $G_0\sim\phi^2$ [3,12]. At any stage in the relaxation dynamics of a melt of identical star polymers, therefore, when a segment s is currently relaxing for the first time, the effective entanglement molecular weight is $M_e(s)=M_{e0}/(1-s/L)$. To recompute the relaxation times $\tau(s)$ we consider the activated diffusion in a hierarchical way: to retract from s to $s+\Delta s$, the attempt frequency is $\tau(s)^{-1}$ and the barrier to diffusion is exp $\{(-1/kT)$ $[U(s+\Delta s); M_e(s)) - U(s; M_e(s))$]} where the notation for U indicated that the running value of the entanglement molecular weight is kept. Taking the limit of Δs small gives the differential equation [9]

$$\frac{\partial \ln \tau(s)}{\partial s} = \frac{15}{4}\frac{M_a}{M_{e0}}\frac{s}{L^2}(1-\frac{s}{L}) \tag{18}$$

where the last term arises from the "dynamic dilution" of the tube. Integration of this equation leads to an effective renormalisation of the potential $U(s)$ which is now a cubic in s

$$U_{eff}(s) = kT\frac{15}{4}\frac{M_a}{M_e}\left(\frac{s^2}{2L^2} - \frac{s^3}{3L^3}\right) \tag{19}$$

The terminal time and viscosity are dominated, as we saw above, by the potential at complete retraction $U(0)$. This is now given by $15/24(M_a/M_e)$, in much closer agreement with experiments. The formula for the relaxation modulus also needs modifying since each element of chain ds contributing to the stress relaxation now does so in an environment diluted by (1-s/L), so picks up this factor [9]. Now the predicted shape of the relaxation spectrum is better fitted by the modified distribution of relaxation times implied by the cubic potential $U_{eff}(s)$ [9], except at high frequencies where the assumption of activated diffusion breaks down (it may easily be verified that $U_{eff}(s)<kT$ for the fraction $(M_e/M_a)^{1/2}$ of chain nearest the chain end).

3.2.2 *Parameter-Free Treatment of Star Polymer Melts*

The approximate treatment described above does remarkably well in accounting for the linear rheology of star polymer melts. In fact it has been remarked that the case for the tube model has its real strength in the results for star polymers rather than for linear chains, where the problems of constraint release and breathing modes are harder to account for (see chapter of M. Rubinstein, this volume). However, there are still some outstanding issues and questions [13]:

(1) What are the proper prefactors to the expression for $\tau(s)$ using $U_{eff}(s)$?
(2) How sensitive is the result to the exact dilution behaviour of the entanglement network?
(3) At short times, what is the effect of non-activated Rouse motion on the relaxation of segments near the end of the star arms?

In particular it was conjectured that the terminal relaxation of star polymers might be the most sensitive test of the "dilution exponent" β in $G_0 \sim \phi^\beta$. The favoured candidates seem to be the close values of 2 and 7/3 (see [12] and chapter of M. Rubinstein). Very recent work [14] has answered all these questions, and in particular has given a cross-over formula for $\tau(s)$ incorporating the associated effects, so valid for all timescales longer than the Rouse time of an entanglement segment τ_e:

$$\tau(x) \cong \tau_e \left(\frac{M_a}{M_e}\right)^{3/2} \left(\frac{2\pi^5}{15}\right)^{1/2} \frac{\exp[U_{eff}(xL)]}{x\left[(1-x)^{2\alpha} + \left(\left(\frac{4M_a}{15M_e}\right)(1+\alpha)\right)^{\frac{2\alpha}{\alpha+1}} \Gamma\left(\frac{1}{\alpha+1}\right)^{-2}\right]^{1/2}} \quad (20)$$

where $x=s/L$ and $\beta=1+\alpha$. The prediction for the relaxation spectrum of a monodisperse star melt can be made without any free parameters at all if G_o and τ_e are taken from data on linear polymers in the low and high frequency range respectively. When this is done the shape and range of relaxation times (via the rheological function $G''(\omega)$) is very well accounted for (see figure 5). To achieve a quantitative fit at the level of prefactors, shifts of about 1.6 in the modulus and 2 in the monomeric friction factor are required. These shifts seem to be independent of polymer chemistry, and may result from the level at which constraint-release has been treated via the dilution approximation.

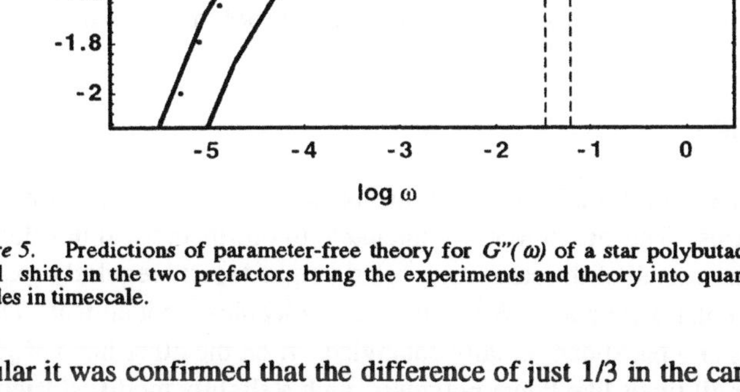

Figure 5. Predictions of parameter-free theory for $G''(\omega)$ of a star polybutadiene from reference [15]. Small shifts in the two prefactors bring the experiments and theory into quantitative agreement over 5 decades in timescale.

In particular it was confirmed that the difference of just 1/3 in the candidate values for the dilution exponent can alter the terminal time of a moderately-entangled star polymer by over an order of magnitude. Within the range of reported values of M_e for the range of chemistries examined [16], the value of 7/3 is preferred.

3.2.3 *Single Chain Structure Factor for Star-Polymer Dynamics*

The dynamic structure factor for scattering from a single chain as defined in equation (10) had a simple long-time form for the case of linear reptating polymers (12), due to the homogenous nature of the chain's orientational relaxation. The special entangled dynamics of star polymers, however, lends a little more structure to $S(k,t)$ [17]. The full result is relatively complex, but differs from the approximation for linear polymers (12) in that the scattering within the range $R_g^{-1} < k < a^{-1}$ does not follow the same time-dependence, but decays more rapidly with increasing k. This is due to two effects: the exponentially slower relaxation of deeper parts of the chain, which dominate the low-k region, and also the continued contribution of relaxed outer parts of the chain to the low-k scattering. In particular, pairs of monomers nearby on the outer parts of the star arm, which contribute at early times to high-k scattering, contribute to lower-k scattering at longer times, when they have become part of a retracted and renewed portion of chain. The interference in the scattering amplitude is then between the former and present pieces of chain (the two copies of the outer parts of the star arm in figure 4). The scattering function can be approximated by:

$$S_{star}(k,t) \cong S_{star}(k,0)\left(\frac{t}{\tau_R}\right)^{-z(k)} \tag{21}$$

for $\tau_R < t < \tau_{max}$. The scale-dependent dynamic exponent $z(k)$ is an approximately linear function of k, decreasing from 1 to 0.2 over the k-range $3R_g^{-1}$ to R_g^{-1}.

4. More Complex Topologies

The recognition of the two fundamental mechanisms of reptation and retraction for linear and branched entangled polymers respectively allows the theoretical of the linear rheology of more complex polymers. The essential tool is the "renormalisation" of the dynamics on a hierarchy of timescales, as for the case of star polymers. It is important to stress that experimental checks on well-controlled architectures of higher complexity are still very few due to the difficulty of synthesis, but the case of comb-polymers is an example where good data exits [18].

4.1 COMB POLYMERS

These polymers posses a high-molecular-weight "backbone" from which are attached a number, q, of side arms. The structure most amenable to our theoretical development are those in which the arms are all of the same molecular weight M_a and regularly spaced along the backbone of molecular weight M_b. We assume all molecules are identical. The case of interest has both arms and backbone initially entangled - if on the other hand $M_a/M_e<1$, the backbones behave for all times like linear polymers, with a slightly modified friction factor and entanglement length since the arms just act as diluent. However for $M_a/M_e>1$ the behaviour is predicted to be quite different [19]. Now the backbones are quite immobile for early times since the side arms behave just like star arms, and do not release the diffusion of their branching points until the timescale for full retraction. On the other hand the retraction of the side arms is *slower* than in a melt of stars of identical M_a because of the presence of the backbone material, which acts as permanent network for the side arms throughout their retraction. This reduces the extremely efficient process of dynamic dilution considered in the previous section, so that the effective potential for relaxation times for the arms is now:

$$U_{eff}(s) = kT\frac{15}{4}\frac{M_a}{M_e}\left(\frac{s^2}{2L^2} - \phi_{arm}\frac{s^3}{3L^3}\right) \tag{22}$$

where ϕ_{arm} is the volume fraction of arm-material $qM_a/(qM_a+M_b)$. This interpolated behaviour for retracting arms, intermediate between the star-network and star-melt cases also arises in a blend of star polymers with long linear chains [20].

Now at times $t>\tau_{arm}$, the longest retraction time of the side arms, the backbone is free to move, controlled by the effective frictional drag of the branch-points, which typically far outweighs the monomeric frictional drag from the backbone itself. Each branch point takes a random step of size a with a rate τ_{arm}^{-1} so the effective friction constant of the branch point $\zeta_{br}\sim kT\tau_{arm}/a^2$. At the timescales for backbone motion, all the arm material has disentangled (typically at much shorter timescales) so, just as in the mutual dilution of star polymers, acts as unentangled solvent for the backbone. The melt thus behaves like a melt of *linear* polymers at long times, with a degree of entanglement per chain of $M_b(1-\phi_{arm})/M_e$ and a total frictional drag of $q\zeta_{br}$. In this way it is possible for a highly branched polymer to behave like a linear polymer for some range of timescales. Whether the long-time dynamics of the backbones is effectively entangled or not depends on the renormalised number of entanglement strands, which is reduced as the amount of material in the side arms increases so swelling the value of M_e. Thus even if the backbones are originally entangled, the final motion may be a slowed-down Rouse motion if mutual entanglements between backbones are not sufficient to force them to reptate. Both cases of renormalised Rouse and renormalised reptation have been observed in model materials [18].

The simplest case of comb polymer is the H-shaped structure in which two side arms of equal length are grafted onto each end of a linear "cross-bar" [21]. In this case the backbones may reptate, but the reptation time is proportional to the *square* of M_b rather than the cube, because the drag is dominated by the dumbbell-like frictional branch points at the chain ends [22, 23].

Experiments on both combs and H-polymers show rheological spectra with two distinct regions at frequencies typical of entangled modes: a high-frequency "shoulder" attributable to the arm retractions, and a low-frequency region which may be of the form $\omega^{1/2}$ in the case of unentangled backbones, or exhibit a characteristic time via a maximum in $G''(\omega)$ for entangled backbones (see figure 6).

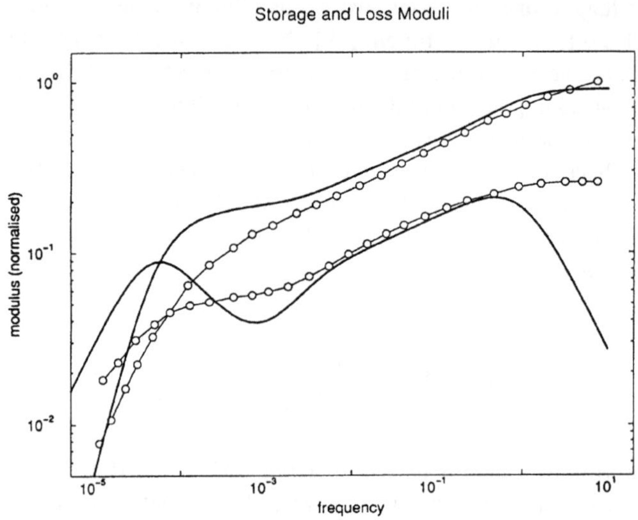

Figure 6. The rheological functions $G'(\omega)$ and $G''(\omega)$ for a H-shaped PI of arm molecular weight 35k and backbone 130k [23]. The high-frequency arm retraction modes can be seen together with a low-frequency feature due to the "cross-bar" dynamics. The smooth curves are the predictions of a model which takes eqn. (22) as the basis for the arm retraction times and a Doi-Edwards reptation spectrum for the backbone eqn. (9). The reptation time is correctly predicted, as the spectrum from the arm modes, though the low frequency form is more polydisperse than the simple theory predicts.

Combining the predictions for the reptation time of the H cross-bar and the effective modulus when the arms are acting as solvent gives a prediction for the scaling of the viscosity of a melt of H-polymers on their structural parameters:

$$\eta_H \cong G_0 \tau_0 (1-\phi_{arm})^3 \left(\frac{M_b}{M_e}\right)^2 \exp[-2\nu \frac{M_a}{M_e}(\frac{1}{2}-\frac{\phi_{arm}}{3})] \tag{23}$$

4.2 TREE POLYMERS

If combs represent one extreme of the topological family of branched polymers, then another extreme is given by the case of tree-polymers, which retain a branched structure at all timescales. The study of tree-like branched architectures is also motivated by the important commercial polymer Low Density Polyethylene (LDPE), which has remarkable rheological properties making it suitable for many processing operations [24].

4.2.1. *Cayley Tree*

A useful "toy" theoretical model which captures the essential features of self-entangled trees is the monodisperse Cayley tree in which each chain segment branches with a fixed functionality z at each of its ends, except those at the extremity of the molecule (see figure 7). Smaller versions of these structures, too low in molecular weight to be entangled, have been synthesised and are usually referred-to as "dendrimers"[25].

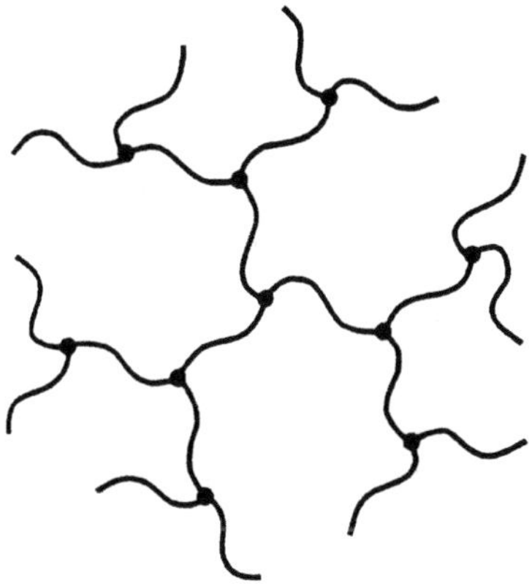

Figure 7. The Cayley-tree model structure

The application of tube-theory incorporating a "dynamical dilution" model for constraint-release proceeds as for the comb-polymers [26], but in this case there are as many distinct stages in the dynamic renormalisation of the molecular structure as there are levels of branching in the molecule. When there are many levels, a further approximation can be taken by working in discrete, rather than continuous, timescales - the relaxation times of the successive layers of branch points, τ_i. Considering the entangled segment connecting the *ith* to the *(i+1)*th level of the molecule as a star-arm retracting in an entanglement network whose concentration is at all times given by the concentration of *unrelaxed* segments, gives a recursion-relation for the τ_i:

$$\tau_{i+1} \approx \tau_i \exp\left(\frac{2\nu M_x}{M_e}\left[\frac{(z-1)^i - 1}{(z-1)^N - 1}\right]\right) \tag{24}$$

where M_x is the molecular weight between the branch points and N here the number of levels in the Cayley tree. The constant ν is 15/8 as for star polymers. The z-dependent function in the square brackets of eqn. (24) is just the effective concentration of entangled segments C_i operative at level i which dilutes the entanglement molecular weight just as for star polymers,

The hierarchy of relaxation times τ_i as a function of level i can be inverted to give an approximate form for $G(t)$ (since $G(t) \sim C_i^2(t=\tau_i)$), which in the physical case of z=2 becomes:

$$G(t) = 4G_0[\ln\left(\frac{\tau_{\max}}{t}\right)]^{\theta} \tag{25}$$

where τ_{max} is the longest entangled relaxation time of the tree ($\sim \tau_0 \exp(2\nu M_\xi/M_e)$). θ is a "topological exponent" which in the case of the Cayley tree is 2, but assumes other values for more irregular trees (see below).

A feature of theories for tree-like polymers is the "disentanglement transition", which occurs when the tube dilation becomes faster than the arm-retraction within it. In fact this will happen even for simple star polymers, but very close to the terminal time itself when very little orientation remains in the polymers. In tree-like polymers, it is possible that several levels of molecule near the core are not effectively entangled, and instead relax via Rouse dynamics (though slowed down by the underlying entangled dynamics of the effective solvent) or in extreme cases by Zimm dynamics, when the surroundings fail to screen the hydrodynamic interactions between the core sections of the molecules (see chapter by M. Rubinstein, this volume).

4.2.2 *Mean-field Gelation Ensemble*

Perhaps a more appealing model for a randomly branched structure such as LDPE is given by a stochastic tree in which branch points have a probability p of occurring at the end of chain segments, and ($1-p$) of termination [27]. This model, which leads to a naturally-polydisperse set of branched structures even when the segments between branch points are monodisperse (and of molecular weight M_x) is also known as the Flory-Stockmayer gelation ensemble [28, 29]. It models a cross-linking ensemble of chains, but without loops, and has a gelation transition at $p=p_c=(z-1)^{-1}$, when a polymer of infinite molecular weight first appears.

To model a polydisperse branched polymer melt, it suffices to keep p below p_c and to generate the concentrations of segments at each level of the disentanglement hierarchy. The term "seniority" has been coined for the effective topological level of a segment - within retraction dynamics all segments of equal seniority relax on the same timescale. Conversely the proportion of segments at all higher seniorities controls the effective concentration of the entanglement network on that timescale.

The result of treating the Flory-Stockmayer ensemble at the same level of approximation as that of the Cayley tree sketched in the previous section, is that a very similar relaxation spectrum to (25) is predicted, but with a value for the exponent θ of 4 rather then 2 [27]. A disentanglement transition is also found, as well as the appealing feature that at each seniority a fixed fraction of the molecules still entangled actually renormalise into *linear* polymers (like the comb topology) and relax by reptation rather than by further fluctuation.

The rheology of cross-linking polymer solutions and melts at the gel-point is a wide field of great interest [30] beyond the scope of this chapter. However there is clear experimental evidence on the few entangled systems studied that they differ qualitatively from the universal features observed for unentangled systems at the gel point. In particular, though at the gel-point when $G(t) \sim t^{-u}$, unentangled systems follow a universal value of u=0.67, entangled systems follow an approximate power law in which $u=u(M_x/M_e)$ and takes much smaller values than 0.7.

5. Response to Large Deformations and Flows

It became clear in the early development of the tube model that it provided a means of calculating the response of entangled polymers to large deformations as well as small [3]. In particular the same step-strain experiment used to determine $G(t)$ directly in shear is straightforward to extend to large shear strains γ. In many cases of such experiments on polymer melts both linear and branched, monodisperse and polydisperse, the experimental strain-dependent relaxation function $G(t, \gamma)$ may be written:

$$G(t, \gamma) = G(t)h(\gamma) \tag{26}$$

for all times $t > \tau_k$, a material time constant [31]. So for the longest relaxation times of the material, the non-linear viscoelastic response factorises into a time-dependence identical to that of linear deformations, and a strain-dependence $h(\gamma)$, which is known as the *damping function*. In particular, linear polymer melts have a rather weak non-linear elasticity, in which the effective modulus falls rapidly with strain in a universal way.

5.1 RETRACTION ON STEP-STRAIN IN THE TUBE MODEL

When non-linear deformations are made, we need to make additional assumptions within the tube model on how the tube itself deforms with the bulk strain. The simplest, and original assumptions are that:

(i) The tube contour (primitive path) deforms affinely.
(ii) The tube diameter remains unchanged.

although other assumptions are possible [32, 33] (and see below). The first may be justified on the grounds that by definition the tube diameter sets the scale below which non-affine deformation of chains dominates - this just needs to extend into the nonlinear region. The second assumption is less compelling, but is motivated by the observation that the local configuration of chains is not greatly different from that at equilibrium even at large strains due to the fact that $N_e >> 1$.

When these assumptions are made, an additional mechanism for stress-relaxation arises - that of *retraction*. (see figure 8). When a chain is embedded randomly in a deforming incompressible medium, the contour length of both chain and tube is initially increased by the rapid strain. However, when the strain stops, the stretched chain now supports a greater curvilinear tension than the equilibrium tension f_{eq} still applied to the chain ends by their entropic exploration of the surrounding entanglement network. So the chains retract back along the deformed tubes until their curvilinear tension returns to equilibrium, which will be achieved when they regain their equilibrium contour length. This process happens much faster than reptation for long chains because no diffusion of the centre of mass of the molecule is required. The retraction dynamics is rather achieved through curvilinear Rouse motion (local friction on a connected linear object) and will decay as a Rouse stress relaxation - $t^{-1/2}$ cut-off at the Rouse time τ_R.

We now use this to calculate the stress in the melt *after* the retraction has occurred.

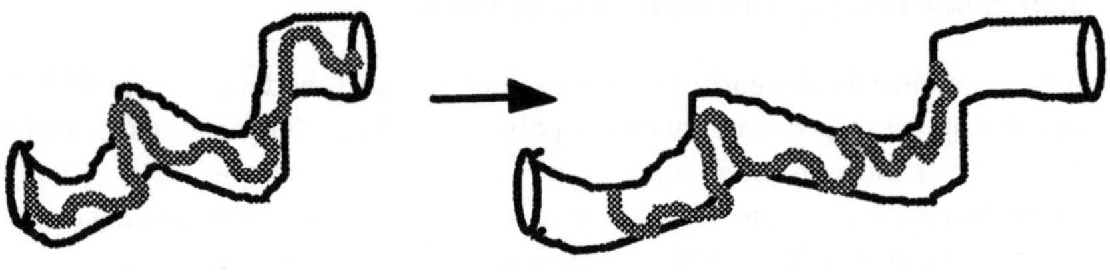

Figure 8. Chain retraction in a tube under non-linear strain.

The deformation is described by the tensor **E** defined so that an arbitrary vector **v** in the material is deformed affinely into the vector **E.v**. For example, in simple shear of shear strain γ, and in uniaxial extension of strain ε, the tensor **E** takes the forms:

$$\begin{pmatrix} 1 & \gamma & 0 \\ 0 & 1 & 0 \\ 0 & 0 & 1 \end{pmatrix} \qquad \begin{pmatrix} \varepsilon & 0 & 0 \\ 0 & \varepsilon^{-1/2} & 0 \\ 0 & 0 & \varepsilon^{-1/2} \end{pmatrix}$$

When applying the formula (3) to calculate the stress the length of an entanglement strand is increased by a factor $|\mathbf{E.u}|$ where **u** is the unit vector along the strand. This needs to be inserted inside the average over all initial orientations **u**. The unit vector itself is deformed to $\mathbf{E.u}/|\mathbf{E.u}|$. The other modification is to the concentration of entanglement strands c_{Ne}. The retraction process means that a the same proportion of original entanglement strands are lost as the length retracted, so c_{Ne} is reduced by the factor $1/\langle|\mathbf{E.u}|\rangle$, where we may pre-average because a well-entangled chain explores all orientations in its tube segments.

The final result for the stress-tensor is that

$$\sigma = kT\, c_{eq} \mathbf{Q}(\mathbf{E}) \tag{27}$$

where the geometric tensor **Q** is defined by the orientational averages over the unit vector **u**:

$$\mathbf{Q} = \frac{1}{\langle|\mathbf{E.u}|\rangle}\left\langle \frac{\mathbf{E.u}\;\mathbf{E.u}}{|\mathbf{E.u}|} \right\rangle \tag{28}$$

The averages <...> may be written in spherical polar co-ordinates as angular integrals, which have simple numerical, and in some cases analytic, forms.

5.1.1 *Properties of the Q-tensor and Consequences*

In the two classic viscometric deformations of simple shear and extension, the appropriate components of **Q** have very different behaviour. For small shear strains, the shear stress depends on the component Q_{xy}, which has the linear asymptotic form $4\gamma/15$. This is the origin of the constant ν in the tube potential of section 3 above. For large strains, however, $Q_{xy}\sim\gamma^{-1}$, so is very shear-thinning. Physically this comes from the entanglement loss on retraction and the alignment of the remaining tube segments into the plane of the shear. In extension the relevant experimental component is the difference $Q_{xx}-Q_{yy}$. This is linear in extensional strain ε at low strain, but this time asymptotes to a constant (15/4) at high strain.

The behaviour of the **Q**-tensor in shear gives directly the damping function from its definition in (26) so that $h(\gamma)=(15/4)Q_{xy}(\gamma)/\gamma$. This function is plotted in the usual convention in figure 9. It is very close to results on monodisperse entangled solutions and melts [34].

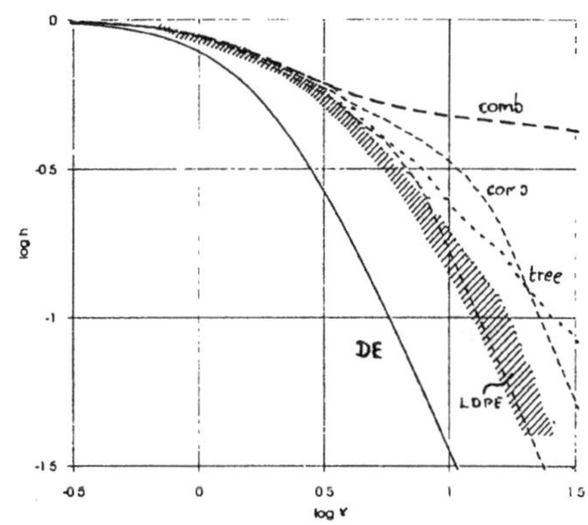

Figure 9. Damping functions in shear from the tube model for linear polymers (lowest curve) and various branched architectures.

5.1.2 *Damping Functions for Branched Polymers*

The process of chain retraction can be applied to more complex topologies of entangled polymers under the same additional assumptions discussed for linear polymers [35]. In the case of star polymers retraction may proceed just as for linear chains, and the damping function is expected to be of the universal "Doi-Edwards" type. This has been confirmed experimentally [31]. However for polymers with higher levels of branching the situation is different. In spite of being stretched along its contour by a large bulk strain, a segment one level further into a branched molecule than the outermost arms cannot in general retract, because this would mean drawing those outer segments into its tube. It may only do this when its tension exceeds the sum of the tensions in the impeding arms - a criterion only met beyond a critical strain proportional to the number of arms attached at the outermost branch point.

In a manner akin to the hierarchical relaxations with time in a highly branched polymer, this new form of retraction via "branch point withdrawal" also happens hierarchically, though as a function of strain rather than time. Segments two levels into the molecule may only retract when their tension exceeds that of the first level, and so on. As a consequence of this balance of entropic tensions, the strain at which any segment withdraws its outer branch point (and so the functional form it contributes to the strain dependence) just depends on the number of free ends at the edge of the molecule it is connected to. This statistic has been termed the "priority" distribution, in analogy with the "seniority" distribution which controls the relaxation times. In general a branched molecule has different seniority and priority distributions - a knowledge of the former is required to predict the linear stress-relaxation function, of the latter to predict the non-linear strain response [35].

Predictions of the damping function (defined now as the effective modulus after all rapid (Rouse) retractions have occurred) for both Cayley tree and comb architectures are shown alongside the Doi-Edwards function in figure 9. It is immediately apparent that the damping functions are only "universal" for a particular topology of polymer, and may eventually be useful in characterising branched melts. Results on LDPE are also shown, indicating that the structure is more likely to be tree-like than comb-like within this picture. Though very appealing, this is probably pushing the tube model as far as one ought to, if not a little further.

5.1.3 *Strain-dependence of the Tube*

It is of interest to think about relaxing the assumptions (i) and particularly (ii) of this section, although hard experimental tests for specific assumptions of the deformation of the tube constraint itself can never be confined to rheology alone, but will involve at least careful analysis of neutron-scattering experiments [33, 36].

Reasons have been advanced for both an increase and a decrease of the tube diameter with strain. A justification of the former view might be the retraction process itself [37]. If it acts in a similar way to the dynamic dilution process and the effective concentration of entanglement network follows the retraction then $c_{eff} \sim \langle |\mathbf{E.u}| \rangle$ so that $a \sim \langle |\mathbf{E.u}| \rangle^{1/2}$. On the other hand one might guess that at large strains the tube deforms at constant tube volume La^2. The tube length must increase as $\langle |\mathbf{E.u}| \rangle$, so from this effect $a \sim \langle |\mathbf{E.u}| \rangle^{-1/2}$. Indeed, Marrucci has recently proposed that both these effects exist and remain unnoticed because they cancel [38]!

5.2 CONSTITUTIVE EQUATIONS FOR CONTINUOUS FLOW

The step-strain experiments discussed above furnish the simplest example of a strong flow. Many other flows are of experimental importance: start-up of steady shear, start-up of extensional flow and reversing step strains, to give a few examples. Indeed the development of phenomenological constitutive equations to systematise the wealth of behaviour of polymeric liquids in general flows has been something of an industry over the past forty years [39]. It is important to note that it is not possible to derive a constitutive equation from the tube model in closed form without making rather uncontrolled mathematical approximations - indeed this suggests that the space of non-Newtonian constitutive behaviour is rather larger than that spanned by compact sets of differential or integral equations. However, the behaviour of the tube model for linear polymers in steady shear flow has recently become very important in the study of "living" polymers, and in particular wormlike surfactant micelles [40] (and see chapter by M. Cates, this volume). A further challenging area is in the special elongational behaviour of highly branched polymers [24].

5.2.1. *Linear Polymers in Continuous Flow*

The essential picture of stress arising from occupied and deformed tube segments carries over naturally to the case of continuous flow. Tube segments are born at chain ends, are carried by diffusion along the chain during which time they are stretched and oriented, and die when they are passed again by a chain end. Thus to calculate the stress a knowledge of the survival probability function $p(s,t)$ just as in equation (4) is required in the form of $p(s,t,t')$, the probability that a segment born at time t' still exists at time t at contour co-ordinate s along the occupying chain. The equation (4) for p is now modified by the *convection* of the chain past the tube segment by the flow as well as by the curvilinear diffusion. We assume that the flow rates are less than the inverse retraction time so that the chain length is constant. If $v(s)$ is the relative velocity of tube and chain (due to the extension of the tube by the flow past the constant-length chain), then (4) becomes

$$\left[\frac{\partial}{\partial t} - D_C \frac{\partial^2}{\partial s^2} + v(s,t)\frac{\partial}{\partial s}\right] p(s,t,t') = 0 \qquad (29).$$

where $v(s,t)$ arises from the extension rates of all the segments from the centre of the chain to the point s. If any segment has an orientation $\mathbf{u}$ then its local rate of increase in length if deformed affinely in the flow is $\mathbf{K{:}uu}$ where $\mathbf{K}$ is the velocity gradient tensor. So $v(s,t)$ in (29) is itself given by:

$$v(s,t) = \int_0^s ds'\mathbf{K}(t) : \int \mathbf{u}(s)\mathbf{u}(s)\, f(\mathbf{u},s,t)\, d^2\mathbf{u} \qquad (30)$$

The probability distribution $f(\mathbf{u}, s,t)$ for the local segment distribution is related to the contribution from the ensemble of segments at s to the total stress. It is calculated self -consistently from the survival probability function by letting the surviving tube segments be deformed by the total deformation tensor over their lifetimes, $\mathbf{E}(t,t')$:

$$f(\mathbf{u},s,t) = \int_{-\infty}^{t} dt' \int d^2\mathbf{u}'\, \delta(\mathbf{u}-\mathbf{u}')\, \mathbf{E}(t,t').\mathbf{u}' \frac{\partial}{\partial t'} p(s,t,t') \qquad (31)$$

The stress tensor can be written now as:

$$\sigma(t) = \frac{15}{4} G_0 \frac{1}{L} \int_{-L/2}^{L/2} ds \int d^2\mathbf{u}\ \mathbf{uu}\ f(\mathbf{u}, s, t) \qquad (32).$$

This full set of self-consistent equations is clearly a very difficult task to solve even numerically. However, good approximations of closed integral type have been proposed. These essentially ignore the s-dependence of the survival and orientation functions, which makes them a *physically* appealing approach in the case of wormlike surfactants (see chapter by M. Cates, this volume). For ordinary monodisperse polymers the following approximate integral constitutive equation results:

$$\sigma = \frac{15}{4} G_0 \int_{-\infty}^{t} dt' \left(\frac{\partial}{\partial t'} \mu_{DE}(t - t') \right) \mathbf{Q}(\mathbf{E}(t, t')) \qquad (33)$$

The remarkable feature shared by this constitutive set and its approximations is that the stress in steady shear is non-monotonic. This has been used to explore the unstable and inhomogenous flows in both Pouseuille ("spurt effect") [41] and Couette ("shear-banding") [40, 42] flows. The non-monotonic behaviour in ordinary polymer melts is still debated, but it now seems clear that this phenomenon does exist in entangled solutions of living polymers. The consequences of such a constitutive instability for an apparent multiplicity of possible bulk flow fields under the same imposed conditions still represents a severe current challenge to theory [43].

5.2.2 *Constitutive Equations for Branched Polymers*

A current challenge for molecular rheology is the continuous flow of highly branched polymers [24, 39]. The central issue is that in both uniaxial and planar extensional flows, LDPE is strain-hardening (stress grows faster than linearly with strain over some range of strain and time in a time-dependent flow), while retaining a softening characteristic in shear. It has proved impossible to fit simultaneously all the flow behaviour of LDPE with existing phenomenological constitutive equations, even qualitatively. However, the insights of the special molecular features of long-chain-branched polymers under high strain, and in particular the delayed retraction and branch-point withdrawal processes, make a molecular approach doubly appealing.

Very recently the continuous flow theory of this section and the theory for branched molecules in section 4 have been combined in a constitutive theory for a class of model branched polymers related to the H-structure, but now with an arbitrary number of arms, q, attached at each end of a cross-bar [44]. For a wide range of flow rates, the cross-bar may be oriented and stretched, while the dangling arms remain isotropic due to their much faster relaxation dynamics. Thus important contributions to the stress arise only from the cross-bars. However the arms play an important role: when the cross-bar becomes highly stretched as in an extensional flow, branch-point withdrawal diminishes the size of the frictional "blob" of dangling arms, which in turn reduces very rapidly the effective frictional drag of the branch points. The result, after some simplification, is that the average stretch of the cross-bar, $\lambda(t)$, and its averaged orientation-tensor, $\mathbf{S}(t)$, each follow coupled dynamic equations, with a characteristic relaxation time for each: τ_s and τ_b respectively [22]. These may be thought-of as the Rouse and reptation time for the cross-bars as renormalised linear polymers.

$$\frac{\partial}{\partial t}\lambda = \lambda(\mathbf{K}:\mathbf{S}) - \frac{1}{\tau_s}(\lambda - 1)$$

$$\mathbf{S}(t) = \int_{-\infty}^{t} \frac{dt'}{\tau_b(t')} \exp\{-\int_{t'}^{t} \frac{dt''}{\tau_b(t'')}\} \mathbf{Q}(t,t') \tag{34}$$

The relaxation times vary with time itself only when the backbone becomes full-stretched ($\lambda=q$), and then in such a way as to maintain this maximum stretch until the flow no-longer tends to stretch the molecules further. The history of the relaxation time τ_b τneeds to be taken into account in the integral part of the dynamical equations, just as for wormlike micelles ([40] and chapter by M. Cates, this volume). The stress itself is a function of both molecular variables:

$$\sigma(t) = G_0\, \phi_b^2\, \lambda^2(t)\, \mathbf{S}(t) \tag{35}$$

with corrections arising from the configuration of the dangling arms. In (35) we have the pleasing structure of a stress calculated at each time from a number of molecular variables, each following their own (coupled) dynamics. This feature gives such a constitutive formulation the same qualitative features of extensional and shear flows of LDPE (see Figure 10), though a full test awaits experiments on well-characterised architectures.

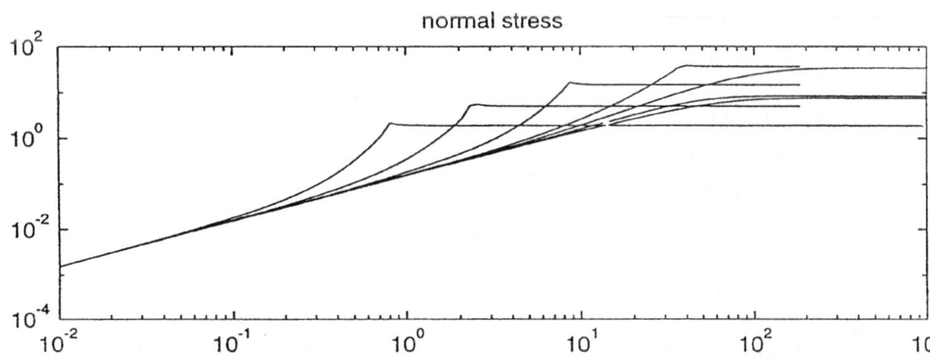

Figure 10. Predicted "startup-flow" in extension for the constitutive equation derived for a "pom-pom" polymer with q=5. Extensional stress divided by the extensional strain-rate is plotted (so that early time response is rate-independent) against time after initiation of the flow for a range of dimensionless rates from 0.001 to 3. Both axes are logarithmic.

6. Conclusions

The molecular approach to the dynamics of entangled polymers outlined in this and the chapter by M. Rubinstein has proved remarkably fruitful, and has accounted for a very wide range of phenomena, some initially very puzzling. There are still many open challenges to theory, however, of which a few might be the following questions:

(i) Why does the tube-model do better for the linear rheology of branched polymers than linear?
(ii) Is the prediction of a stress-maximum in the shear flow of linear (and some branched) polymers pathological in the case of melts, and if so what is the correct molecular physics that is missing?
(iii) In which cases is the dilution-approach to constraint-release valid, and how does it operate in non-linear flows?
(iv) Is the extension to non-linear flows of branched polymers outlined here valid and what experimental tests of it are the best to devise?

The role of numerical simulation has not been mentioned so far in this chapter, but it is bound to be important in at least two ways. Already simulation at a molecular, albeit coarse-grained, level has picked-up the presence of entanglement constraints on an ensemble of Rouse chains [45]. At a quite different level is the need for *flow-solvers* capable of numerical prediction of flows in complex geometries arising from molecular constitutive equations. In this area there is at present an exciting confluence of an established area of applied mathematics with statistical physics [46, 47].

References

1. Ferry, J.D. (1986) *Viscoelastic Properties of Polymers*, Wiley.
2. Fetters, L.J., Kiss, A.D., Pearson, D.S., Quack, G.F. and Vitus, F.J. (1993) *Macromolecules* **26**, 647-654.
3. Doi, M. and Edwards, S.F. (1986) *The Theory of Polymer Dynamics*, Oxford.
4. Higgins, J.S. and Benoit, H. (1995) *Neutron Scattering of Polymers*, Oxford.
5. Higgins, J.S. and Roots, J.E. (1985) *J. Chem. Soc., Faraday Trans. II* **81**, 757-767; Richter, D. *et al.* (1990) *Phys Rev. Lett.* **64**, 1389-1392.
6. Utracki, L.A. and Roovers, J. (1973) *Macromolecules* **6**, 366.
7. Doi, M. and Kuzuu, N.Y. (1980) *J. Polym. Sci., Polym. Lett. Ed.*, **18**, 775.
8. Pearson, D.S. and Helfand, E. (1984) *Macromolecules*, **19**, 888.
9. Ball, R.C. and McLeish, T.C.B. (1989) *Macromolecules*, **22**, 1911.
10. Marrucci, G. (1985) *J. Polym. Sci. Polym. Phys. Edn.*, **23**, 159.
11. Viovy, J.L., Rubinstein, M. and Colby, R.H. (1991) *Macromolecules*, **24**, 3587-3596.
12. Colby, R.H. and Rubinstein, M. (1990) *Macromolecules*, **23**, 2753.
13. Milner, S.T. and Nicolai, T., discussion contribution during NATO ASI, Cambridge 1996.
14. Milner, S.T. and McLeish, T.C.B. (1996) submitted to *Macromolecules*.
15. Adams, C., Hutchings, L.R., Klein, P.G., McLeish, T.C.B., and Richrds, R.W. (1996) *Macromolecules*, **29**, 5171-5722.
16. Fetters, L.J., Lohse, D.J., Richter, D., Witten, T.A and Zirkel, A. (1994) *Macromolecules*, **27**, 4639.
17. Clarke, N. and McLeish, T.C.B. (1993) *Macromolecules*, **26**, 5264-5266.
18. Roovers, J. and Graessley, W.W. (1981) *Macromolecules*, **14**, 766.
19. Yurasova,T.A., McLeish, T.C.B. and Semenov, A.N. (1994) *Macromolecules*, **27**, 7205-7211.
20. McLeish, T.C.B. and O'Connor, K.P. (1993) *Polymer*, **34**, 2998-3003.
21. Roovers, J. (1984), *Macromolecules*, **17**, 1196.
22. McLeish, T.C.B. (1988), *Macromolecules*, **21**, 1062-1070.

23. Bick, D.K., Johnson, J.M., Hakiki, A., Groves, D.J., Young R.N. and McLeish T.C.B., (1996) in preparation.
24. Meissner, J. (1975) *Pure Appl. Chem.*, **42**, 553.
25. Boris,, D. and Rubinstein, M.,(1996) *Macromolecules*, **29**, 7251-7268.
26. McLeish, T.C.B. (1988) *Europhys. Lett.*, **6**, 511-516.
27. Rubinstein, M., Zurek, S., McLeish, T.C.B. and Ball, R.C. (1990), *J. Phys. France*, **51**, 757.
28. Flory, P.J. (1953) *Principles of Polymer Chemistry*, Cornell University Press, Ithaca NY.
29. Stockmayer, W.H. (1944) *J. Chem. Phys.*, **38**, 355.
30. Lairez, D., Adam, M., Emery, J.R. and Durand, D. (1992) *Macromolecules*, **25**, 286-289 and references therein.
31. Osaki, K. (1993) *Rheologica Acta*, **32**, 429-437.
32. Wagner, M. (1994) *J. Rheol.*, **38**, 655-679.
33. Straube, E., Urban, V., Pyckhout-Hintzen, W., Richter, D. and Glinka, C.J. (1995) *Phys. Rev. Lett.*, **74**, 4464-4467.
34. Osaki, K., Nakamura, H. and Kurata, M. (1982) *Macromolecules*, **15**, 1068.
35. Bick, D.K. and McLeish, T.C.B. (1996) *Phys. Rev. Lett.*, **76**, 2587.
36. Read, D.R. and McLeish, T.C.B. (1996) in preparation.
37. Viovy, J.L. (1985) *J. Physique Lett.*, **46**, 847.
38. Marrucci, G. (1996) Communication during, Newton Institute workshop, Cambridge.
39. Larson, R.G. (1988) *Constitutive Equations for Polymer Melts and Solutions*, Butterworth
40. Spenley, N.A., Cates, M.E. and McLeish, T.C.B. (1993) *Phys. Rev. Lett.* , **71**, 939-942.
41. McLeish, T.C.B. and Ball, R.C. (1986) *J. Polym. Sci. Polym. Phys. Edn.*, **24**, 1735; McLeish,T.C.B. (1897) *J. Polym. Sci. Polym. Phys. Edn.*, **25**, 2253.
42. Berret, J.-F., Porte, G. and Decruppe, J.-P. (1996) *J. Phys. E.*, submitted.
43. Olmsted, P.D.(1996) *Phys. Rev. Lett.*, submitted.
44. McLeish, T.C.B. and Larson, R.G. (1996), in preparation.
45. Kremer, K. and Grest, G. (1995) in *Monte Carlo and MD Simulations in Polymer Science*, ed. K. Binder, (Oxford)
46. Oettinger; H,C, (1995) *Stochastic Processes in Polymer Fluids*, Springer (Berlin).
47. Bishko, G., Harlen, O. and McLeish, T.C.B. (1996) in preparation.

SHEAR-ENHANCED DIFFUSION IN CHARGE-STABILISED COLLOIDS

SRIRAM RAMASWAMY
Department of Physics
Indian Institute of Science, Bangalore 560 012, India
sriram@physics.iisc.ernet.in

Abstract. The generalised Langevin equation method for the dynamics of interacting colloids presented in my previous lecture is extended here to the case of a sheared suspension. A calculation of shear-dependent diffusivities using these methods is found to account for puzzling observations in experiments and simulations. The limitations of the method are discussed, and important unresolved questions presented. This lecture summarises work done in collaboration with A.V. Indrani [1].

1. Introduction

1.1. SUMMARY OF EXPERIMENTS AND SIMULATIONS

As I said in my previous lecture, the slow dynamics and low rigidity of colloidal suspensions of latex spheres make them an ideal laboratory for the study of nonequilibrium and nonlinear phenomena at very modest shear-rates $\dot{\gamma} \sim 10 - 100$ Hz. An early example of such an effect is the observation of strongly anisotropic spatial correlations in a steadily sheared suspension [2]. A phenomenon which caused some surprise when it was first reported [3] is the enhancement of diffusion by shear flow in a strongly charge-stabilised suspension. The measurement was done by forced Rayleigh scattering: the interference pattern between two "write" laser beams is used to modulate the concentration field of the suspension spatially at a specified wavevector. The "write" beams are switched off, and the relaxation of the pattern to equilibrium monitored by diffracting another "read" laser beam off the pattern and watching the the decay of the Bragg peaks. This gives directly the relaxation rate of concentration fluctuations at the imposed wavevector.

T McLeish (ed), Theoretical Challenges in the Dynamics of Complex Fluids, 115-125

Nonequilibrium Brownian dynamics simulations without hydrodynamic interaction [4, 5], have also seen this shear-enhanced diffusion. A theory of this phenomenon, which was lacking until recently [1], is the subject of this chapter. The method is an extension, to the case of a sheared suspension, of the ideas developed in the context of dynamics at equilibrium in the previous lecture. Apart from the intrinsic interest in the phenomenon, this chapter together with the one by S. Milner in this volume should serve as an introduction to the methods of time-dependent statistical mechanics in nonequilibrium steady states. The work I discuss here was also done in collaboration with A.V. Indrani [1].

The experiments [3] and simulations [4, 5] were done in oscillatory shear flow with velocity field

$$\mathbf{v}(\mathbf{r}) = \dot{\gamma}(y)\ cos(\omega_0 t)\ y\ \hat{\mathbf{x}} \tag{1}$$

Note, however, that the behaviour in steady and oscillatory shear flow will be identical as long as we consider oscillation frequencies much smaller than the shear rate. Our theory will be done explicitly for steady shear. In the laboratory experiment [3], $\dot{\gamma}(y)$ in (1) varied linearly with y, corresponding to plane Poiseuille flow and self-diffusion coefficients D_i, $i = x, z$ along x and z were measured. The Brownian simulations measured D_y and D_z; [4] used the plane Couette flow geometry, $\dot{\gamma}$ = constant, while [5] considered both Couette and Poiseuille flows. Both simulations and experiments found an excess diffusivity ΔD_i, relative to that in the absence of shear, that went to zero in the noninteracting (large concentration of ionic impurities) limit. This is of course in *addition* to, and *not* to be confused with, the well-known Taylor dispersion for the displacement along the flow direction $\hat{\mathbf{x}}$ [6], a purely kinematic effect which arises because the thermal wandering of the y–coordinate of a particle leads it into regions of ever–larger x–velocity. It is easy to show, looking at a single random walker in a background with x–velocity varying linearly with y, that this latter effect gives $x^2 \sim \dot{\gamma}^2 t^3$, and leaves y and z unaffected. The data of [3] seems to fit $\Delta D_i \propto |\dot{\gamma}|$ while the simulations, suggest $\dot{\gamma}^{1/2}$ at intermediate shear–rates, an apparent but ultimately not very serious disagreement about which I will comment at the end of this section. In addition, [5] found a quadratic growth for much lower shear-rates, while [4] noted a distinct saturation for the largest $\dot{\gamma}$, and both report strong shear-induced enhancement of correlations and a slowing-down of diffusion along the gradient direction y at very strong shear.

1.2. RESULTS

Using a mode–coupling treatment [7] of the type discussed in the previous chapter, now adapted to the case of a sheared suspension, we [1] calculated the shear-rate dependence of the self-diffusivity. Note that we *ignore* self-consistency effects: this reduces the problem to evaluating an integral rather than solving an integral equation. This was done purely because the anisotropic nature of correlations under shear makes the integral equations involved much uglier, but it does mean some loss of detailed predictive power. As before, we neglect the hydrodynamic interaction (HI), but this should be acceptable not only for the simulations but, at moderate shear-rates, for the experiments as well, since the latter are in the dilute, large screening-length regime [3]. Recall (see Brady's lecture in this volume) that HI effects are proportional to powers of the ratio of particle size to inter-particle distance, which is about 1/5 here.

Fig. 1 shows the self–diffusivities, with Taylor dispersion subtracted out, plotted against the dimensionless shear-rate (or Péclet number) $\gamma \equiv \dot{\gamma}\sigma^2/6D_0$. Gratifyingly, three regimes are clearly seen: (i) an initial quadratic growth, followed by (ii) an appreciable range of $\gamma^{1/2}$, crossing over to (iii) a saturation, for which analytical arguments for a hard-sphere suspension give the form $D_i = D_0 - const.\gamma^{-1/2}$ where D_0 is the diffusivity in the absence of interactions. The behaviour is found to be *isotropic* until the largest shear–rates, in agreement with [3, 4]. That this is an *interaction* effect is clear from the figure, since the effect of shear on self–diffusion is greater for more strongly interacting colloids, as observed by [3, 4]. We are thus able to account largely for the behaviour observed in the simulations [5, 4]. What is more, a re-examination of the experimental data of [3] suggests strongly that what was fitted to a linear behaviour is better described as a crossover from quadratic growth at low shear-rates to sublinear behaviour at larger shear-rates. A good quantitative comparison would, however, require more data.

2. Theory

2.1. QUALITATIVE ARGUMENTS

It is relatively easy to understand qualitatively why shear should enhance diffusion in a collection of interacting Brownian particles. Recall from the previous lecture that the self–diffusion coefficient of the *unsheared* colloidal particles, at temperature T, is equal to T/Γ where the friction coefficient Γ has two contributions: a "bare" part Γ_0, from the solvent, and a correction $\Delta\Gamma$, from interparticle interactions. The important contribution to $\Delta\Gamma$ in a suspension with liquid-like correlations is from concentration fluctuations with wavenumber $q \sim q_0$ not far from the peak of the static structure factor. Although a sheared suspension is not at thermal equilibrium, the relation between friction coefficient and diffusivity can be used if by "friction" we mean the decay–rate of the velocity autocorrelation function (VACF) and

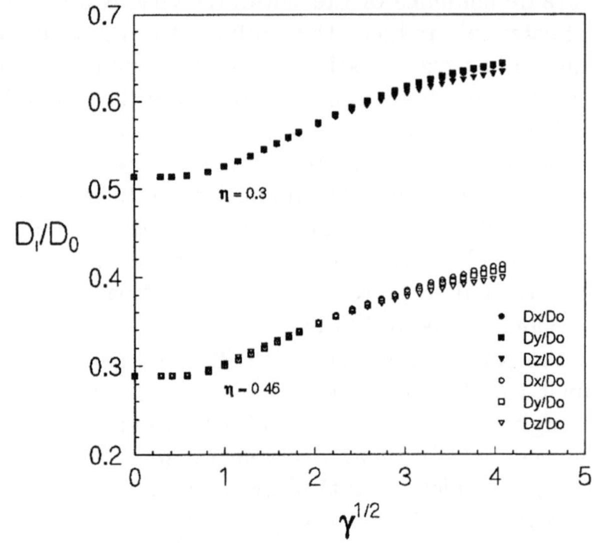

Figure 1. Self–diffusion coefficients D_i/D_0 under shear (for $i = x, y, z$), *vs.* $\gamma^{1/2}$ (where γ is the dimensionless shear–rate), for packing–fractions $\eta =$ 0.3, and $\eta =$ 0.46. The initial γ^2 growth followed by $\gamma^{1/2}$ are clearly seen, and there is some evidence for the onset of saturation.

not necessarily the inverse of a mobility. With this caution [8], let us ask what shear does to the system. The primary physical consequence of imposing a flow with shear-rate $\dot{\gamma}$ on the suspension is that a new timescale enters the problem. Modes with relaxation rate larger than $\dot{\gamma}$ relax before shear has affected them appreciably, and thus behave much as they would at thermal equilibrium. Modes with relaxation rate *smaller* than $\dot{\gamma}$, however, are highly sheared before they can relax by diffusion, and their contribution to $\Delta\Gamma$ effectively eliminated. The effective friction thus decreases as $\dot{\gamma}$ increases. As $\dot{\gamma} \to \infty$, all modes are cut out, and Γ flattens out to its bare value Γ_0 after passing through some kind of crossover region. Since the diffusivity goes as the inverse of Γ, the above argument explains qualitatively the shapes of the curves in Fig. 1.

2.2. EQUATIONS OF MOTION

We begin as in the previous lecture, and with the same notation, with the continuity equation for the self-number density $n_s(\mathbf{r}, t)$ [9], a Langevin equation for the self-momentum density $\mathbf{g}_s(\mathbf{r}, t) \equiv m n_s \mathbf{v}_s$ (which defines $\mathbf{v}_s$, the self-velocity field of a ***tagged particle*** of mass m) and a fluctuating-diffusion equation for the collective density $n(\mathbf{r}, t)$. Now, however, we must describe a colloidal suspension in a steady, uniform shear flow $\mathbf{v} = \dot{\gamma} y \hat{\mathbf{x}}$ [(1), with $\dot{\gamma}(y) =$ constant, and $\omega_0 = 0$]. The friction will thus try to make the self-velocity relax to the background profile (1), not to zero, and the collective diffusion equation will now contain the effects of advection by the mean flow. Since we persist in neglecting the hydrodynamic interaction, the velocity field will always be kept at its mean profile value, with no fluctuations. The identity

$$\frac{\partial n_s}{\partial t} + \nabla.(n_s \mathbf{v}_s) = 0 \tag{2}$$

is the statement of conservation of tagged particle(s);

$$\frac{\partial \mathbf{g}_s}{\partial t} + \Gamma_0(\mathbf{v}_s - \dot{\gamma} y \hat{\mathbf{x}}) + n_s \nabla \frac{\delta F_s}{\delta n_s} = \mathbf{f}(\mathbf{r}, t) \tag{3}$$

balances friction (between the particles and the mean solvent flow) against the thermodynamic driving force arising from the density functional defined in my first lecture; and

$$\frac{\partial n_{\mathbf{q}}(t)}{\partial t} - \dot{\gamma} q_x \frac{\partial n_{\mathbf{q}}}{\partial q_y} + \tau_q^{-1} n_{\mathbf{q}} = \zeta_{\mathbf{q}}(t) \tag{4}$$

is the sheared fluctuating-diffusion equation [10] (see also chapter by S. Milner in this volume), where

$$\tau_q^{-1} = D_0 q^2 / S_0(q). \tag{5}$$

Here we define $S_0(q) = [1 - n_0 c(q)]^{-1}$, the ***equilibrium*** static structure factor of the suspension, to distinguish it from the observed steady-state sheared structure factor $S(\mathbf{q}; \dot{\gamma})$, and $\mathbf{f}$ and $\zeta_{\mathbf{q}}$ are Gaussian thermal noises with variances proportional to γ_0 and $D_0 q^2$ respectively, where $D_0 = T/\Gamma_0$ is the bare self–diffusivity.

2.3. MODE-COUPLING TREATMENT UNDER SHEAR

We must now generalise our definition of the "self-energy" for the friction coefficient to the sheared case. The first step is to define

$$\mathbf{u}_s(\mathbf{r}, t) = \mathbf{v}_s(\mathbf{r}, t) - \dot{\gamma}\; y\; \hat{\mathbf{x}}, \tag{6}$$

the self–velocity field relative to the local background shear–flow, thus subtracting out the effects of Taylor dispersion. This allows us to extract the interaction contribution to diffusivities *along* the flow as well as normal to it. Information about self-diffusion is contained in the correlation function of the u_i fields,

$$C_{ui}(\mathbf{k}, z; \dot{\gamma}) = \int_0^\infty \mathrm{d}t\; e^{izt} \int \mathrm{d}^3 r e^{-i\mathbf{k}\cdot\mathbf{r}} \left\langle u_{si}(\mathbf{0}, 0) u_{si}(\mathbf{r}, t) \right\rangle \tag{7}$$

(with no sum over repeated indices implied), Fourier–transformed in space and Laplace–transformed in time. The relation between velocity correlations and self-diffusivity in a shear flow can then be written as

$$D_i(\dot{\gamma}) = \lim_{z\to 0} \lim_{k\to 0} C_{ui}(\mathbf{k}, z; \dot{\gamma}) \tag{8}$$

where the limit exists because Taylor dispersion has been removed. Since the mean flow has been subtracted, the noninteracting limit of $C_{ui}(\mathbf{k}, z$ is the same as at equilibrium. It is therefore natural to define a self–energy $\Sigma_i(\mathbf{k}, z; \dot{\gamma})$ by

$$C_{ui}(\mathbf{k}, z; \dot{\gamma}) = \frac{T}{z + \Gamma_0 \left[1 + \Sigma_i(\mathbf{k}, z; \dot{\gamma})\right]} \tag{9}$$

Then

$$\frac{D_i(\dot{\gamma})}{D_0} = \frac{1}{1 + \Sigma_i(\dot{\gamma})} \tag{10}$$

where $\Sigma_i(\dot{\gamma}) = \Sigma_i(\mathbf{k} = 0, z = 0; \dot{\gamma})$. In the same Gaussian decoupling approximation as was used in my earlier lecture, Σ_i can be expressed as the product of a self–density and a collective–density correlation function. In detail, starting from (2) and (3), one finds

$$\begin{aligned} \Sigma_i(\dot{\gamma}) &= \int \frac{\mathrm{d}^3 q}{(2\pi)^3} \int_0^\infty \mathrm{d}t\; (q_i^2 + \dot{\gamma} t q_x q_y \delta_{iy})\, c(q) \\ &\times\; c(|\mathbf{q} + \dot{\gamma} t q_x \hat{\mathbf{y}}|) C_s(\mathbf{q}, t; \dot{\gamma}) C_c(\mathbf{q}, t; \dot{\gamma}) \end{aligned} \tag{11}$$

where $c(q)$ is the direct pair correlation function at *equilibrium*, and C_s and C_c are the "bare" correlation functions of the self–density (n_s) and the collective density (n) respectively, ***in the sheared steady state***. These can be calculated from the linearised versions of (2), (3), and (4) as I show below. Note first [11] that uniform shear flow has a kind of translation-invariance along the gradient direction: a translation merely moves you to a different local velocity, which can be compensated by a Galilean boost. Operationally this means that equal-time correlations are translation-invariant in the usual sense, while for time-displaced correlations the result is more complicated, *viz.*, $\langle \delta n(\mathbf{q},t)\delta n(\mathbf{q}',0)\rangle \propto C_c(\mathbf{q},t;\dot\epsilon)\delta(\mathbf{q}+\mathbf{q}'+\dot\epsilon t q_x\hat{\mathbf{y}})$, (and likewise for n_s). Indeed the preceding expressions define what is meant by C_c and C_s. To calculate them, eliminate the self-velocity from (2) by solving for it in terms of n_s in (3) in a linearised approximation. Multiply the resulting equation and (4) by the appropriate densities and average over the noise. This leads to

$$C_s(\mathbf{q},t;\dot\gamma) = e^{-t(D_0 q^2 - \dot\gamma q_x \frac{\partial}{\partial q_y})} 1 \tag{12}$$

and

$$C_c(\mathbf{q},t;\dot\gamma) = e^{-t\left(\frac{D_0 q^2}{S_0(q)} - \dot\gamma q_x \frac{\partial}{\partial q_y}\right)} S(\mathbf{q};\dot\gamma) \tag{13}$$

We can see in (12) and (13) what shear does to the fluctuations in the system: a wavevector $\mathbf{q}$ at time $t=0$ is advected in time to $\mathbf{q}(t) = \mathbf{q} - \dot\gamma t q_x \hat{\mathbf{y}}$. It is this advective transport of modes in wavevector space which in general increases the magnitude of the wavenumber and hastens their decay.

We evaluate the sheared static structure factor [10] $S(\mathbf{q};\dot\gamma)$ in (13) as follows. First multiply (4) by $n_{-\mathbf{q}}$ and average over the noise. This leads to

$$\left(2\tau_q^{-1} - \dot\gamma q_x \frac{\partial}{\partial q_y}\right) S(\mathbf{q};\dot\gamma) = 2 D_0 q^2 \tag{14}$$

where τ_q^{-1} has been defined above. Equation (14) is then integrated directly, leading to

$$S(\mathbf{q};\dot\gamma) = \int_0^\infty \mathrm{d}\lambda e^{-\lambda\left(\frac{D_0 q^2}{S_0(q)} - \frac{1}{2}\dot\gamma q_x \frac{\partial}{\partial q_y}\right)} D_0 q^2. \tag{15}$$

The exponential decay factors appearing in (12), (13) and (15) can be evaluated exactly, using the following identity [11]

$$e^{-\tau\left(h(\mathbf{q}) - b\frac{\partial}{\partial q_y}\right)} = e^{-\int_0^\tau \mathrm{d}\alpha\, h(\tilde{\mathbf{q}})}\; e^{b\tau \frac{\partial}{\partial q_y}} \tag{16}$$

for any function $h(\mathbf{q})$ and constant b, where ($\tilde{\mathbf{q}} = \mathbf{q} + b\alpha\,\hat{\mathbf{q_y}}$) is a shifted wave-vector. For application of the method to polymers in shear see the chapter by S. Milner in this volume. We must then integrate (11) numerically, using equations (12) to (16) to obtain an explicit expression for the shear–dependent diffusivity. As stated at the beginning of this article, this has been done ignoring "self-consistency", i.e., the correlation functions that go into the integral are calculated from the *linearised* theory. We scale all lengths by the mean interparticle distance σ, and all times by $\tau_0 = \sigma^2/6D_0$, the time that a non-interacting particle takes to diffuse a distance σ, and plot our results as functions of the dimensionless shear–rate $\gamma = \dot{\gamma}\tau_0$.

Before describing the numerical calculation, let me make some general observations about the structure of $\Sigma_i(\dot{\gamma})$. (i) C_s and C_c decay faster in time for $\dot{\gamma} \neq 0$ than for $\dot{\gamma} = 0$, so that one expects $\Sigma_i(\dot{\gamma} \neq 0) < \Sigma_i(\dot{\gamma} = 0)$. (ii) If the diffusivity were analytic near vanishing shear-rate, one would expect an initial quadratic growth. The diffusive nature of the dynamics of the concentration fields, however, implies [12] certain non-analyticities for $\dot{\gamma} \to 0$. Nonetheless, the major contribution to $\Sigma_i(\dot{\gamma})$ is from nonzero wavenumbers $q \sim q_0$ where the integrand (controlled by a combination of $S_0(q)$, the vertex $|c(q)|^2 q^2$, and the self–relaxation time $1/D_0 q^2$) is largest. The contribution of density modes near $q = 0$ is suppressed by both phase–space factors and the vanishing of the vertex. Thus, except at the very smallest $\dot{\gamma}$, a quadratic initial growth is still to be expected, and this is indeed what is seen, in both our calculated curve and [5]. (iii) We can do a rough asymptotic analysis on (11) for $\dot{\gamma} \to \infty$, where, from (15), $S(\mathbf{q}; \dot{\gamma}) \to 1$, and C_s as well as C_c decay as $e^{-D_0 q^2 (t + \dot{\gamma}^2 t^3)}$ for most directions of $\mathbf{q}$. The self-energy therefore involves an integral of the form $\int_0^\infty \mathrm{d}t \int_0^\infty \mathrm{d}q\, q^4 (c(q))^2 e^{-q^2(t+\dot{\gamma}^2 t^3)} \equiv I(\dot{\gamma})$ which can be rewritten in terms of a scaling function H as

$$I(\dot{\gamma}) = \int_0^\infty \mathrm{d}q\, q^2\, (c(q))^2\, H(\dot{\gamma}/q^2). \tag{17}$$

Now

$$H(x) \longrightarrow \left\{ \begin{array}{l} 1, x \to 0 \\ const.x^{-2/3}, x \to \infty \end{array} \right\}, \tag{18}$$

so that the dominant contribution to the integral for $\dot{\gamma} \to \infty$ is from *large* q, where $H \simeq 1$, and $I(\dot{\gamma}) \sim \int_{\dot{\gamma}^{1/2}}^\infty \mathrm{d}q\, q^2\, (c(q))^2$. The large-shear-rate behaviour of the diffusivity is thus determined by the way in which $c(q)$ goes to zero for large q. For hard spheres [13] $c(q) \sim q^{-2}$ for $q \to \infty$. Even for particles interacting via a screened Coulomb potential this form, though obviously not the true asymptotic behaviour, seems to fit for a range of wavenumbers beyond the peak. *Assuming* this form for the large q behaviour, it is clear that (17) and (18) lead to $\Sigma_i(\dot{\gamma}) \sim \dot{\gamma}^{-1/2}$ for shear rates such that the corresponding wavenumbers are in the q^{-2} regime. We can therefore expect an appreciable range over which $D_i(\dot{\gamma}) \sim \dot{\gamma}^{1/2}$ before saturating to a constant. Our detailed calculation, whose results are summarized in Fig.1, is in agreement with these trends. As inputs to our numerical calculation, we use the readily available Percus - Yevick [14] analytical expressions for $c(q)$ and $S_0(q)$ for hard spheres. This is quite a reasonable procedure, since the structure and dynamics of a charge-stabilised system with a given screening length are [15] very similar to those for a hard-sphere system with about the same sphere-radius. The general structure of the results we get in this manner should thus apply to charge-stabilised colloids as well. We have evaluated (11) for packing fractions $\eta = 0.3$ and $\eta = 0.46$, where the maximum heights of $S_0(q)$ are about 1.5 and 2.6 and the zero–shear–rate diffusivities about one–half and one–third of the Stokes–Einstein value, respectively. Difficulties with convergence of our integration schemes prevent us from going beyond $\gamma \simeq 15$. Our calculations lead to the results [1] stated in the Introduction and presented in Fig. 1.

3. Conclusion

3.1. DISCUSSION OF RESULTS

Our results for D_i/D_0 (at $\eta = 0.46$, for example) are qualitatively similar to what is seen in the experiments and simulations at similar values of zero–shear–rate diffusivities. The lack of quantitative agreement can also be understood: in our treatment the tagged particle moves through a medium of *freely diffusing* particles, whose self-density field relaxes at a rate given by the *bare* D_0q^2. Shear will thus start to cut off the contribution of these modes to the friction when $\dot{\gamma} \sim D_0{q_0}^2$, where q_0 is the peak position of the structure factor. By contrast, in the experiments and simulations, or in a self–consistent description, the relevant rate is set by the *renormalised* diffusivity, so that the diffusivity *vs.* shear-rate curve should start to turn over at a shear-rate much lower than that given by our "non–self–consistent" calculation. A self-consistent calculation in the presence of shear will largely remedy this discrepancy, but is numerically rather involved. In fact, our calculation suggests a universal relation between suitably non–dimensionalised excess self–diffusivity $(\frac{D_i(\dot{\gamma})-D_i(\dot{\gamma}=0)}{D_i(\dot{\gamma}=\infty)-D_i(\dot{\gamma}=0)})$ and rescaled shear–rate $(\gamma = \dot{\gamma}\tau$, where τ is the ***observed*** rather than bare interparticle diffusion time, in agreement with a speculation in [4].

3.2. CHALLENGES

Improving the present theory to give detailed quantitative agreement with experiment is a relatively minor aim. The major drawbacks of our treatment, which if remedied would yield new physics, are (i) the neglect of the hydrodynamic interaction, (ii) the assumption that the non-hydrodynamic forces are obtained, even far from equilibrium, from an equilibrium free-energy functional, (iii) the use of equilibrium thermal noise to generate fluctuations, and (iv) the inability to account for shear-induced enhancement of correlations along the gradient direction [4, 5]. Two recent papers [12, 16] remedy in part the issue raised in (i). The criticism (ii) requires a reworking of the density functional approach, for steady-state as well as time-displaced correlations, starting from Langevin equations for N interacting Brownian particles in a shear flow. I know of no such treatment, even without the hydrodynamic interaction. Regarding (iii), it is well-known that Stokesian dynamics simulations of non-Brownian hard spheres (see chapter by J. Brady in this volume) [17, 18] find a purely hydrodynamic contribution to the diffusivity of sheared suspensions at large shear-rates, scaling as $\dot{\gamma}a^2$ where a is the radius of the colloidal particle. This means that there is a non-thermal source of effectively random motion for strongly sheared suspensions. A study of the statistics of this nonequilibrium noise, and how to include it in coarse-grained stochastic models of the dynamics at large shear-rates, is a challenge. It is possible that this hydrodynamic dispersion is what makes the data of [3] look close to linear. Turning to (iv), while the analytical results of [16] find incipient ordering along the gradient direction in a treatment in which the hydrodynamic interaction plays a crucial role, this does not explain the simulations, which are simply Brownian dynamics without solvent. There is a potential technical problem in their treatment: their definition of thermodynamic force is obtained from a quadratic approximation to the free-energy functional, which gives a nonlinearity in their expression for the force-density even in the ideal gas limit.

Finally, an understanding of correlations in a sheared suspension is crucial to a complete theory of shear-induced melting and re-entrance of colloidal crystals [19, 20, 21, 22, 23, 24], which remains one of the most fascinating unsolved problems of nonequilibrium statistical mechanics.

References

1. A.V. Indrani and S. Ramaswamy, Phys. Rev. E**52**, 6492 (1995).
2. N.A. Clark and B.J. Ackerson, Phys. Rev. Lett. **44**, 1005 (1980).
3. X. Qiu, H.D. Ou-Yang, D.J. Pine, and P.M. Chaikin, Phys. Rev. Lett. **61**, 2554 (1988).
4. W. Xue and G.S. Grest, Phys. Rev. A **40**, 1709 (1989).
5. J. Chakrabarti, A.K. Sood and H.R. Krishnamurthy, Phys. Rev. E **50**, R 3326 (1994).
6. G.I.Taylor, Proc. Roy. Soc. London Ser. A **219**, 186 (1954); *ibid.* **223**, 446 (1954).
7. W. Hess and R. Klein, Adv. in Phys. **32**, 173 (1983).
8. G. Szamel, J. Bławzdziewicz and J.A. Leegwater, Phys. Rev. A **45**, R2173 (1992).
9. T.R.Kirkpatrick and J.C.Nieuwoudt, Phys. Rev. A **33**, 2651 (1986); *ibid.*, 2658 (1986).
10. D. Ronis, Phys. Rev. A **29**, 1453 (1984).
11. A. Onuki and K. Kawasaki, Ann. Phys. **121**, 456 (1979).

12. J.F. Morris and J.F. Brady, J. Fluid. Mech. **312**, 223 (1996).
13. N.K. Ailawadi, Phys. Rep. **57**, 241 (1980).
14. D. Henderson and E.W. Grundke, J. Chem. Phys. **63**, 601 (1975).
15. E.G.D. Cohen and I.M. de Schepper, J. Stat. Phys. **63**, 241 (1991).
16. B. Morin and D. Ronis, Phys. Rev. E **54**, 576 (1996).
17. G. Bossis and J.F. Brady, in *Hydrodynamics of Dispersed Media*, edited by J.P. Hulin, A.M. Cazabat, E. Guyon, and F. Carmona (North-Holland, Amsterdam, 1990) p. 119.
18. P.J. Mitchell, D.M. Heyes, and J.R. Melrose, J. Chem. Soc. — Faraday Trans. **91**, 1975 (1995).
19. B.J. Ackerson and N.A. Clark, Phys. Rev. Lett. **46**, 123 (1981).
20. H.M. Lindsay and P.M. Chaikin, J. Physique Colloq. **46**, C3-269 (1985).
21. S. Ramaswamy and S.R. Renn, Phys. Rev. Lett. **56**, 945 (1986).
22. B. Bagchi and D. Thirumalai, Phys. rev. A **37**, 2530 (1988).
23. M.J. Stevens, M.O. Robbins, and J.F. Belak, Phys. Rev. Lett. **66** , 3004 (1995).
24. R. Lahiri and S. Ramaswamy, Phys. Rev. Lett. **73** , 1043 (1994).

EFFECTS OF SHEAR FLOW ON SEMIDILUTE POLYMER SOLUTIONS

SCOTT T. MILNER
Exxon Research & Engineering,
Route 22 East, Annandale NJ 08801, USA

Abstract. The effects of flow on concentration variations in semidilute polymer solutions leads to a rich variety of experimental phenomena and predicted observations. Examples I will consider here include 1) coupling of stress and concentration fluctuations observed in dynamic light scattering; 2) large shear-induced concentration fluctuations, including possible shear-induced shifts in the spinodal for polymer-solvent demixing; and 3) polymer migration in inhomogenous shear flows.

1. Introduction.

The outline of this lecture is as follows. Section 1 presents physical arguments for the "two-fluid" model of polymer solution hydrodynamics. As a first application, I summarize the consequences of that model for coupled dynamical fluctuations of stress and concentration in quiescent polymer solutions. In Section 2, I consider a semidilute polymer solution in a marginal solvent under shear; within the "adiabatic" approximation in which dynamics of stress relaxation is neglected, I discuss the basic mechanism for shear-enhanced concentration fluctuations. I then describe the how the stress relaxation time cuts off this mechanism at short distances and times, leading to a characteristic length scale observed in light-scattering under shear. In Section 3, I discuss the shift of the demixing spinodal at second order in shear rate. (For background on spinodals, see the chapter by M. Doi in this volume.)

The theoretical results I present here [1] [2] draw on the work of Fredrickson and Helfand [3], Doi [4], and Onuki [5], and Brochard and de Gennes [6]. The experiments I describe were made by Pine et al. [7] and Hashimoto

T McLeish (ed), Theoretical Challenges in the Dynamics of Complex Fluids, 127-140.

et al. [8] (static light scattering under shear), and Adam and Delsanti [9] (dynamic light scattering from solutions).

2. Two-Fluid Model.

To a physicist, hydrodynamics is a precise and well-controlled theory, valid in the limit of low frequencies and long wavelengths when local equilibrium is reached, and only a very limited set of degrees of freedom, namely the conserved quantities (concentrations $\phi(\boldsymbol{r})$, momentum density $\boldsymbol{g}(\boldsymbol{r})$, energy density) and broken-symmetry variables (simple displacements for crystals, rotations for ferromagnets,...), are still out of equilibrium. Very general arguments exist that prescribe the form of the hydrodynamic equations, based on conservation laws, Galilean invariance, and the Onsager relation between dissipative coefficients [10]. Unfortunately, for liquids containing complex structures, which because of their size typically have long relaxation times, the window of frequency over which rigorous hydrodynamics is valid is pathetically small. And, to someone interested in "complex fluids", rigorous hydrodynamics is boring; all fluids are Newtonian in the hydrodynamic limit.

To expand the regime where a generalized hydrodynamic theory is valid, we are obliged to add to the list of non-equilibrium variables those which although not conserved or associated with a broken symmetry, have for other reasons slow dynamics. If there are only a few such degrees of freedom, we may have a valid description, while avoiding writing a set of 6×10^{23} equations. In semidilute polymer solutions, the entanglement of the chains with each other leads to an arbitrarily long timescale for stress relaxation via reptation, even though the stress tensor is not a conserved quantity. This suggests adding the stress tensor $\boldsymbol{\sigma}(\boldsymbol{r})$ to the set of dynamical variables.

Having said that, the question remains what equations to write for the coupled dynamics of concentration and stress. Here, I appeal to physical arguments to write the "two-fluid" model for a polymer solution. The name refers to the idea that there is a separate balance of forces on the polymer and on the solvent. First, a (large) drag force $-\Gamma(\phi)(\boldsymbol{v}_p - \boldsymbol{v}_s)$ acts on the polymer concentration (and its negative acting on the solvent) if the polymer is in motion relative to the solvent. By analogy with flow in porous media (or dimensional analysis), we expect $\Gamma(\phi)$ to be of order $\eta_0/\xi(\phi)^2$, where ξ is the hydrodynamic correlation length.

To specify the two-fluid model, we must decide for each of the forces acting in the solution whether it acts "directly" on the polymer (or on the solvent), or only influences the polymer indirectly via the drag force. We expect that 1) the hydrostatic pressure p acts on both solvent and polymer alike (to enforce incompressibility); 2) the osmotic pressure acts directly on the polymer concentration (mediated by monomer-monomer repulsion); 3) the solvent viscous stress $\boldsymbol{\sigma}^{(s)}$ acts on the solvent (which we will shortly neglect anyway); and 4) the polymer elastic stress $\boldsymbol{\sigma}^{(e)}$, representing the pull of the entropic springs acting across surfaces in the fluid, acts directly on the polymer. Thus we are led to write

$$0 = -\Gamma(\phi)(\boldsymbol{v}_p - \boldsymbol{v}_s) + \nabla \cdot (\boldsymbol{\sigma}^{(o)} + \boldsymbol{\sigma}^{(e)}) - \phi \nabla p \tag{1}$$

$$0 = -\Gamma(\phi)(\boldsymbol{v}_s - \boldsymbol{v}_p) + \nabla \cdot \boldsymbol{\sigma}^{(s)} - \nabla p \tag{2}$$

Here $\sigma_{ij}^{(o)}$ is the "osmotic stress". In the absence of gradient terms, $\sigma_{ij}^{(o)} = \pi(\phi)\delta_{ij}$, where $\pi(\phi) = \phi\partial f/\partial\phi - f$ is the osmotic pressure, and f is the free energy density. More generally, $\nabla \cdot \boldsymbol{\sigma}^{(o)}$ is a force density, and is equal to $\phi\nabla\delta F/\delta\phi = \phi\nabla\mu$ (monomers per volume times thermodynamic force per monomer) including gradient terms.

The velocity $\boldsymbol{v}_p$ enters the conservation law for ϕ in the usual way, with a concentration current $\phi\boldsymbol{v}_p$, as

$$\begin{aligned} 0 = & \quad \dot{\phi} + \nabla \cdot (\phi \boldsymbol{v}_p) & (3) \\ = & \quad \dot{\phi} + \nabla \cdot (\phi \boldsymbol{v}) + \nabla \cdot \zeta^{-1} \nabla \cdot (\boldsymbol{\sigma}^{(0)} + \boldsymbol{\sigma}^{(e)}) & (4) \end{aligned}$$

where the mobility ζ is defined by $\phi\zeta = \Gamma$. Because ϕ is typically small (a few percent), we approximate the fluid velocity $\boldsymbol{v}$ by $\boldsymbol{v}_s$, and drop other terms of order $O(\phi)$ where convenient.

The coupling of concentration to gradients of the chemical potential gives rise to conventional diffusion. If we expand around the uniform concentration, $\delta\mu = \chi^{-1}\delta\phi$ (which defines the osmotic susceptibility χ^{-1}) and we have a diffusion coefficient $D = \phi\zeta^{-1}\chi^{-1}$. The second source of polymer concentration current is novel: it represents stress-driven concentration currents. To see this term in action, consider an initial condition in which a polymer solution of uniform concentration contains chains that are stretched along the x-direction by an amount that varies as $\cos qx$. In those regions where the chains are more stretched, they will tend to contract, increasing the local polymer concentration and driving solvent into the regions where the chains are less stretched.

Thus inhomogenous polymer stresses can drive concentration currents. Of course, the resulting inhomogenous concentration will eventually diffuse away. Thinking about this simple example, we already see an important length scale emerge. The polymers relax their stress in some finite time τ (the stress relaxation time), but the time for diffusion to smooth out the resulting concentration variation is $(Dq^2)^{-1}$. For large q, diffusion is faster than stress relaxation, and the concentration tends to stay uniform.

Equating the rates τ^{-1} and Dq^2 determines a length scale q^*, at which the chain contraction begins to be ineffective at producing a concentration variation. This length scale can be large compared to the radius of gyration of a chain. To see this, note that D is the collective diffusion constant, which (in good solvents) scales as $D \sim kT/(\eta_0 \xi)$, where ξ is the solution correlation length. This is much larger than the self-diffusion constant D_{self}, which scales [11] as $D(\phi/\phi^*)^{-5/2}$. Since a chain diffuses its own radius in the time τ, the length $\sqrt{D\tau}$ scales as $R(\phi/\phi^*)^{5/4}$, which for entangled solutions can be large compared to R.

So far, we have said nothing about the equation of motion for the stress tensor, other than that it relaxes in a reptation time τ. Writing sensible evolution equations for the stress in a polymer solution is the province of rheologists; books have been written about the subject [12]. Fortunately, we have only modest requirements for our stress evolution equation. For present purposes, the Maxwell model suffices:

$$\dot{\boldsymbol{\sigma}}^{(e)} + v \cdot \nabla \boldsymbol{\sigma}^{(e)} = \boldsymbol{K}_p \cdot \boldsymbol{\sigma}^{(e)} + \boldsymbol{\sigma}^{(e)} \cdot \boldsymbol{K}_p^T - \tau^{-1}(\boldsymbol{\sigma}^{(e)} - G\mathbf{1}) \tag{5}$$

The left-hand side is the usual convective time derivative; spatial inhomogenieties of the stress tensor are moved along by the velocity. Naturally, the appropriate velocity to advect the polymer stress tensor is $\boldsymbol{v}_p$.

The stress relaxes with a single relaxation time to an isotropic tensor, the magnitude of which corresponds to the thermally generated tension of entangled chains seeking to reptate out of their tubes [13]. The stress is built up by convection, by the polymer velocity gradient $(K_p)_{ij} = \nabla_j (v_p)_i$. Thus concentration perturbations are able to produce inhomogenous elastic stresses; as the concentration diffuses away, a nonuniform velocity $\boldsymbol{v}_p$ is established, which tends to build up the stress via the terms involving $\boldsymbol{K}$.

For the Rouse model, the convective terms in (5) are exactly correct; the adjacent beads on a Rouse chain defining a spring move relative to each other in the velocity gradient. For more realistic models of viscoelastic polymer solutions, the terms building up the stress would become non-affine at larger stresses or strain rates, leading to shear-thinning of the solution viscosity.

The simplest application of the coupled equations of motion (4) and (5) is the relaxation of small fluctuations around a uniform quiescent solution. We expand in perturbations $\delta\phi$ and $\delta\boldsymbol{\sigma}$, and Fourier transform the equations to obtain

$$\dot{\phi} + Dq^2\phi - \Gamma^{-1} q^2 \sigma_{qq}^{(e)} = 0 \tag{6}$$

$$\dot{\sigma}_{qq}^{(e)} - 2Dq^2\phi + (\tau^{-1} + G\Gamma^{-1} q^2)\sigma_{qq}^{(e)} = 0 \tag{7}$$

The two cross-couplings correspond to the physical effects described above: generation of concentration variations by the relaxation of inho-

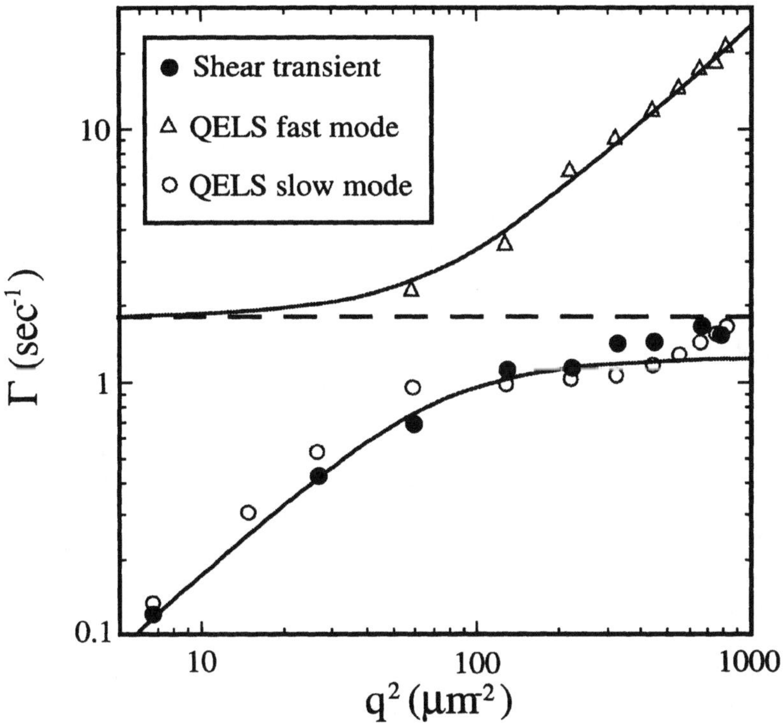

Figure 1. The eigenmodes of the two-fluid model are mixed by the dynamical coupling; the effect is strongest at q^*. Quasielastic light scattering experiments reveal these two modes [data from Ref. 7].

mogenous stress via chain relaxation, and the generation of stress through convection by an inhomogenous polymer velocity as a concentration variation diffuses away. In the absence of cross-coupling, stress variations would relax at a rate τ^{-1} independent of q, and concentration variations would relax diffusively at a rate Dq^2. As a result of the cross-couplings, the eigenmodes of this 2x2 system are mixed, with a pair of relaxation rates that are not simple functions of q^2. (See Fig. 1.)

The "mode mixing" is strongest at a wavenumber of order q^*, at which the two separate relaxation rates are comparable. This behavior is evident in dynamic light-scattering studies of semidilute polymer solutions, in which the time autocorrelation function of scattered light intensity decays not as a single exponential, but as a weighted sum of exponentials with decay rates given by the eigenvalues of (7).

3. Shear-induced fluctuations.

It has been known experimentally for a long time that entangled polymers in marginal solvents develop large concentration fluctuations when sheared. For instance, a 5 percent solution of 3×10^6 molecular weight polystyrene in dioctyl phthalate flowing through a 1mm capillary tube at 1 cm/sec will become cloudy and white in appearance [7]. The effects are largest in the general vicinity of the polymer-solvent critical point (though the temperature may be many degrees from T_c), on the entangled (high-concentration) side.

This shear-induced turbidity was commonly thought to indicate phase separation of the polymer and solvent, until light-scattering measurements [7] [8] probed the spatial structure of the concentration fluctuations. It was found at moderate shear rates (to be defined more precisely below) that the scattering was peaked not at small wavenumbers, as would be usual for an approaching spinodal or for incipient phase separation, but at a finite wavevector corresponding to a length scale of about a micron.

In a solution under shear flow, all directions are inequivalent, so it is important to ask in which direction of $\boldsymbol{q}$ the peak in $S(\boldsymbol{q})$ lies. We take as a coordinate system $\hat{\boldsymbol{x}}$ in the velocity direction, $\hat{\boldsymbol{y}}$ in the gradient direction, and $\hat{\boldsymbol{z}}$ in the vorticity direction. The peak in the structure factor for low shear was in the $\hat{\boldsymbol{x}} + \hat{\boldsymbol{y}}$ direction, i.e. in the first (and third) quadrant; as the shear rate was increased, the peak progressively migrated towards the $\hat{\boldsymbol{x}}$ direction, finally ending up in the $\hat{\boldsymbol{x}} - \hat{\boldsymbol{y}}$ direction (fourth and second quadrants). (See Fig. 2).

Note that $\hat{\boldsymbol{x}} + \hat{\boldsymbol{y}}$ is a surprising direction in $\boldsymbol{q}$-space to see peaks in sheared fluids. A naive argument says that the shear flow affinely deforms equilibrium concentration fluctuations, which may be thought of as round blobs of size ξ. This leads to deformed blobs slightly thicker along the $\hat{\boldsymbol{x}} + \hat{\boldsymbol{y}}$ direction (and thinner along $\hat{\boldsymbol{x}} - \hat{\boldsymbol{y}}$, with a form factor that correspondingly falls off faster along $\hat{\boldsymbol{x}} + \hat{\boldsymbol{y}}$. This argument thus leads to *weaker* scattering at a given wavenumber in the first and third quadrants; also, the argument continues to give the peak in $S(\boldsymbol{q})$ at $\boldsymbol{q} = 0$.

We can understand the scattering experiments in terms of Eqns. (4) and (5) in finite shear. As a first approach, we make an adiabatic approximation for the polymer elastic stress [3]; namely, that it relaxes quickly to a level consistent with the local concentration and shear rate. For small shear (no normal stress) this is simply

$$\boldsymbol{\sigma}^{(e)} \approx \eta(\phi)\dot{\gamma}(\hat{\boldsymbol{x}}\hat{\boldsymbol{y}} + \hat{\boldsymbol{y}}\hat{\boldsymbol{x}}) \tag{8}$$

where $\eta = G\tau$ is the solution viscosity. Note that $\eta(\phi)$ is concentration dependent, which will be essential in what follows.

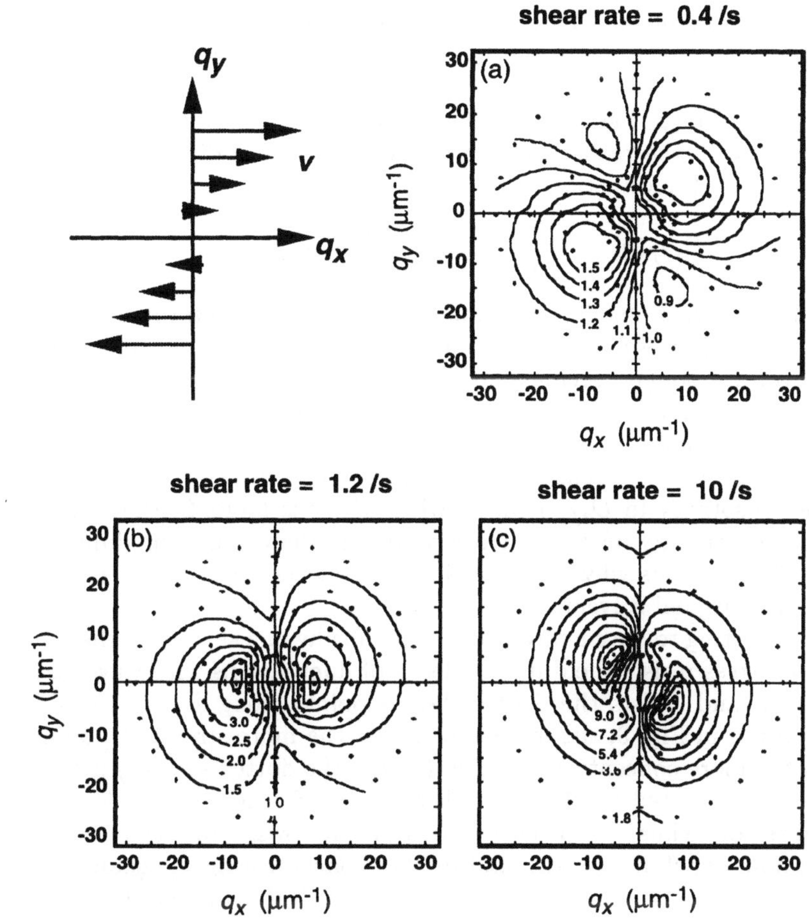

Figure 2. Contour plot of steady-state structure factor in the shear plane, at various shear rates, for 4 percent 1.8×10^6 polystyrene in dioctyl phthalate [from Ref. 7].

If we use the adiabatic stress in Eqn. (4), expand to linear order in $\delta\phi$ about a uniform concentration, and take the spatial Fourier transform, we obtain

$$\dot{\phi} + (Dq^2 - 2\dot{\gamma}\eta'\zeta^{-1}q_x q_y)\phi = f_\phi \tag{9}$$

We have added a random noise source $f_\phi(q,t)$ to this equation, to represent the effect of thermal fluctuations. Since the microscopic origin of the noise is the motion of fast degrees of freedom left out of the hydrodynamic description,.which are in local equilibrium, we expect the noise statistics to be unaffected by the shear flow. Thus we take the noise amplitude to be given by the usual fluctuation-dissipation theorem [10], i.e., $[f_\phi(\boldsymbol{q},t)f_\phi(-\boldsymbol{q},t')] = 2\zeta q^2\delta(t-t')$.

Equation (9) looks like a diffusion equation in shear flow (the convection term gives rise to the rather ugly second term), with a diffusion constant that depends on the direction of $\boldsymbol{q}$:

$$D_{eff}(\hat{\boldsymbol{q}}) = D\left(1 - 2\dot{\gamma}\tau\frac{\partial \ln\eta}{\partial\ln\phi}\frac{G}{\chi^{-1}\phi_0^2}\hat{q}_x\hat{q}_y\right) \tag{10}$$

Note that $D_{eff}(\boldsymbol{q})$ is smaller for $\boldsymbol{q}$ pointing along the $\hat{\boldsymbol{x}}+\hat{\boldsymbol{y}}$ direction, and larger for $\boldsymbol{q}$ along $\hat{\boldsymbol{x}}-\hat{\boldsymbol{y}}$. Thus fluctuations born with wavevectors in that direction will live longer, giving a larger steady-state value of $S(\boldsymbol{q};\dot{\gamma})$ in the first and third quadrant, and contrariwise for wavevectors in the second and fourth quadrants, in agreement with the light-scattering experiments.

An estimate of the strength of the effect is likewise contained in Eqn. (10). The anisotropy in D_{eff} is large when 1) $\dot{\gamma}\tau$ is large, 2) the viscosity $\eta(\phi)$ depends strongly on concentration, and 3) the osmotic modulus χ^{-1} is small. This explains why the shear-induced effects are seen in entangled solutions in marginal solvents.

It is comforting to have a physical picture of why in a shear flow the polymer elastic stress resists the diffusive decay of fluctuations with wavevector along $\hat{\boldsymbol{x}}+\hat{\boldsymbol{y}}$, and assists the decay when $\boldsymbol{q}$ points along $\hat{\boldsymbol{x}}-\hat{\boldsymbol{y}}$. Consider the fluctuation in Fig. 3; the force exerted by the more-stretched polymer chains in the more concentrated region (across the dotted surface) is larger than that that exerted by the fewer, less-stretched chains in the less concentrated region. Thus, the surface tends to move toward the region of high concentration, expelling solvent, and further concentrating the chains. Diffusion opposes this, of course, but the rate at which this fluctuation relaxes is reduced relative to a quiescent solution.

The expression for the steady-state structure factor from Eqn. (9) is somewhat involved, because of the nasty convective term, but the physics is relatively simple. 1) Fluctuations are born at all wavevectors, with a birth rate given by the noise amplitude. 2) By the action of the convective term, a wavelet born with wavevector $\boldsymbol{q}$ can be regarded as "travelling in $\boldsymbol{q}$-space" according to

$$\dot{\boldsymbol{q}} = -\dot{\gamma}q_x\hat{\boldsymbol{y}} \tag{11}$$

3) As the wavelet moves along, its amplitude decays due to diffusion with a rate $D_{eff}(\boldsymbol{q})q^2$. The structure factor is a steady-state average over this process of birth, convection, and decay by diffusion. This is the physical content of the formula

$$S(\boldsymbol{q}_0) = \chi\int_{-\infty}^{0} dt' 2Dq^2(t')\exp\left(-2\int_{t'}^{0} D_{eff}(\hat{\boldsymbol{q}}(t''))q^2(t'')dt''\right) \tag{12}$$

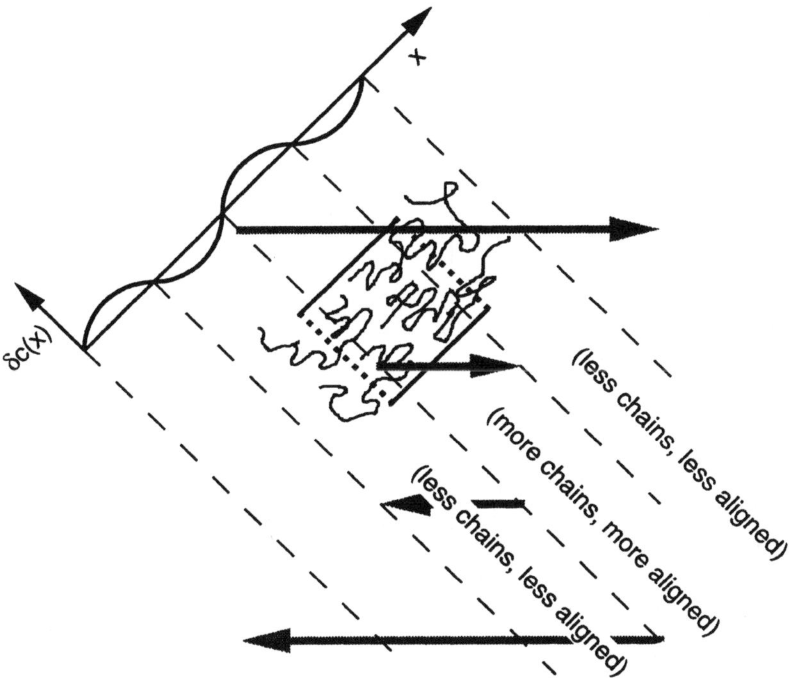

Figure 3. The larger tension of the more-stretched, more-numerous chains in the high concentration region tend to pull the boundary (dotted line) inwards, resisting diffusive decay of the fluctuation.

The adiabatic approximation captures much of the physics of the shear-induced concentration fluctuations, but fails to give a peak; D_{eff} is not a function of $|q|$. The peak originates from the breakdown of the adiabatic approximation. If the concentration increases somewhere because of a fluctuation, the stress takes a time τ to respond, i.e. to relax to the value that the adiabatic approximation assumed was reached instantly. If the concentration fluctuation is of short wavelength, it may diffuse away before the stress has time to respond. Evidently, this reduces the effectiveness of the mechanism that led to the anisotropy in $D_{eff}(\hat{q})$. The wavenumber where stress dynamics becomes important is just the q^* we encountered before.

To compute the steady-state structure factor from Eqns. (4) and (5) is somewhat of an ordeal. [2] The basic idea is to 1) expand to linear order in $\delta\phi$ and $\delta\sigma$ about a state of uniform concentration under steady shear, and 2) to work only to first order in shear rate. This results in linear equations

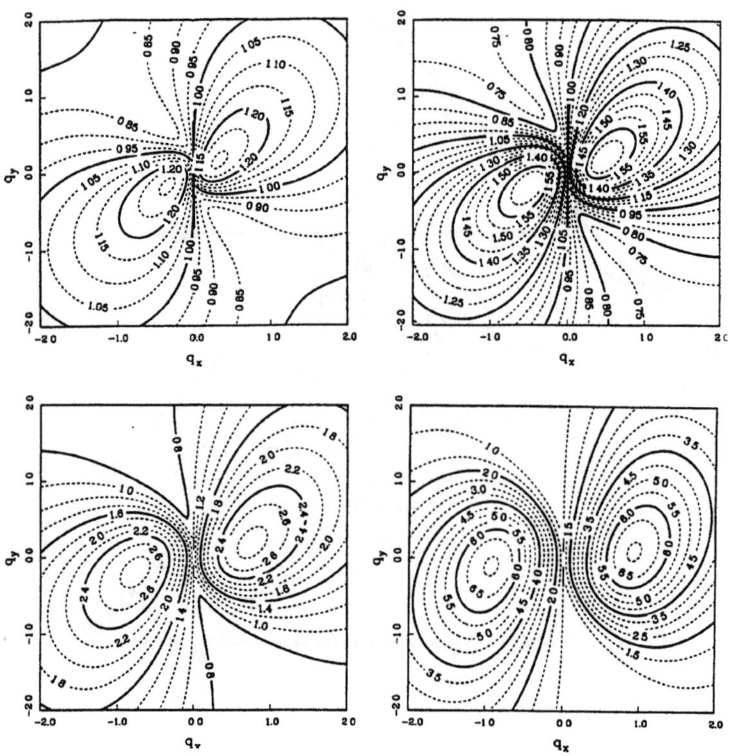

Figure 4. Contour plots (from Ref. 1) of numerically calculated $S(q)$ for parameters corresponding to experiments of Ref. 7.

coupling the dynamics of $\delta\phi(\boldsymbol{q})$ with only one component of the stress tensor, namely $\hat{\boldsymbol{q}} \cdot \delta\boldsymbol{\sigma} \cdot \hat{\hat{\boldsymbol{q}}}$, where $\hat{\hat{\boldsymbol{q}}}$ is at right angles in the shear plane to $\hat{\boldsymbol{q}}$.

The structure factor can then be calculated by an appropriate generalization of the "history integral" Eqn. (12). Resulting plots of $S(\boldsymbol{q};\dot{\gamma})$ show peaks of a reasonable width, position, and amplitude compared to light-scattering experiments at modest shear rates. Even within the $O(\dot{\gamma})$ theory, the peak migrates towards the $\hat{\boldsymbol{x}}$-axis, due to the effect of convection (which carries enhanced fluctuations in that direction before they expire by diffusion). (See Fig. 4.)

4. Shifts in the Spinodal.

So far, we have not discussed whether or not the polymer-solvent demixing transition is actually shifted by the shear flow. A simpler question is whether shear shifts the spinodal, the temperature at which long-wavelength concentration fluctuations grow spontaneously. This question can only be asked for concentration fluctuations with wavevectors perpendicular to the velocity direction.

Such fluctuations are not distorted by the shear flow, and we can watch them for an arbitrarily long time to see if they grow or not. Then convective terms drop out of Eqn. (4), and because we seek the point at which fluctuations just begin to grow, the adiabatic approximation for the stress is justified. Working at second order in shear rate, a general constitutive equation for the polymer elastic stress is

$$\begin{aligned}\sigma^{(e)} &\approx \eta(\phi)\dot{\gamma}(\hat{\boldsymbol{x}}\hat{\boldsymbol{y}}+\hat{\boldsymbol{y}}\hat{\boldsymbol{x}})+\dot{\gamma}^2(\psi_x(\phi)\hat{\boldsymbol{x}}\hat{\boldsymbol{x}}+\psi_y(\phi)\hat{\boldsymbol{y}}\hat{\boldsymbol{y}}+\psi_z(\phi)\hat{\boldsymbol{z}}\hat{\boldsymbol{z}})\\ &= \eta(\phi)\dot{\gamma}(\hat{\boldsymbol{x}}\hat{\boldsymbol{y}}+\hat{\boldsymbol{y}}\hat{\boldsymbol{x}})+\dot{\gamma}^2[(2n_1/3+n_2/3)\hat{\boldsymbol{x}}\hat{\boldsymbol{x}}+\\ &\quad +(-n_1/3+n_2/3)\hat{\boldsymbol{y}}\hat{\boldsymbol{y}}+(-n_1/3-2n_2/3)\hat{\boldsymbol{z}}\hat{\boldsymbol{z}}]\end{aligned} \tag{13}$$

The second line above expresses the normal stress coefficients $\{\psi_x, \psi_y, \psi_z\}$ in terms of normal stress difference coefficients $\{n_1, n_2\}$, *assuming* that there is no contribution to the trace of the elastic stress tensor at $O(\dot{\gamma}^2)$. A constant-trace stress tensor is a consequence of tube models for the stress. Onuki [5] has argued for a term in the osmotic pressure (interchangeable with the isotropic part of the elastic stress) arising from a contribution to the free energy density of

$$f_{el} = (\eta\dot{\gamma})^2/(2G(\phi)) \tag{14}$$

which is meant to be the steady-state elastic free energy at lowest order in shear rate.

We then impose a small periodic variation in concentration $\delta\phi$ with wavevector either along the $\hat{\boldsymbol{y}}$ (gradient) or $\hat{\boldsymbol{z}}$ (vorticity) direction. If the concentration varies along $\hat{\boldsymbol{y}}$, so must the shear rate $\dot{\gamma}$, so that the shear stress $\eta(\phi)\dot{\gamma}$ is constant; if the concentration varies along $\hat{\boldsymbol{z}}$, the shear rate itself is constant. We then use Eqn. (13) in Eqn. (4) expanded to $O(\delta\phi)$ to find the effective diffusion constant $D_{eff}(\hat{\boldsymbol{q}})$:

$$D_{eff}(\hat{\boldsymbol{y}}) = D - C_G(C_G-1)\Gamma^{-1}\frac{(\eta\dot{\gamma})^2}{2G} + 2C_\tau\psi_y\dot{\gamma}^2\Gamma^{-1} \tag{15}$$

for wavevector along $\hat{\boldsymbol{y}}$, and

$$D_{eff}(\hat{\boldsymbol{z}}) = D + 2(C_G-1)(2C_\tau + C_G)\Gamma^{-1}\frac{(\eta\dot{\gamma})^2}{2G} - C_{\psi_z}\psi_z\dot{\gamma}^2\Gamma^{-1} \tag{16}$$

for wavevector along $\hat{\boldsymbol{z}}$. In the above, $C_G = \partial\log G(\phi)/\partial\log\phi$ and similarly for C_τ.

To reach the above equation, we have assumed 1) η scales as $G(\phi)\tau(\phi)$, and 2) ψ_i scales as $G(\phi)\tau(\phi)^2$, where we expect the log derivatives to satisfy $C_G > 1$ and $C_\tau > 0$. Now note that $N_1 >> N_2$, so that $\psi_y < 0$ and $\psi_z < 0$. Then, both terms in Eqns. (15) and (16) have the same effect: the spinodal is enhanced along the velocity direction and suppressed along the vorticity direction. This is in agreement with light-scattering experiments at higher shear rate.

5. Polymer Migration.

Another simple application of Eqn. (4) and Eqn. (13) is to polymer solutions in inhomogenous steady flows [4]. Consider a plate-plate rheometer, in which shear rates and normal stresses vary with radial position ρ. This leads to radial stress-driven currents, which in steady state balance against diffusion currents driven by concentration gradients. Setting $\hat{\rho} \cdot v_p = 0$ implies

$$0 = \frac{\partial \sigma_{\rho\rho}}{\partial \rho} + (\sigma_{\rho\rho} - \sigma_{\theta\theta})/\rho - \chi^{-1}\frac{\partial \phi}{\partial \rho} \tag{17}$$

Using the second-order constitutive equation Eqn. (13), recalling that $\dot{\gamma}(\rho) \propto \rho$, and assuming $N_1 >> N_2$ leads to a concentration variation

$$\frac{\delta\phi(\rho)}{\phi_0} = \frac{-5n_1\dot{\gamma}(\rho)^2}{6\chi^{-1}\phi_0} \tag{18}$$

The concentration is highest in the center. This is a large effect under basically the same conditions (strong shear, marginal solvent) that gave rise to large concentration fluctuations in shear.

Note that if the plate-plate rheometer is replaced by a cone-plate setup, which has constant shear, there is still an effect:

$$\frac{\delta\phi(\rho)}{\phi_0} = \frac{-n_1\dot{\gamma}^2 \log(\rho/\rho_0)}{\chi^{-1}\phi_0} \tag{19}$$

No effect in a cone-plate setup would be predicted by a theory which supposed that the trace of the stress tensor was affected by shear, i.e., that the osmotic pressure was shear-rate dependent [5].

Such experiments have not to my knowledge been performed, perhaps because of the long equilibration time, set by the collective diffusion constant and the dimensions of the rheometer. To estimate this, take a diffusion constant $D \sim kT/(\eta_0\xi)$ with $\xi = 100\mathring{A}$ and $\eta = 1$ centipoise, and a plate radius of 1 cm, and the resulting timescale is about 10^5 seconds, or about 10 days.

To shorten the timescale, one could imagine a cylindrical Couette geometry with a modulated gap, perhaps on the scale of 1 mm. A similar

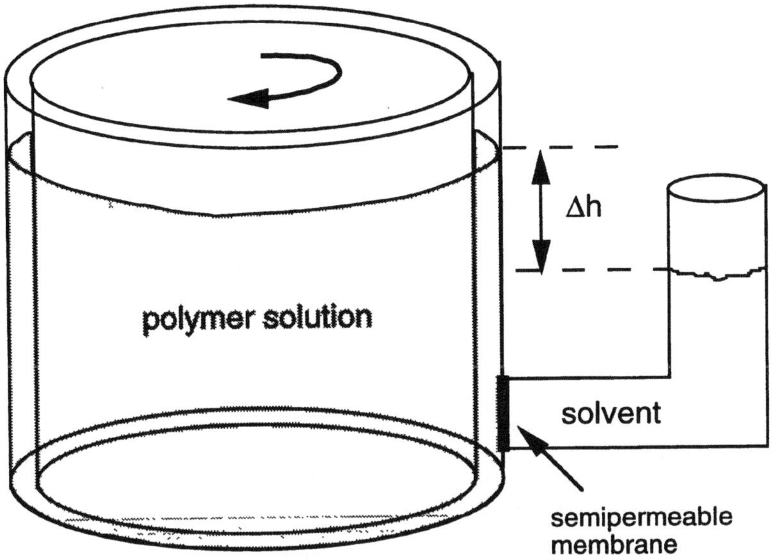

Figure 5. Does the level of solvent in the osmometer rise, fall, or remain the same when the solution is sheared?

effect to the above would be predicted, with concentration higher in the large-gap region, but with an equilibration 100 times shorter than above, i.e., 1000 sec. The concentration variation could be observed by index of refraction changes, preferably after stopping the shear and waiting a short time for the shear-induced concentration fluctuations to decay away.

Another worthwhile experiment yet to be performed would ask directly whether or not there is a contribution to osmotic pressure arising from the "elastic free energy" of Eqn. (14). One could imagine an osmometer connected to a Couette shear cell (Fig. 5); the question would be, does the level of solvent change when the solution is sheared?

Conclusions.

The dynamic coupling between concentration and stress in an entangled polymer solution, proposed some time ago by Brochard and de Gennes in connection with dynamic light scattering experiments of Adam and Delsanti, has dramatic consequences when such solutions are subjected to a shear flow. Strong anisotropic scattering from shear-induced concentration fluctuations result, all the main features of which are accounted for in a minimal dynamical theory. Some unanswered questions remain, including whether and how the first-order phase boundary for polymer-solvent demixing is shifted by shear. A proper dynamical definition of coexisting phases could be given in terms of a shear-stabilized stationary interface between polymer-rich and polymer-poor phases. The nightmarish possibility exists that such a shear-stabilized interface would be hydrodynamically unstable, tending to ripple and break up in shear flow, rendering experimental observations of the coexistence curve difficult at best.

References

1. S. T. Milner, Phys. Rev. Lett. **66**, 1477 (1991).
2. S. T. Milner, Phys. Rev. E **48**, 3674 (1993).
3. E. Helfand and G. H. Fredrickson, Phys. Rev. Lett. **62**, 2468 (1989).
4. M. Doi, in "Dynamics and Patterns in Complex Fluids: New Aspects of the Physics-Chemistry Interface", edited by A. Onuki and K. Kawasaki, Springer Proccedings in Physics vol. 52 (Springer-Verlag: Berlin, 1990).
5. A. Onuki, J. Phys. Soc. Japan **59**, 3427 (1990).
6. F. Brochard and P.-G. de Gennes, Macromolecules **10**, 1157 (1977).
7. X.-L. Wu, D. J. Pine, and P. K. Dixon, Phys. Rev. Lett. **66**, 2408 (1991).
8. T. Hashimoto, T. Takebe, and K. Fujioka, in Ref. 4.
9. M. Adam and M. Delsanti, Macromolecules **18**, 1760 (1985).
10. D. Forster, "Hydrodynamic Fluctuations, Broken Symmetry, and Correlation Functions" (Benjamin/Cummings: Reading, MA 1978).
11. P.-G. de Gennes, "Scaling Concepts in Polymer Physics" (Cornell University Press: Ithaca, NY, 1979), Chapter III.
12. R. G. Larson, "Constitutive Equations for Polymner Melts and Solutions", (Butterworths: Boston, 1988).
13. M. Doi and S. F. Edwards, "The Theory of Polymer Dynamics", (Clarendon: Oxford, 1986).

RHEOLOGY OF LIQUID CRYSTALLINE POLYMERS

Linear and Nonlinear Behaviour of the Nematic Phase

G. MARRUCCI

University of Naples, Department of Chemical Engineering
Piazzale Tecchio, 80125 Napoli, Italy

1. Introduction

The nematic phase of main-chain liquid crystalline polymers (LCPs in the following) is to be considered a complex fluid for more than one reason. To begin with, similarly to nematics of low molecular weight, LCPs are locally anisotropic, as their (more or less rodlike) molecules want to stay roughly parallel to one another. The average common direction is described by a unit vector **n**, called *director*. The anisotropy is *per se* a factor of complexity: the viscosity, for example, is no longer a uniquely defined quantity, as the rate of energy dissipation during steady flow will be different, depending on how the director is oriented with respect to the velocity gradient.

Secondly, the director will generally vary in space. Gradients in the director field generate an elastic response, named after Frank, which is a long range effect, as it arises from *distortions* of the "crystalline" order. Frank elasticity, which again is present both in LCPs and in small molecule nematics (see contribution of R. Larson, this volume), should not be confused with the ordinary entropic elasticity of polymeric substances. The latter is also to be expected in LCPs, however, as they are polymers after all, and as such relax much more slowly than small molecules. Thus, effects due to entropic elasticity are bound to become important in fast flows.

Finally, it should be appreciated that in most cases LCPs form *polydomains*. By this word it is meant that the distortions of the director field previously mentioned permeate the sample to such an extent that the orientation can be considered roughly uniform only over very small distances, of the order of a few μm or less. The average size of these "domains", indicated as *a* in the following, also defines the density of *defects* or *disclinations*. These are points or, more

T McLeish (ed), Theoretical Challenges in the Dynamics of Complex Fluids, 141-158 frequently, lines at which the director field becomes singular. Should all these defects be eliminated, the sample would become a monodomain (a perfect nematic "crystal"). Now, if defects are more easily eliminated in small molecule nematics, they tend to be extremely persistent in LCPs, where they are also readily generated during most flows. The fact that LCPs form polydomains more often than not makes them strikingly similar in behaviour, at least in some respects, to other complex fluids where the domain structure arises for different reasons, like phase separation. Notice further that, because of the polydomain structure, the material can appear nearly isotropic at a large scale in spite of the anisotropy which prevails at the scale of a single domain.

In the following, the main theoretical aspects of LCP rheology are reviewed, with an emphasis on the open problems. As usual, it is useful to distinguish between linear and nonlinear response, although, as we shall see, the distinction is less clearcut in the present case. Before starting with flow, the main concepts describing order and orientation in nematic phases are briefly revisited.

2. Order and orientation

In the Introduction, the local director **n** has been (vaguely) defined as the common direction of the (nearly) rodlike molecules. On the other hand, it is to be expected that thermal motion makes *individual* orientations of rodlike molecules (or of molecular segments, for the case of semi-flexible chains) to fluctuate in both space and time. The magnitude of these orientation fluctuations defines the "degree of order" of the nematic phase. To a smaller (larger) spread of molecular orientations obviously corresponds a more (less) ordered system. Under equilibrium conditions, the degree of order of a uniaxial nematic is described through a scalar order parameter S, ranging from 0 to 1, where 0 corresponds to the isotropic phase, and 1 to a "perfect" nematic with all molecules strictly straight and parallel.

Both the average orientation **n** and the order parameter S depend on individual molecular (or segmental) orientations. Indicating the individual orientation through a unit vector **u**, the orientation distribution can be described by a function $\psi(\mathbf{u})$ which gives the probability density of finding a molecule (or a molecular segment) oriented at **u**. Because **u** and −**u** are equivalent, all odd moments of the orientation distribution are nil. Hence, the simplest nonzero moment is the 2nd moment, i.e., the integral ($d\Omega$ is the infinitesimal solid angle about **u**)

$$\mathbf{S} = \langle \mathbf{uu} \rangle = \int_{4\pi} d\Omega\, \mathbf{uu}\,\psi(\mathbf{u}) \qquad (1)$$

For whatever $\psi(\mathbf{u})$, tensor **S** is symmetric and positive definite, hence it can be described geometrically by an ellipsoid. Also, the three principal values of **S** are linked by the condition tr**S** = $\langle \mathbf{u}\cdot\mathbf{u}\rangle = 1$.

Under equilibrium conditions (no flow), and outside defect cores (see later), the orientation distribution of a uniaxial nematic is axially symmetric about the director **n**, and so is the ellipsoid **S**. Under these conditions two of the three principal values coincide, and since the sum of all three is fixed (tr**S**=1), *a single scalar value* is required to complete the description of the ellipsoid. Indeed, the order parameter S describes the "slenderness" of the ellipsoid through the relationship

$$S = \frac{3}{2}S_{ZZ} - \frac{1}{2} \quad ; \quad S_{ZZ} = \left\langle (\mathbf{u}\cdot\mathbf{n})^2 \right\rangle = \left\langle \cos^2\varphi \right\rangle \tag{2}$$

Thus isotropy (S=0, S_{zz}=1/3) corresponds to a sphere, while positive S values approaching 1 correspond to increasingly prolate (slender) ellipsoids. Negative S values (corresponding to oblate ellipsoids) can also exist under special circumstances, all the way down to S = −1/2 (i.e., S_{zz}= 0) where the oblate ellipsoid reduces to a disk of zero thickness.

Although both the director **n** and the order parameter S are "properties" of the molecular orientation distribution $\psi(\mathbf{u})$, only S is a thermodynamic quantity. Thus, while S is rather insensitive to external fields, the director **n** can readily be changed by weak forces, like magnetic fields, wall effects, flow, etc. Conversely, S is modified by temperature and, in the lyotropic case, by concentration.

In terms of the ellipsoid representing the tensor **S** we may say that, while its orientation is readily modified by weak forces, the ellipsoid "shape" is a robust thermodynamic quantity. Nevertheless, also the shape gets modified by strong fields, for example in fast flows. The ellipsoid then not only changes its slenderness, but even loses the uniaxial symmetry in the general case, so that the single scalar S, and the single vector **n**, become insufficient for the description of such a biaxial situation induced by flow. In the next section, we shall see that even in the static, equilibrium situation shape changes of the ellipsoid can occur in close proximity of defects.

Before closing this section it is worth emphasising that, so far, our discussion of order and orientation refers either to a spatially uniform situation or, otherwise, to *local* values. In the latter case, averages such as those appearing in Eq. (1) and Eq. (2) must be interpreted as ensemble averages made at a "point" in the sample. However, in case of nonuniformities, particularly for the polydomain situation, it may be useful also to consider the so called "mesoscopic" averages.

These are defined over a sample volume which, though possibly still quite small with respect to the apparatus dimension, nevertheless contains many domains. Thus it may occur that an anisotropic nematic phase, locally endowed with a well defined nonzero order parameter, looks virtually isotropic at the mesoscopic level because of the polydomain texture. A clear distinction between local and mesoscopic averages is very important, as thermodynamics (or other strong forces) determine the former, while in most cases our measuring techniques are sensitive to the latter. Typically, main-chain LCPs have order parameters as large as 0.8 (NMR determined); yet, the birefringence of a bulk phase will reveal *mesoscopic order parameters* which are much smaller, or even nearly zero.

3. Distortions and defects

As mentioned in the Introduction, gradients of the director field are called "distortions" and are dealt with by means of Frank elasticity. Frank considered that the energy of the distorted "crystal" must be an even function of $\nabla\mathbf{n}$. Then, by assuming a quadratic dependence (which certainly applies, asymptotically, to small distortions) he derived that the most general form can be written as the sum of three terms, corresponding to the basic distortions of splay, twist, and bend. The free energy density expression is

$$F = \frac{1}{2}K_1(\nabla \cdot \mathbf{n})^2 + \frac{1}{2}K_2(\mathbf{n} \cdot \nabla \times \mathbf{n})^2 + \frac{1}{2}K_3(\mathbf{n} \times \nabla \times \mathbf{n})^2 \qquad (3)$$

where K_1, K_2, K_3 are the elastic constants (dimensions of a force) for splay, twist, and bend, respectively. Minimisation of the integral of F over the volume of the sample (with given **n** at the boundary) will then determine the equilibrium director field. Such a procedure is in good agreement with the observations in defect free samples.

Unfortunately, in the presence of defects the method breaks down. Indeed, since $\nabla\mathbf{n}$ diverges while approaching a defect point or line, so does F from Eq. (3), and in the case of lines even the volume integral of F becomes infinite (logarithmically), thus frustrating any minimisation procedure. The physical interpretation of this apparent mathematical paradox is simple, however. The fact that $\nabla\mathbf{n}$ diverges while approaching a defect obviously implies that the distortions can no longer be considered small; hence Eq. (3) should not be expected to remain valid in close proximity of defects. We find here an example (in statics!) of a transition from linear (a quadratic expression for the energy is synonymous for a linear force) to nonlinear behaviour.

In other words, while approaching the centre of a defect, the topology of the orientation field overcomes thermodynamics, so to speak, and forces the order parameter (which, at a constant temperature, stays constant throughout the nematic field) to change, locally, all along the thickness of the defect "core" (as if along a temperature gradient). More precisely, the whole shape of the ellipsoid is expected to change, also loosing uniaxiality in general.

What is important to realise is that, although vector **n** is singular at the defect centre, the field of tensor **S** (i.e., of the ellipsoid) is expected to remain continuous throughout the defect core. Thus, a mathematical treatment of defects based on **S** instead of **n** becomes possible. Already in 1969, de Gennes [1] proposed an expression for the free energy density in the form of a Landau expansion in S. Although originally proposed for the isotropic phase of nematic materials, de Gennes' expression has been subsequently used by many authors for the nematic phase as well. In that expansion, the terms quadratic in $\nabla \mathbf{S}$ are two (and only two), giving rise to the following form for the distortion energy density of a nematic phase

$$F = \frac{1}{2} a_1 (\nabla \mathrm{S})^2 + \frac{1}{2} a_2 (\nabla \cdot \mathrm{S})^2 \tag{4}$$

where a_1 and a_2 are material constants, and the quantity indicated as $(\nabla \mathrm{S})^2$ is the scalar which, in indicial notation, is written as $\partial_\alpha S_{\beta\gamma} \partial_\alpha S_{\beta\gamma}$.

Although Eq. (4) runs smoothly through defects, it poses nevertheless a conceptual problem. Away from defects, it does not degenerate properly, as it should, into Frank elasticity. Suffice it to observe that from two independent constants, a_1 and a_2 , it is impossible to obtain three independent ones. Indeed, from Eq. (4), in the limit of uniaxiality and constant S, one obtains for the Frank constants the result

$$\begin{aligned} K_1 = K_3 &= S^2 (2a_1 + a_2) \\ K_2 &= 2S^2 a_1 \end{aligned} \tag{5}$$

where the condition $K_1 = K_3$ is not in agreement with observations. Suggestions were therefore advanced that a_1 and a_2 should not be taken as constants but as polynomials of S [2], or that higher order moments should be included in the Landau-de Gennes expansion [3]. Either of these suggestions brings the number of the constitutive constants in the theory to a value which is essentially arbitrarily large so that, if the problem of proper degeneration into Frank elasticity is amply solved, the practical use of the theory is virtually nil since it becomes impossible to determine all those constants.

The truth of the matter is that, whenever one is faced by a *nonlinear constitutive problem*, continuum theories can no longer be profitably used, at least not exclusively, and some guidance from molecular considerations must be obtained. In the case at hand, the rod-like molecular model has been used [4] to understand why a theory containing only the gradient of the second-order moment **S** is certainly insufficient to predict three different constants of Frank elasticity, and to obtain some definite prediction. In the rest of this section, that result is briefly presented in intuitive terms.

Let us start by recalling that a nematic phase is generated by a mean-field orientation potential, the simplest example of which is the Maier-Saupe potential. This can be written in the form

$$V(\mathbf{u}) = -kTU\mathbf{S}:\mathbf{uu} \qquad (6)$$

where kT is Boltzmann constant times absolute temperature, and U is the non-dimensional intensity of the interaction, itself a function of temperature and concentration. Recalling that $\mathbf{S} = \langle\mathbf{uu}\rangle$ (cf Eq. 1), the mean-field nature of the potential is apparent. Indeed, should **S** reduce to a sphere, then the potential becomes ineffective (V reduces to a constant) no matter how large is U. Conversely, for a prolate ellipsoid, the potential attains a minimum when **u** is along the axis of the ellipsoid, i.e., each rod-like molecule is driven by the potential to align to **n** (the more so the larger the product US).

Equation (6) applies to a spatially uniform situation. Now consider the case where spatial gradients are present, and therefore the orientation potential V(**u** , **r**) varies with **r**. Equation (6) must now be rewritten as

$$V(\mathbf{u},\mathbf{r}) = -kTU\langle\{\mathbf{u}'\mathbf{u}'\}\rangle:\mathbf{uu} \qquad (7)$$

where the symbol $\langle\{\ldots\}\rangle$ is meant to indicate the ensemble average over a suitable "interaction region" about the point **r**, and **u**' is used inside the average to avoid confusion with the orientation **u** of the test molecule on which the potential is acting.

For not too large gradients, i.e., such that an expansion can be used, one might perhaps expect that

$$\langle\{\mathbf{u}'\mathbf{u}'\}\rangle \cong \mathbf{S} + l^2\nabla^2\mathbf{S} \qquad (8)$$

where l is a characteristic interaction length. Equation (8) is wrong, however (it would lead to $K_1 = K_2 = K_3$). To understand why, consider Figure 1 where the test rod is shown together with its "interaction region". Clearly, by changing the orientation **u** of the test rod, the interaction region also changes. As a consequence the average to be used in Eq. (7) *must depend on* ***u***. Detailed calculations reported in [4] show that, instead of Eq. (8), the correct expansion has the form

$$\langle\{\mathbf{u}'\mathbf{u}'\}\rangle \cong \mathbf{S} + l^2\nabla^2\mathbf{S} + L^2 u_\alpha u_\beta \partial_\alpha \partial_\beta \mathbf{S} \tag{9}$$

where L is the molecular length (see Figure 1). When Eq. (9) is replaced in the expression for the potential, Eq. (7), and the ensemble average of V is taken to obtain the free energy density, the last term in Eq. (9) generates the fourth-order moment $\langle \mathbf{uuuu} \rangle$. This leads "naturally" to three different, system dependent, Frank elasticity constants [4].

The same theory (or any similar one, for that matter, based on some different molecular model) can be used to investigate the core structure of defects. Only the simplest case of the spherically symmetric hedgehog defect has so far been considered with this approach, however, and not even completely [5]. Other ways of investigating defect cores were either calculations based on continuum assumptions [6, 7], or a few Monte Carlo simulations (cf, e.g., [8, 9]). There seems to be room for further theoretical investigation of defects in nematics.

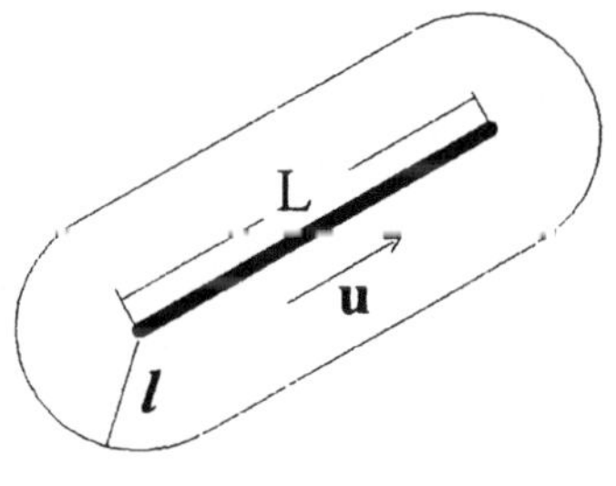

Figure 1. The interaction region of a rodlike molecule. The length l characterises the interaction distance, while L is molecular length. By changing the orientation ***u****, the rod "explores" a different region.*

4. Leslie-Ericksen theory of nemato-dynamics

We now go back to linear behaviour, but in the dynamic situation, which is described by the Leslie-Ericksen continuum theory (cf e.g. [10]). The starting point is the expression for the viscous stress, which is

$$\sigma = \alpha_1 \mathbf{D}:\mathbf{nnnn} + \alpha_2 \mathbf{nN} + \alpha_3 \mathbf{Nn} + \alpha_4 \mathbf{D} + \alpha_5 \mathbf{nn} \cdot \mathbf{D} + \alpha_6 \mathbf{D} \cdot \mathbf{nn} \qquad (10)$$

where the α's are Leslie coefficients (dimensions of a viscosity), $\mathbf{D}$ is the symmetric part of the velocity gradient $\mathbf{K}$, and $\mathbf{N}$ is the vector defined as

$$\mathbf{N} = \dot{\mathbf{n}} - \frac{1}{2}\left(\mathbf{K} - \mathbf{K}^{\mathrm{T}}\right) \cdot \mathbf{n} \qquad (11)$$

where $\dot{\mathbf{n}}$ is the rate of change of the director $\mathbf{n}$, and $\mathbf{K}^{\mathrm{T}}$ is the transpose of $\mathbf{K}$.

Equation (10) is the most general expression, properly frame invariant, linking *linearly* σ to $\mathbf{K}$. Conceptually, it plays the same role of the equation $\sigma = 2\eta\mathbf{D}$, holding for isotropic fluids with a (constant) viscosity η. Nematics require six viscosity coefficients in the description of the linear response because of the role of $\mathbf{n}$. The viscosity of a nematic phase depends on how the director is oriented with respect to the velocity gradient $\mathbf{K}$, and, through $\dot{\mathbf{n}}$, on how fast it is rotating. The range of validity of the linear viscous response conceptually coincides with that of Frank elasticity. Both hold true as long as the "shape" of the ellipsoid $\mathbf{S}$ is not modified, that is, as long as the order parameter remains constant. In the bulk of the nematic phase (i.e., away from walls and defect cores), S remains constant in slow flows, whereas deviations must be expected in fast flows, especially in LCPs. Such nonlinearities will be discussed in the next section.

To proceed with the linear response, it must be noted that the stress σ in liquid crystals is a non-symmetric tensor in the general case. This occurs because of the possibility of exerting torques on anisotropic materials, either externally through a magnetic field, say, or internally (between different points in the material) through distortions of the $\mathbf{n}$-field and the associated Frank elasticity. In the torque balance equation, viscous forces contribute the following term

$$\mathbf{T} = \left[\left(\alpha_3 - \alpha_2\right)\mathbf{N} + \left(\alpha_3 + \alpha_2\right)\mathbf{D} \cdot \mathbf{n}\right] \times \mathbf{n} \qquad (12)$$

where only two of the Leslie coefficients appear.

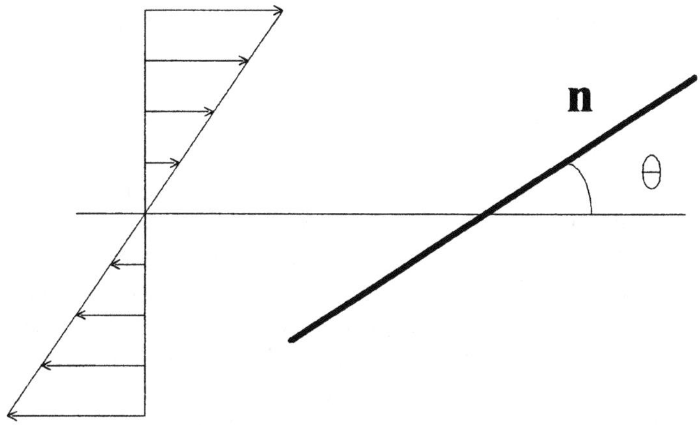

Figure 2. Shear flow of a nematic. In the shear plane, the director **n** *forms the angle* θ *with the shear direction.*

In view of its special importance, let us consider in some detail the case in which viscous forces dominate so that the torque balance equation reduces to **T**= **0** (director inertia seems negligible in all cases). Then, for a given velocity gradient **K**, Eq. (12) with **T**= **0** becomes a direct link between $\dot{\mathbf{n}}$ and **n**, allowing for an explicit calculation of how the director evolves in time. For a shear flow, Eq. (12) gives

$$\dot{\theta} = -\dot{\gamma}\,\frac{\sin^2\theta - (\alpha_3/\alpha_2)\cos^2\theta}{1-\alpha_3/\alpha_2} \tag{13}$$

where the angle θ defines the director in the plane of shear (see Figure 2), and $\dot{\gamma}$ is the shear rate.

Two quite distinct behaviours are described by Eq. (13) depending on the sign of the ratio $\beta=\alpha_3/\alpha_2$. If β is negative, dθ/dt never changes sign with changing θ. Hence the director keeps rotating indefinitely in the shear flow. Materials having a negative ratio β of Leslie coefficients are called *tumbling nematics*.

Conversely, if β is positive, Eq. (13) shows that the director approaches the stable orientation $\tan^{-1}\sqrt{(\beta)}$, called Leslie angle. We then speak of *flow-aligning nematics*.

Leslie-Ericksen theory is completed by a specification of the elastic contributions to both the stress tensor and the torque. These contributions are obtained by taking suitable derivatives (cf, e.g., [10]) of the Frank expression for the free energy density, Eq. (3). Thus, in the general case where the theory is applicable, problems of nemato-dynamics present themselves as follows. With the usual assumption of incompressibility, the unknowns are the vector fields of the velocity $\mathbf{v}$ and of the director $\mathbf{n}$, plus the scalar field of the pressure p. The available equations are the vectorial equations describing, respectively, the force balance (in which $\nabla \cdot \sigma$ and ∇p will appear) and the torque balance, plus the scalar equation $\nabla \cdot \mathbf{v} = 0$ describing mass conservation. Thus, there are as many equations as unknown, and the problem is closed.

A last comment on linearity is now in order. Although Leslie-Ericksen constitutive equation is linear in the velocity gradient, all non-trivial problems of nemato-dynamics are nonlinear. In particular, the apparent viscosity of a nematic slab sheared between two platens may well change with changing platen velocity. Similarly, the apparent viscosity of a nematic flowing in a capillary will generally depend on flow rate. Such nonlinearities are due to coupling effects between flow and orientation. We recall that, through Eq. (10), the viscosity depends on $\mathbf{n}$. On the other hand, through the torque balance, the $\mathbf{n}$ field will generally depend on velocity. In spite of these complications, which can become prohibitive in some cases, it remains true that the physical status of the theory is simple. It corresponds to situations where the shape of the ellipsoid **S**, as determined by thermodynamics, remains essentially unaltered in spite of flow.

5. Nonlinear behaviour of LCPs in fast flows

We have previously stated that nonlinear constitutive problems cannot be effectively dealt with without the help of molecular models. The simplest molecular model, appropriate to LCPs, is that of a rod endowed with some mean-field orientation potential. The dynamical equation for such a model was written in the form of a Smoluchowski equation by Hess [11] and, independently, by Doi [12]. The distribution function of rod orientations $\psi(\mathbf{u})$ (already encountered in Section 2) becomes time dependent in dynamics, and obeys the evolution equation

$$\frac{\partial \psi}{\partial t} = \nabla_{\mathbf{u}} \cdot D\left(\nabla_{\mathbf{u}} \psi + \psi \nabla_{\mathbf{u}} V/kT\right) - \nabla_{\mathbf{u}} \cdot (\psi \dot{\mathbf{u}}) \tag{14}$$

where $\nabla_{\mathbf{u}}$ indicates the gradient operation in **u**–space, i.e., on the unit sphere, D is the rotational diffusivity of the rod-like molecules, V is the mean-field orientation potential (already encountered in Section 3), and $\dot{\mathbf{u}}$ is the rate of change of rod orientation due to flow, that is

$$\dot{\mathbf{u}} = \mathbf{K} \cdot \mathbf{u} - \mathbf{K}{:}\mathbf{uuu} \tag{15}$$

in which the last term automatically ensures that **u** remains of unit length. Of course, the dynamical equation also describes the equilibrium condition as a special case. This is obtained by setting to zero the term in the first bracket, which gives the Boltzmann distribution $\psi \propto \exp(-V/kT)$.

If Eq. (14) is used in the limit of slow flows it is expected to reproduce Leslie-Ericksen theory with some definite predictions for the Leslie coefficients (though not for the Frank constants because Eq. (14) is written for the spatially uniform case). Notice that Eq. (14) is nonlinear because ψ also appears within the mean-field potential V. The limit of slow flows is obtained when the magnitude of the velocity gradient **K** is much smaller than the rotational diffusivity D, so that ψ will deviate from the Boltzmann distribution only by small terms of order $|\mathbf{K}|/D$ and Eq. (14) can be linearized with respect to that deviation. In order to reproduce the non-symmetric stress of Leslie-Ericksen theory, to the mean-field potential V an external (magnetic, say) potential must also be added.

These calculations for the linear limit (which remain of a significant complexity) were carried out independently by Semenov [13] and by Kuzuu and Doi [14]. It is interesting to report what their predictions are for the important ratio of Leslie coefficients $\beta=\alpha_3/\alpha_2$. In his calculations, Semenov used for V the Maier-Saupe expression of Eq. (6). With such a form for the mean-field potential, the order parameter in the nematic at the isotropic-nematic transition is $S = 0.43$. Semenov found that in the interval $0.43 < S < 0.53$ the β ratio is positive, hence the nematic phase is flow aligning in that range. Above $S = 0.53$, and up to $S = 1$, β becomes negative and the nematic is of the tumbling type. Kuzuu and Doi used, instead of Maier-Saupe potential, the Onsager excluded volume potential, according to which the order parameter at the isotropic-nematic transition is already as large as 0.8. Their calculations predict a negative β value throughout the available nematic range $0.8 < S < 1$. Considering that LCPs typically show large values of the order parameter, the message emerging from both these calculations is that LCPs should be *tumbling nematics*. The message was not immediately accepted, however, also because β, though negative in sign, comes out a very small number, and it was deemed possible that other effects neglected in the model might easily determine a re-crossing of zero, so overturning the prediction.

Tumbling was confirmed by the nonlinear calculations, however, which were first performed on the two-dimensional version of Eq. (14) [15], and soon after on the full three-dimensional equation [16]. It was shown that the peculiar rheological phenomenon of negative normal stresses, observed for many lyotropic LCP systems, could in fact be explained as a nonlinear effect marking the dynamic transition from tumbling to flow-aligning behaviour, taking place with increasing shear rate. In other words, it was demonstrated that LCPs showing the effect *must be tumbling nematics at low shear rates.*

A simple, intuitive explanation of the sign changes of normal stresses is based upon the molecular "elasticity" of the ellipsoid **S** [15]. First consider an isotropic phase, where the ellipsoid is in fact a sphere at equilibrium. As a consequence of a shear flow, the sphere deforms into an ellipsoid which, with increasing shear rate, further elongates along the shear direction. The elasticity of this ellipsoid which, if left alone, would "bounce back" to the spherical shape justifies why isotropic phases show normal stresses the sign of which is always positive. (By convention, positive means a traction in the shear direction or, equivalently, a compression normal to it.)

Now consider a nematic phase where the equilibrium **S** is an ellipsoid of given "slenderness" S. We have good reasons to believe that this ellipsoid is oriented by the shear flow, for the most part (though not exclusively) along the shear direction. Nonlinear effects can now deform the ellipsoid in two ways. The ellipsoid can become either slimmer or fatter with respect to equilibrium. In the former case, the situation is not qualitatively different from the case of the sphere. Hence positive normal stresses will be found in such a case. Conversely, when the ellipsoid becomes "fatter" than the equilibrium one the elasticity is in the opposite direction, and negative normal stresses are generated. Both possibilities do occur in fact with changing the shear rate in the nonlinear range.

Also the predictions of Eq. (14) on the behaviour of the second normal stress difference in a shear flow (i.e., between the direction of the velocity gradient and that of the vorticity) are in excellent qualitative agreement with data [17]. Similarly confirmed is the prediction that, at large enough shear rates, LCPs become well aligned to the shear direction [18]. In the range of high shear rates, even fine details of the transient response which is observed after stepwise changes in the shear rate seem to be qualitatively accounted for by the rod-like molecular theory [19].

6. Nonlinear behaviour of LCPs in slow flows

At this point one might believe that the discussion of the dynamic situation is essentially over. One might be tempted to conclude that the whole of LCP dynamic behaviour is described, to a good approximation at least, by the rod model, i.e. by Eq. (14). Indeed, at high shear rates, that equation predicts all the rich phenomenology which is observed experimentally, and, at low shear rates, it degenerates properly into the linear limit described by the Leslie-Ericksen continuum theory. That conclusion would be too optimistic, however, for the good reason that the polydomain structure of LCPs causes them to disagree with Leslie-Ericksen theory in slow flows, at least at the mesoscopic level. The experiments show that *LCPs behave nonlinearly even in slow flows*.

Polydomains have a high defect density, and we know from Section 3 that linearity gets lost within defect cores. So we should not be surprised that, although the flow is slow, linearity is not regained by the polydomain. We can nevertheless surmise that locally, within each domain, Leslie-Ericksen theory applies to the defectless nematic. This idea can be profitably exploited to deal somehow, and in a simple way, with the dynamics of the polydomain.

To start with, we can apply dimensional analysis to the steady state situation of the polydomain undergoing a shear flow to try and determine the size of the domains [20]. The physical picture is as follows. On the one hand tumbling creates spatial windings of the director [21] thus tending to decrease domain size, the more so the faster the flow, i.e., the stronger the viscous forces. On the other hand, Frank elasticity drives the system towards an increase of the domain size due to healing and disappearance of defects. At least, this is what is observed in the quiescent polydomain if the defect density is initially very high. In the dynamical situation, a balance of the opposing effects will be attained, on average, at some appropriate domain size. Now, if the Frank elastic constants are collectively indicated as K, the Leslie coefficients as η, and $\dot{\gamma}$ is the shear rate, the following length can be formed from dimensional analysis

$$a = \sqrt{K/(\eta\dot{\gamma})} \tag{16}$$

Thus, if no other physical quantities are relevant, Eq. (16) provides the scaling for the domain size. An inverse square root law with shear rate for the so called structure refining seems to be well obeyed, at least as long as a is much larger than molecular length.

If we consider the stress σ as a function of the same variables, $\sigma(K,\eta,\dot{\gamma},a)$, dimensional analysis allows us to write

$$\frac{\sigma}{\eta\dot{\gamma}} = f\left(\eta\dot{\gamma}a^2/K\right) \tag{17}$$

where the nondimensional argument of the function has the form of the Ericksen number. However, the characteristic length a is not imposed externally. In fact, in view of Eq. (16) this *internal* Ericksen number stays constant when $\dot{\gamma}$ is varied. Therefore, Eq. (17) allows us to conclude that also the group $\sigma/\eta\dot{\gamma}$ remains constant with varying $\dot{\gamma}$ or, in words, that the *stress is linear with shear rate*. One should be aware, however, that this "linearity" is very peculiar. Indeed, if on one hand the apparent viscosity stays constant (which is typical linear behaviour), it is also true that *normal stresses come out proportional to shear rate* (to the absolute value of shear rate, to be precise, since normal stresses cannot change sign upon flow reversal). This behaviour of normal stresses of LCPs in shear is highly anomalous. For ordinary polymers, the linear range of viscoelasticity is characterised by a *quadratic* dependence of normal stresses on shear rate.

The transient behaviour of LCPs in slow flows has been modelled by Larson and Doi [22], who proposed a constitutive equation. Independently of the details of that equation, which also incorporates phenomenological assumptions, what is certainly universal is the extension of the dimensional analysis to the transient situation, which fully explains peculiar scaling laws, some of which had been found experimentally, and constituted another LCP puzzle for some time. It had been observed by various authors that the curves of stress vs. time during relaxation, or those of recovered strain vs. time in recoil experiments, would scale with $\dot{\gamma}t$, where $\dot{\gamma}$ is the shear rate of the pre-existing steady shear flow. Also the transient stress response during flow reversal would scale in the same way, i.e., with strain units $\dot{\gamma}t$ rather than time t.

All such behaviour, and more, is consistent with simple dimensional analysis. Indeed, if the transient experiment is characterised by a single value of $\dot{\gamma}$, extension of Eq. (17) to the transient situation (and recalling that a must not be listed among the independent variables of the dimensional analysis because it is self-determined by the system dynamics) gives

$$\frac{\sigma}{\eta\dot{\gamma}} = f(\dot{\gamma}\, t) \tag{18}$$

Experiments obeying Eq. (18) are the already mentioned stress relaxation, and flow reversals. A similar equation is obeyed by recoil experiments. If, on the other hand, *two* values of shear rate are involved in the experiment, such as in stepwise changes of shear rate (up or down), then Eq. (18) is replaced by

$$\frac{\sigma}{\eta\dot{\gamma}_1} = f\left(\dot{\gamma}_1 t, \frac{\dot{\gamma}_2}{\dot{\gamma}_1}\right) \tag{19}$$

where $\dot{\gamma}_1$ and $\dot{\gamma}_2$ are the shear rate values before and after the stepwise change.

The functions which describe these stepwise changes in shear rate are shown by experiment to have the form of damped oscillations. This behaviour is interpreted as being representative of winding and unwinding of the director field of the tumbling nematic, i.e., as a manifestation of Frank elasticity [23] (see also R. Larson, this volume). It is noteworthy that, although such an interpretation seems to describe correctly the physical situation, the Frank elastic constants K do not appear explicitly in the scaling laws of Eqs (18) and (19). This is due to the fact that Frank elasticity *per se* does not define a characteristic time for the material, nor a characteristic length, but only a combination of them. Indeed, given K and η, what remains fixed is the ratio of a square length to a time. Thus, as long as the characteristic length is not imposed externally through a dominance of the wall conditions (i.e., as long as the material response is independent of sample thickness), the length and time scales of the observed phenomena remain interdependent. If a shear rate is imposed, hence a characteristic time is assigned, then the domain size remains determined (cf Eq. 16), and the stress with it (cf Eq. 17). If the shear rate is changed, since the domain size self-adjusts to the new situation, the transient response of stress must obey the peculiar scaling of Eqs (18) or (19).

The just mentioned observed independence on sample thickness of the LCP response (with the exception of very thin samples), and the consequent scaling of Eqs (16-19), is a central property of the polydomain, which markedly distinguish LCPs from ordinary defectless nematics obeying Leslie-Ericksen theory. It so appears that defects play an important role in LCPs, and conjectures have been made on what such a role might be [24, 25], but the question remains open. One further mystery of the nonlinear behaviour of LCPs at low shear rates is the shear thinning region which exists for many LCP systems at very low shear rates (so called Region I [26]). Here again speculations have been advanced [27] but no conclusive interpretation was reached.

7. Conclusions

We wish to conclude by summarising the challenges which face the theorist, and more generally the scientist, who wants to continue unravelling the complex dynamic behaviour of LCPs. Though certainly interconnected, the following four main challenges can be distinguished.

To start with, we need to know more about the static structure of defect cores. There is some information, but surely not enough. Moreover, the existing calculations and/or simulations do not refer specifically to polymeric nematics. In view of their molecular anisotropy, LCPs are expected to show a large separation among the values of the three elastic constants, with the twist constant being the lowest and the splay one the largest. The structure of defect cores for large values of the molecular anisotropy is not known.

The structure of defects during flow, specifically in the region of constant viscosity of the polydomain, is also unknown. Defect formation during flow is currently being investigated through simulations in reduced dimensionalities [28]. There appears to be a long way to go, however, before a clear understanding of the polydomain structure and behaviour is available. Among other things, several optical results (cf e.g. [29]) await detailed interpretation.

The shear thinning behaviour at very low shear rates (Region I of the viscosity curve) also poses a challenge. It is not even understood whether or not such a behaviour is universal. For a few lyotropic LCPs which do not show the effect at the lowest shear rate accessible to the experiments it might simply be a matter of inaccessibility of the relevant region. Indeed, for the same polymers, by increasing the concentration the effect suddenly appears [26].

Finally we should mention that most theoretical results were obtained by means of the rigid rodlike molecular model, whereas the molecules of many main-chain LCP systems are in fact somewhat flexible one way or the other. The effect of partial flexibility on the characteristic dynamic transitions is still essentially unknown, and only preliminary results (sometimes conflicting with one another) were obtained with models which somehow relax the rigidity constraint [30-32].

8. References

1. de Gennes, P.G. (1969) Phenomenology of short-range-order effects in the isotropic phase of nematic materials, *Physics Letters* **30A**, 454-455.
2. Poniewierski, A., and Sluckin T. J. (1985) On the free energy density of non-uniform nematics, *Molecular Physics* **55**, 1113-1127.
3. Rosciszewski, K. (1983) The nematic elastic constants from the Ginzburg-Landau Hamiltonian of the nematic to isotropic phase transition, *Acta Physica Polonica* **A64**, 277-283.
4. Marrucci, G., and Greco, F. (1991) The elastic constants of Maier-Saupe rodlike molecule nematics, *Molecular Crystals Liquid Crystals* **206**, 17-30.
5. Greco, F., and Marrucci, G. (1992) Molecular structure of the hedgehog point defect in nematics, *Molecular Crystals Liquid Crystals* **210**, 129-141.
6. Fan, C. (1971) Disclination lines in liquid crystals, *Physics Letters* **34A**, 335-336.
7. Schopohl, N., and Sluckin, T.J. (1987) Defect core structure in nematic liquid crystals, *Physical Review Letters* **59**, 2582-2584.
8. Hudson, S.D., and Larson, R.G. (1993) Monte Carlo simulation of a disclination core in nematic solutions of rodlike molecules, *Physical Review Letters* **70**, 2916-2919.
9. Chiccoli, C., Pasini, P., Semeria, F., Sluckin, T.J., and Zannoni, C. (1995) Monte Carlo simulation of the hedgehog defect core in spin systems, *J. Physique II France* **5**, 427-436.
10. de Gennes, P.G. (1974) *The Physics of Liquid Crystals*, Clarendon Press, Oxford.
11. Hess, S. (1976) Fokker-Planck-equation approach to flow alignment in liquid crystals, *Z. Naturforsch.*, **31a**, 1034-1037.
12. Doi, M. (1981) Molecular dynamics and rheological properties of concentrated solutions of rodlike polymers in isotropic and liquid crystalline phases, *J. Polymer Science: Polymer Physics Ed.* **19**, 229-243.
13. Semenov, A.N. (1983) Rheological properties of a liquid-crystal solution of rodlike molecules, *Soviet Physics JETP* **58**, 321-326.
14. Kuzuu, N., and Doi, M. (1984) Constitutive equation for nematic liquid crystals under weak velocity gradient derived from a molecular kinetic equation. II. Leslie coefficients for rodlike polymers, *J. Physical Society Japan.* **53**, 1033-1040.
15. Marrucci, G., and Maffettone, P.L. (1989) Description of the liquid-crystalline phase of rodlike polymers at high shear rates, *Macromolecules*, **22**, 4076-4082.
16. Larson, R.G. (1990) Arrested tumbling in shearing flows of liquid crystal polymers, *Macromolecules* **23**, 3983-3992.
17. Magda, J.J., Baek, S.G., De Vries, K.L., and Larson, R.G. (1991) Shear flows of liquid crystal polymers: Measurements of the second normal stress difference and the Doi molecular theory, *Macromolecules* **24**, 4460-4468.
18. Hongladarom, K., Burghardt, W.R., Baek, S.G., Cementwala, S., and Magda, J. J. (1993) Molecular alignment of polymer liquid crystals in shear flows. Spectrographic birefringence technique, steady-state orientation, and normal stress behavior in poly(benzylglutamate) solutions, *Macromolecules* **26**, 772-784.
19. Maffettone, P.L., Marrucci, G., Mortier, M., Moldenaers, P., and Mewis, J. (1994) Dynamic characterization of liquid crystalline polymers under flow-aligning shear conditions, *J. Chemical Physics* **100**, 7736-7743.
20. Marrucci, G. (1985) Rheology of liquid crystalline polymers, *Pure and Applied Chemistry* **57**, 1545-1552.

21. Manneville, P. (1981) The transition to turbulence in nematic liquid crystals, *Molecular Crystals Liquid Crystals* **70**, 223-250.
22. Larson, R.G., and Doi, M. (1991) Mesoscopic domain theory for textured liquid crystalline polymers, *J. Rheology* **35**, 539-563.
23. Burghardt, W.R., and Fuller, G.G. (1990) Transient shear flow of nematic liquid crystals: Manifestation of director tumbling, *J. Rheology* **34**, 959-992.
24. Marrucci, G., and Greco, F. (1992) A molecular approach to the poly-domain structure of LCPs in weak shear flows, *J. Non-Newtonian Fluid Mechanics* **44**, 1-13.
25. Marrucci, G., and Greco, F. (1993) Flow behavior of liquid crystalline polymers, *Advances in Chemical Physics* **86**, 331-404.
26. Onogi, S., and Asada, T. (1980) Rheology and rheo-optics of polymer liquid crystals, in G. Astarita, G. Marrucci, and L. Nicolais (eds.), *Rheology, Volume 1*, Plenum Press, New York, pp. 127-147.
27. Marrucci, G., and Maffettone, P.L. (1993) LCP defect dynamics in shear flows, in C. Carfagna (ed.), *Liquid Crystalline Polymers*, Pergamon, Oxford, pp. 127-131.
28. Maffettone, P.L. (1996), personal communication.
29. Hongladarom, K. and Burghardt, W.R. (1993) Molecular alignment of polymer liquid crystals in shear flows. 2. Transient flow behavior in poly(benzylglutamate) solutions, *Macromolecules* **26**, 785-794.
30. Semenov, A.N. (1990) Dynamics of nematic solutions of persistent macromolecules, in D.R. Oliver (ed.), *Third European Rheology Conference*, Elsevier Applied Science, London, pp. 434-436.
31. Subbotin, A. (1993) Dynamics of slightly flexible rods in the liquid crystalline state, *Macromolecules* **26**, 2562-2565.
32. Greco, F., and Marrucci, G. (submitted) The tumbling or flow-aligning nature of nematics as predicted from the Slightly Bending Rod molecular model, *Liquid Crystals*.

THE SHEAR-FLOW PROPERTIES OF LIQUID CRYSTALS

R.G. LARSON
Bell Labs, Lucent Technologies
700 Mountain Ave., Murray Hill, NJ 07974, USA

AND

P.T. MATHER
U.S.A.F. Phillips Laboratory
10 East Saturn Blvd, CA 93524-7680, USA

1. Introduction

Liquid crystalline phases, such as small-molecule thermotropes, polymeric liquid crystals, block copolymers, and ordered surfactant solutions, show an incredibly rich variety of phenomena when subjected to shearing flows. The reason is that these fluids possess order not only at molecular length scales, but also at mesoscopic and macroscopic length scales, extending up to the size of the sample. Each different length scale responds somewhat differently to shearing deformations; as a result, the degree and type of order induced or destroyed by shearing depends in general on a wide range of factors, including the symmetry of the phase (nematic, smectic, etc.), the molecular relaxation time, the type of the molecule, the proximity of phase transitions, the rate of shear and the magnitude of the imposed strain, the mode of deformation (steady vs. oscillatory shear), etc. We focus here on the 'simplest' cases of small-molecule thermotropic nematic and smectic phases in a shearing flow. Even for these, a wide variety of interesting, and often unexplained, phenomena have been observed. In addition to being interesting and important in their own right, an understanding of the behavior of these 'simple' thermotropic liquid crystalline phases in shearing provides necessary background for understanding the more complex liquid crystalline polymers and ordered surfactant solutions. The behavior of these more complex ordered liquids is described in chapters by Marrucci and Roux, respectively.

T. McLeish (ed.), Theoretical Challenges in the Dynamics of Complex Fluids, 159-180.

The simplest types of liquid crystalline symmetry are the nematic and smectic-A. A simple *nematic* phase is one with uniaxial orientational order at rest, but no positional order. The direction of preferred alignment of the molecules is defined by a unit vector called the director **n**. A *smectic* has, in addition to orientational order, layer-like positional order. For the simplest case of a smectic-A, the positional order consists of a one dimensional periodicity in the centers of mass of the molecules in a direction parallel to that of the preferred orientational order. Thus, in a smectic-A phase the layers are orthogonal to the director.

2. Deborah Number and Ericksen Number

To distinguish various regimes of shearing flow of a liquid crystalline fluid, the shear rate $\dot{\gamma}$ must be made dimensionless using a relevant time scale of the fluid. If the time scale is the molecular relaxation time τ, then the relevant dimensionless quantity is the *Deborah number De*,

$$De = \dot{\gamma}\tau \tag{1}$$

For polymeric nematics, De can easily be of order unity or greater, and the shear rate is then large enough to influence the molecular distribution of orientations; this regime is considered in the chapter by Marrucci. When De is much smaller than unity, then the shear does not affect the orientation at the molecular level; this is the usual case for small-molecule nematics and smectics, which are the focus of this chapter.

The second important dimensionless group is the *Ericksen number Er*, which is defined by

$$Er = \frac{\eta\dot{\gamma}d^2}{K} \tag{2}$$

Here η is a typical viscosity, K is an constant, and d is a length scale of the flow geometry, such as the gap. The Ericksen number is the ratio of the viscous stress, $\eta\dot{\gamma}$, to the *Frank elastic stress*, K/d^2. The Frank elastic stress results from spatial gradients in the director field, where K is a typical elastic constant with units of force [1, 2]. For a small-molecule nematic, $\eta \sim 0.01$ Pa s, and K $\sim 10^{-11}$ N; hence, for d $= 10^{-4}$ m and $\dot{\gamma} = 10\ s^{-1}$, the Ericksen number Er is around 10^3, and the viscous stresses should dominate the elastic stresses, at least until the flow produces spatial gradients that are steep compared to the gap dimension. Spatial gradients often occur during flows of nematics and smectics because the flow competes with a director orientation imposed by the walls of the apparatus. A common boundary condition on the director is *homeotropic* alignment, in which the director is perpendicular to the walls of the apparatus. Under shearing flow, the director in the bulk of the liquid crystal will often rotate towards an

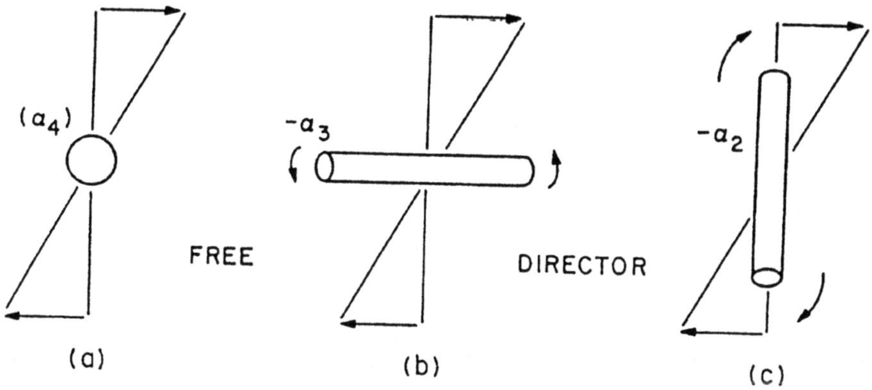

Figure 1. Orientations "a", "b", and "c" correspond to director orientations that are parallel to the (a) vorticity, (b) flow, and (c) velocity gradient directions. The viscosities α_3 and α_2 control the torque exerted on the director in each of these two directions. If $-\alpha_3$ and $-\alpha_2$ are positive in sign, the director rotates in the direction indicated (adapted from Skarp et al. [4].

orientation differing from the homeotropic condition at the walls; at high Er, the homeotropic condition is confined to thin, steep, boundary layers at the walls.

3. Nematics

3.1. TUMBLING VERSUS FLOW ALIGNING

For fluids with nematic symmetry, at low De the orientation of the nematic director can be understood from the Leslie-Ericksen continuum theory [3]. The simplest orientations of the director with respect to the axes of the shearing flow are depicted in Figure 1. In the "a", "b", and "c" orientations, the director is parallel to the vorticity, flow, and velocity gradient directions, respectively. The plane containing directions "b" and "c" (namely the flow and gradient directions) is called the *shearing plane*. Under steady shearing, the direction towards which the nematic director tends to rotate is controlled by the reactive or *tumbling parameter* λ of the Leslie-Ericksen theory. This appears in the equation for the director motion in the absence of Frank elastic torques:

$$\dot{\mathbf{n}} - \mathbf{W} \cdot \mathbf{n} - \lambda(\mathbf{D} \cdot \mathbf{n} - \mathbf{D} : \mathbf{nnn}) = \mathbf{0} \qquad (3)$$

Here $\dot{\mathbf{n}}$ is the substantial time derivative of $\mathbf{n}$, i.e., $\dot{\mathbf{n}} = \partial\mathbf{n}/\partial\mathbf{t} + \mathbf{v}\cdot\nabla\mathbf{n}$, $\mathbf{W}$ is the antisymmetric part of the velocity gradient, $\mathbf{W} = (1/2)(\mathbf{K} - \mathbf{K}^{\mathbf{T}})$, and $\mathbf{D} = (1/2)(\mathbf{K} + \mathbf{K}^{\mathbf{T}})$ is the symmetric part. The tensor $\mathbf{K}$ is the transpose of the velocity gradient, $\mathbf{K} = (\nabla \mathbf{v})^{\mathrm{T}}$. The term $-\mathbf{D} : \mathbf{nnn}$ is included in equation 3 to satisfy the constraint that $\mathbf{n}$ be a unit vector; i.e., $\mathbf{n} \cdot \mathbf{n} = 1$.

The tumbling parameter λ controls the rotation of the director in a fast shearing flow. It is related to the viscosities α_2 and α_3 in Figure 1 by $\lambda = (-\alpha_3 - \alpha_2)/(\alpha_3 - \alpha_2)$. When $\lambda > 1$, the director tends to rotate towards a fixed alignment angle θ_a in the plane of the velocity and the velocity gradient, given by

$$tan(\theta_a) = [\frac{\lambda - 1}{\lambda + 1}]^{1/2} \tag{4}$$

Normally λ is not much greater than unity, and the angle θ_a is then close to zero, which corresponds to the flow direction. The flow direction is called the "b" orientation; see Figure 1. In this case, $\lambda > 1$, the nematic is said to be "flow-aligning".

If $\lambda < 1$, on the other hand, equation 4 has no solutions in the shearing plane, and under fast shearing flow the director rotates or "tumbles" in the shearing plane until defects form and the fluid becomes disordered. The time period P corresponding to a rotation of the director through an angle of π can be calculated from the tumbling parameter λ using

$$P = \frac{2\pi}{\dot{\gamma}(1 - \lambda^2)^{1/2}} \tag{5}$$

The tumbling parameter λ can be estimated analytically by using the following simple approximate form [5-7]:

$$\lambda = \frac{5S_2 + 16S_4 + 14}{35S_2} \tag{6}$$

where S_2 and S_4 are the second and fourth moments of the molecular orientation distribution function at equilibrium (see the contribution of G. Marrucci, this volume). This approximate expression, using the Maier-Saupe theory for S_2 and S_4, agrees reasonably well with measurements of λ for a variety of liquid crystals, as long as there is no transition to a smectic phase near the temperature(s) considered. When a smectic-A phase is nearby, however, as is the case for 4'-n-octyl-4-cyanobiphenyl, or 8CB, then smectic-like fluctuations of the nematic state can significantly reduce λ.

3.2. FAST START-UP OF SHEARING

Consider first a tumbling nematic, such as 8CB, that is suddenly subjected to a torsional shearing flow with homeotropic boundary conditions at a high Ericksen number, Er $> 10^3$. For a tumbling nematic, there is no flow

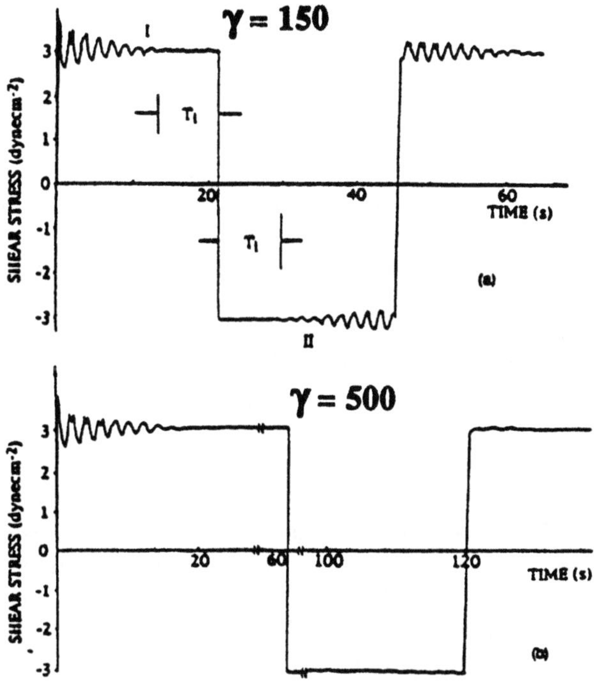

Figure 2. (a) Shear stress as a function of time after start-up of shearing at a shear rate of $\dot{\gamma} = 8\ s^{-1}$ for a tumbling nematic 8CB. A pattern of damped oscillations is evident. After 22 s of shearing, the direction of shearing is reversed, and the oscillations return. After another 22 s of shearing, the direction of shearing is reversed again. (b) The same as in (a), except strain is continued for over 60 s before the shearing direction is reversed. In this case, the oscillations do not return. (From ref. [9]).

alignment angle at which the viscous torques balance, and at a high Ericksen number such as this, Frank elasticity is initially negligible except at thin boundary layers at the walls. Hence, the director in the bulk is forced to rotate through multiple periods of π, as discussed above. Corresponding to these rotations of the director, shear stress measurements show multiple oscillations that slowly dampen with continued shear in cone-and-plate or plate-and-plate geometries at high Er [8, 9]; see Figure 2. The multiple stress oscillations are produced by the tumbling orbits of the director as it "winds up" in the shearing flow at high Er, and are predicted by the

Leslie-Ericksen theory in the limit Er $\to \infty$ when $\lambda < 1$. One reason for damping of these oscillations is discussed below.

The multiple stress oscillations are not present at temperatures near T_{NI} at which 8CB is a flow-aligning nematic. They are also absent for other nematics, such as 5CB (4'-n-pentyl-4-oxycyanobiphenyl), which is a flow-aligning nematic at all temperatures in its nematic range. Flow-aligning nematics with homeotropic anchoring in sudden startup shearing flows have a single stress overshoot, as predicted by the Leslie-Ericksen equations [9]. The single overshoot occurs as the director rotates through an angle intermediate between the homeotropic starting angle ($\theta = \pi/2$), and the final flow alignment angle. Thus, *a pattern of multiple stress oscillations in start-up of shearing flow is a reliable indicator that the nematic is of tumbling character.* In fact, the first observations of multiple overshoots in shearing flows of nematics were made by Moldenaers et al. [10] for the lyotropic polymeric nematic PBLG in metacresol, which provided an important clue that PBLG solutions are tumbling nematics [11].

If shearing of 8CB is not continued too long (less than $\gamma \sim 150$ strain units), flow reversal leads to a *reversal* in the damping of the stress oscillations; see Figure 2. That is, when the direction of the shearing flow is reversed, the oscillations *increase* in amplitude! The time dependence of the stress in the "reverse" shearing is just the opposite of that for the "forward" shearing, and the amplitude of the oscillations in the reverse shearing builds up to that observed at the beginning of the forward shearing when the number of "reverse" strain units equals the number imposed in the "forward" direction. Thus, apparently the reverse shearing corresponds to an "unwinding" of the director pattern produced during the forward shearing. However, if the unidirectional shear is continued for as many as 500 strain units, the damping can no longer be reversed (see Figure 2.)

The damping of the stress oscillations arises from inhomogeneities in the director field, i.e., a gradual loss of spatial coherence across the sample in the phase of the tumbling orbit. Using optical microscopy of 8CB between parallel glass disks in torsional (plate-and-plate) shearing flow, Mather et al. [12] discovered that this dispersion is associated with an inward radial migration of axisymmetric *twist walls.* These walls are boundaries separating regions whose orientation differs by roughly π; they are created periodically at the outer rim of the sample, and propagate inward radially as shearing progresses. Thus, the twist walls separate regions of differing director orientation. The director structure within the twist wall is complex and involves rotation of the director out of the plane of shear.

For plate-and-plate flow, the radial inhomogeneity in the director field which leads to the formation of twist walls is readily understood to be a consequence of the variation in strain across the sample. At a given instant in time, the strain varies linearly with increasing radial distance from the axis of rotation. By accounting for this variation of strain in the limit Er $\rightarrow \infty$, the damping of the stress oscillations in plate-and-plate flow can be accounted for quantitatively [12]. Thus, at least in the case of plate-and-plate flow, the twist walls serve as convenient "markers" showing the existence of radial director inhomogeneities, but they themselves do not contribute significantly to the damping of the stress oscillations.

In the case of cone-plate flow, the strain is uniform throughout the sample, and hence there must be another source of director nonhomogeneity. One source of inhomogeneity in cone-and-plate flow is the radial variation of the gap, which produces a radial variation of the effective Ericksen number (since the Ericksen number is proportional to the square of the gap). For very high Er, this effect should become negligible, but when the director is highly "wound up", the elastic distortions are severe, and the infinite Er limit might not be attained, even when Er is around 10^3 or even higher. It is significant that the rate of damping of the stress oscillations is much more gradual in shearing between a cone and a plate than it is between a plate and a plate.

The reversed damping of the shear stress when the flow direction is reversed can also be explained by twist wall propagation. For, when the flow direction is reversed, the twist walls reverse their propagation direction by migrating outward, and are expelled, one by one, from the edge of the sample. Eventually, when the strain has been fully reversed, the homeotropic monodomain is restored [12]. If the shearing is continued too long in one direction, however, ($\gamma \sim$ 200-500), *disclinations are nucleated,* and these are not expelled by a reversal in flow direction, but only continue to accumulate. Thus, the damping of stress oscillations can only be reversed when the imposed strain is less than the critical value at which disclinations form.

3.3. GRADUAL INCREASE OF SHEARING RATE

If, on the other hand, a homeotropically aligned tumbling nematic is sheared at small Er $= \gamma_1 \mathrm{V d}/K_3$, then the director at the midplane of the sample rotates modestly toward the flow direction, until it reaches a steady state where the viscous forces driving the director rotation are balanced by the Frank stresses; see Figure 3. As Er increases in small increments, so that a steady-state is attained between each increment in Er, the steady-state director at the midplane is rotated to a greater and greater extent; see Figure 3.

In a parallel-disk device, the shear rate, and hence the Ericksen number, increases linearly with radius. If we define the complementary angle $\theta' = \pi/2 - \theta$ to be the angle of deflection away from the homeotropic orienta-

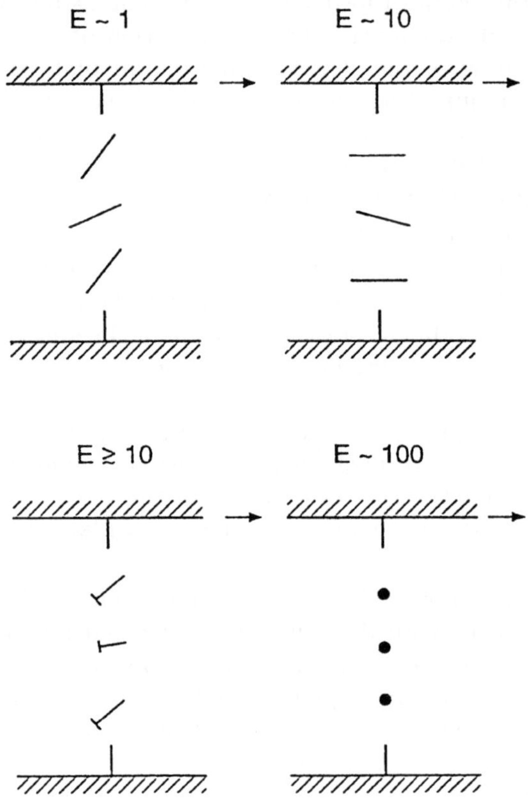

Figure 3. The rotation of an initially homeotropically aligned director field with homeotropic boundary conditions in a shearing flow for which the Ericksen number is gradually increased. At first, the director in the bulk rotates in the plane of shear; at Er $\sim$ 10, the director begins to rotate out of the plane of shear. At higher Ericksen number, Er $\sim$ 100, the director is mostly orthogonal to the plane of shear.

tion, then at the mid-plane θ' increases with increasing radial location r at fixed time. This radially increasing orientation produces a radially increasing *birefringence* which can be observed if the parallel disks are transparent. The multiple, concentric, rings one observes correspond to increasing *orders* of birefringence. A dark *extinction band* appears each time the birefringence, integrated across the gap, equals π times the wavelength of the monochromatic light used to illuminate the sample.

If the nematic were flow-aligning, the director at the midplane would eventually approach the flow-alignment angle and would no longer rotate when the shear rate increases. But for a tumbling nematic, the director can *pass through the flow direction* so that $\theta' > \pi/2$ (or $\theta < 0$). At Er $\sim$ O(10), when θ' is a few degrees to a few tens of degrees greater than 90 degrees, depending on the values of the Leslie viscosities and Frank constants, calculations with the Leslie-Ericksen equations show that *tumbling instabilities* can ensue, in which the midplane director angle θ' abruptly increases [13, 14]. For tumbling nematics, there is also an *out-of-plane* instability [14-18] in which the director begins to rotate toward the *vorticity direction* "a", which is the direction perpendicular to both the velocity and the velocity-gradient directions; see Figure 3. In the parallel-disk flow, this manifests itself as a tipping of the birefringence bands away from the flow direction. At Er $\sim$ O(100), the director becomes nearly parallel to the vorticity direction; see Figure 3. As Er is gradually increased further to around 10^3 in 8CB or PBLG, a *roll-cell instability* occurs. This instability produces a modulation of the director field in the vorticity direction, and leads to a cellular flow, with rolls cells parallel to the flow direction.

All these transitions in the director field have been both predicted from the Leslie-Ericksen theory [19, 20] and observed in both small-molecule and polymeric tumbling nematics, including 8CB [17, 18, 21, 22]. Once the roll cells form in 8CB, *disclinations* are eventually produced, and the flow takes on the irregular time-dependent characteristics of *director turbulence* [17, 23-25]. Similar transitions exist in tumbling polymeric nematics [18]. The disclination lines themselves seem to add little to the viscosity of a nematic, but they do effect the degree and direction of orientation, and in this way affect the stress. The density of disclination lines as a function of shear rate seems to follow a simple scaling law under some conditions [17]. Flow-aligning nematics do not have the out-of-plane and roll-cell instabilities that tumbling nematics have, but nevertheless also form defects at high Ericksen number (Er $> 10^3$) in torsional shearing.

4. Smectics

Since the smectic-A phase is solid-like in one direction and liquid-like in the other two, we expect the rheological properties of a well ordered smectic-A monodomain to depend drastically on the direction of deformation with respect to the orientation of the layers. Because in smectics there is a well defined layer spacing, some modes of elastic deformation permitted to nematics are "forbidden" to smectics, namely twist distortions and director bend (or layer splay). However, director splay (layer bend) is allowed, at a cost in free energy controlled by the splay constant $K_1 \sim 10^{-11}$ N. The layers in a smectic can also be compressed or dilated a bit, at a free energy cost controlled by the compressional/dilational modulus B, which for small-molecule thermotropic smectics is around B $\sim 10^5$ N/m^2 [2].

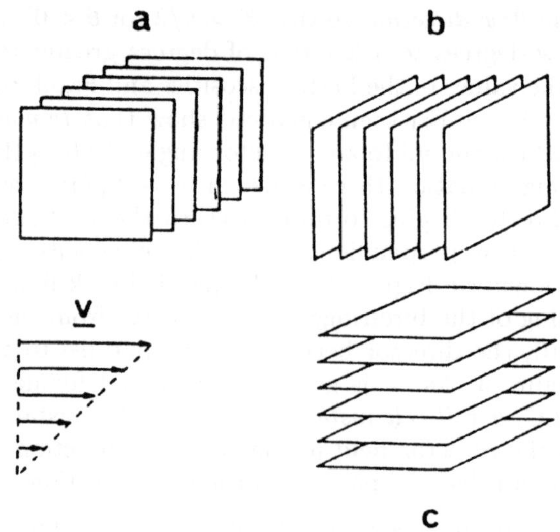

Figure 4. In a smectic-A phase, for which the director is normal to the smectic layer planes, orientations "a", "b", and "c" correspond to director orientations that are parallel to the (a) vorticity, (b) flow, and (c) flow-gradient directions, as in a nematic (adapted from Horn and Kléman [26]).

The three simplest geometries for shearing flow are illustrated in Figure 4, wherein the normal to the layers is oriented parallel to the vorticity, flow, and gradient directions in Figures 4a, b, and c, respectively. The labeling of these orientations is not consistent in the literature, so here we use the letter designations used for oriented nematics (Figure 1).

In orientations "a" and "c" in Figure 4, shearing does not change the layer spacing, at least as long as the sample remains perfectly well oriented and defect free. For these orientations, a liquid-like response is expected, and the smectic is expected to exhibit a largely viscous behavior during shearing, *if the sample remains defect free*. In the "b" orientation, however, uniform shear tends to rotate the layers and change their spacing. Since the layer spacing is a "solid-like" property of the smectic phase, shearing in this orientation should produce a solid-like material response, at least for shearing stresses less than some critical value needed to disrupt the well ordered smectic state.

If the lamellar spacing is increased by an amount $\delta > 2\pi l$, where $l = (K_1/B)^{1/2}$ is the *permeation depth*, which is roughly equal to the layer thickness $\approx$ 5 nm, then the sample will prefer to bend the layers in an *undulational instability* [27, 28] in order to restore the lamellar spacing to its preferred value. The increase in thickness δ required to produce this instability is independent of the initial thickness d of the stack. Hence, for a macroscopic sample of thickness, say, d = 300μ, the *strain* δ/d required to induce the undulational instability is extremely small, $\delta/d \approx 2\pi l/d \approx 10^{-4}$. Thus *smectic monodomains are extremely delicate, and can easily be disrupted by mechanical deformation.*

4.1. SMECTIC DEFECTS

For a dilational strain only 50% larger than that required to produce an undulational instability, the undulations grow large enough to nucleate topological defects. The defects found in this case are *parabolic focal conics,* consisting of a pair of orthogonal parabolic disclination lines with a common focus. When the smectic layers are dilated, a grid of these typically forms. (Example images of such grids are shown in Figures 6 and 7 below.)

Parabolic focal conics are a special case of generic *focal conic defects,* which are composed of layers curved to form toroidal surfaces called *Dupin cyclides.* Each such structure contains a pair of disclination lines, one an ellipse and the other a hyperbola, that share a common focus. In one degenerate case, the ellipse and the hyperbola both become parabolas, forming the parabolic focal conic discussed above. In another a degenerate case, the ellipse becomes a circle and the hyperbola a straight line. Focal conic domains are a predominant type of defect in disrupted smectics, because *they maintain a constant layer spacing throughout the domain.*

Unaligned, disrupted, smectics, such as those prepared by quenching from an elevated temperature into the smectic state, are frequently filled with focal conic domains [29, 30]. Parabolic focal conic domains can be packed to fill space. Another packing, which seems to be prevalent in thicker samples, is the *polygonal focal conic* texture. It is formed by cutting cones from the focal structures, and packing them together into a pyramidal region with the cones sharing a common apex. The base of the pyramid is a polygon. Small regions between cones are filled in by spherical surfaces that are concentric with the apex of the pyramid. By packing pyramids and inverted pyramids of this kind together, large volumes of space can be filled without dilating or rupturing smectic layers, except at disclination lines.

Understanding of the effect of shearing flow on smectics at low Deborah number is much less advanced than for nematics. The following summarizes the few experiments that have been carried out on initially monodomain smectic-A samples in orientations "c", "b", and "a".

4.2. SHEARING FLOW OF ALIGNED SMECTICS

If the layers are in orientation "c" and are perfectly parallel, they can slide smoothly over each other, producing a low-viscosity Newtonian viscous response [26]. However, a dust particle or other imperfection can easily disrupt the layers locally, leaving defects in its wake, as has been beautifully shown by Horn and Kléman [26]. Each moving defect nucleates another, so that a long cascading trail of defects is produced, eventually filling the whole sample with defects. Defects can easily proliferate in thermotropic smectics because of their high layer compression modulus B and small splay-bend constant K_1, which makes the layers more willing to roll up tightly than to compress even in the slightest. Even if dust particles are absent, any nonuniformity in the gap between nominally parallel plates, even one as small as a few layer spacings, is liable to lead to defect formation under steady shear. As we describe below, however, oscillatory shear can drive out the defects, and, in some cases, restore the sample to a state of uniform orientation.

The effect of shear on mondomains of 8CB in orientation "b" in Figure 4 was studied by Marignan et al. [31]. They found that very small shearing strains can introduce defects into the sample. This result is not surprising, since shearing in orientation "b" rotates lamellae and thereby changes their spacing. A shear-induced tilt angle of only 10^{-4} radians with respect to the initial orientation is already enough to produce an array of dislocations; at higher tilt angles, focal conic defects appear.

The effect of shear on smectic monodomains in orientation "a" seems not to have been studied. Like orientation "c," a perfect monodomain in orientation "a" can, in principle, be sheared without rotating the layers. Furthermore, it is believed that orientation "a" is more stable than "c" against layer undulation instabilities, since the layer normal in orientation "a" is orthogonal to both the shear and the shear gradient.

4.3. STEADY SHEARING FLOW OF DEFECT-RIDDEN SAMPLES

The introduction of defects into a smectic sample destroys its fluidity. This contrasts markedly with nematics, for which the presence of defects hardly alters the fluid's viscosity. Of course, this is because in a smectic the direction locally perpendicular to the layers is solid-like, and when defects are present all directions acquire some solid-like character. Horn and Kléman [26] measured shearing stresses in defect-containing smectic samples of 8CB and found that the shear stress σ was given by

$$\sigma = A + B\dot{\gamma} \tag{7}$$

When there are no defects in the sample, A ≈ 0 and B is then the viscosity of the smectic in orientation "b." Both constants A and B increase as the number of defects in the sample increases. A is an elastic "yield stress" of a defect-containing smectic. The magnitude of A can can be estimated from the density of smectic defects as

$$A \sim \frac{K_1}{a^2} \tag{8}$$

where a is the spacing between defects.

Yield-stress behavior is not always seen in the flow of disordered smectics. For higher densities of defects, the shear stress/shear rate curve is sometimes better described by a power law that ends in Newtonian behavior at the highest shear rates [32-34]. In general, however, the rheological behavior of disordered smectics is intermediate between that of a viscous liquid and an elastic solid.

4.4. DYNAMIC MODULUS

The time-dependent rheological properties of disordered smectics are also peculiar. Figure 5 shows apparent frequency-dependent moduli G' and G'' (labeled "MD", for "macroscopically disordered") measured for 8CB after a quench from the isotropic state [35]. (For definitions of G' and G'', see the contributions of T. McLeish and M. Cates in the present volume.) Quenched samples contain a very high density of smectic defects, which produce an elastic-like frequency dependence in G' at low frequencies of oscillation. The apparent moduli G' and G'' in Figure 5 are not true linear viscoelastic moduli, since they were measured in a nonlinear viscoelastic regime in which G' and G'' were strain dependent. In fact, for these smectic materials, no linear regime could be found even for strain amplitudes as low as 1%.

4.5. SHEAR ALIGNMENT

As discovered by Horn and Kléman [26], large-amplitude oscillatory shearing at an amplitude near unity drives out defects and aligns smectic layers. The curves in Figure 5 marked "MO" for "macroscopically ordered" correspond to a sample subjected to 20 minutes of oscillatory shearing at an amplitude of unity and a frequency of 10 s^{-1} in a cone-and-plate instrument. This treatment decreases the low-frequency elastic modulus, and increases the sample's fluidity; see Figure 5. Continued oscillatory shearing can reduce further the number of these defects, leading to an additional decrease in the low frequency elastic modulus G'. Thin samples ($\sim$ 300 μm) sheared between clean glass surfaces that induce homeotropic alignment can be completely freed of defects, and the "b" or "parallel" alignment

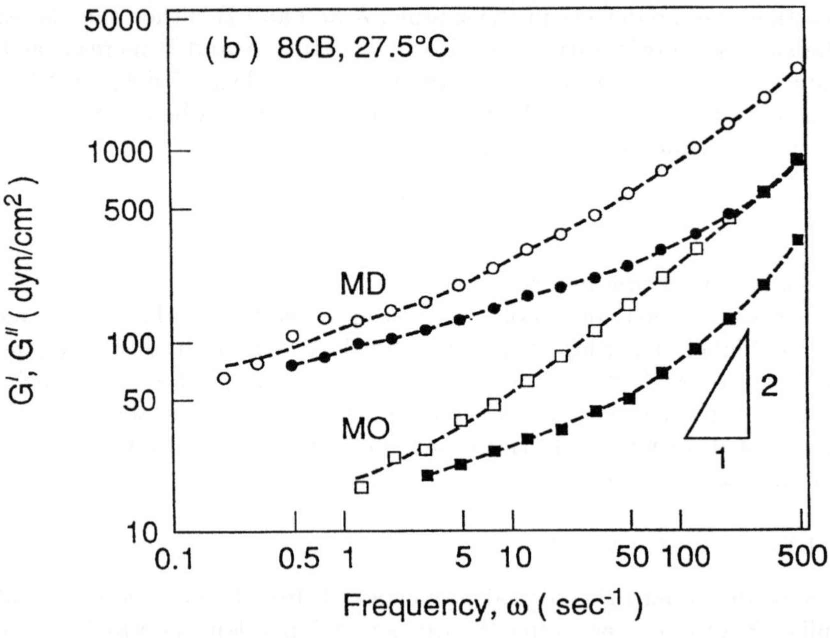

Figure 5. Apparent moduli G' (closed symbols) and G'' (open symbols) as functions of frequency at a strain amplitude of 10% for 8CB in the smectic-A state at 27.5 degrees C. The curves labeled "MD" are for a sample that had been quenched into the smectic-A state by reducing the temperature from the isotropic state. The curves labeled "MO" are for a sample that had been aligned after the quench by large-amplitude oscillatory shear at a frequency of 10 s^{-1} and amplitude of unity for 20 minutes (from ref. [35]).

(the one consistent with homeotropic boundary conditions) is produced [26]. Thicker samples seem never to clear themselves, even after hours of oscillatory shearing. Presumably a perfectly aligned sample would show terminal viscoelastic behavior at low frequencies.

Safinya et al. [36] have shown using x-ray scattering that within a degree or so of T_{AN}, the temperature at which the transition to the nematic state occurs, high shear-rate *steady shearing* of 8CB induces orientation in the perpendicular "a" orientation. At lower temperatures, the orientation is a mixture of "a" - perpendicular, and "c" - parallel, and the sample appears to consist of layers rolled up into tubes that are aligned in the flow direction [33]. (These rolled up layers are apparently analogs of the "onions", or multi-lamellar vescicles, observed by Roux and coworkers in sheared surfactant smectics, and discussed by Roux in this volume.) The "a" orientation is also induced in the nematic state just above the transition at T_{AN}. Intuitively, one expects that in the "c" orientation layer fluctuations will couple to the shearing flow more strongly than they do in the "a" orientation;

hence the "a" orientation is favored when such fluctuations are strong, as they are near T_{AN}. Fluctuations are also evidently responsible for a shear-induced "a" orientation in lamellar-forming block copolymers (which have smectic symmetry) at temperatures just above the transition to the layered state [37]. Goulian and Milner [38] have recently predicted that the "a" orientation should be the stable one in steady shearing of thermotropic smectics even at temperatures well below T_{NA}. However, their analysis ignores defects, which we have seen play a large role in the flow of smectics. Direction "b" seems always to be avoided in small-molecule thermotropic smectics, presumably because in this orientation shearing would change the layer spacing.

5. Visualization Experiments

To study directly the alignment process in smectic liquids, samples of 8CB were observed directly using optical microscopy under an oscillatory shearing flow in a torsional parallel-glass-disk device described elsewhere [39]. The director anchoring condition on each clean, uncoated, glass disk surface was observed to be homeotropic. Hence, in the quiescent nematic state, the sample eventually freed itself of defects, and formed a monodomain with director orientation perpendicular to the glass surfaces. An unaligned, or macroscopically disordered, starting state in the smectic-A phase of 8CB was achieved by starting in the isotropic state just above 40.5 degrees C, and cooling through the nematic state into the smectic-A state at 27.5 degrees C, a few degrees below the transition temperature of 33.5 degrees C. The cooling process typically took around 1 minute or so. This protocol produced macroscopically disordered smectic-A samples, whose texture depended on the sample thickness. Samples 300 μm thick formed parabolic focal conic textures, while samples 1 mm thick had more disorganized textures, with some of the characteristics of polygonal focal conic structures. The samples with parabolic focal conic textures annealed to some extent with time; one could observe patches of homeotropically aligned material growing slowly at the expense of the domains of parabolic focal conic textures. This annealing process could be greatly accelerated by applying an oscillatory shearing flow. For the thicker samples the texture in the quenched quiescent state at 27.5 degrees C remained trapped, and did not anneal appreciably with time unless oscillatory shear was applied.

We first consider the case of oscillatory shear imposed on the "thin" 300 micron-thick sample which was initially filled with parabolic focal conic defects. Figure 6a-b shows a pair of images from such a sample undergoing oscillatory shearing at a frequency of 10 s^{-1} and a strain amplitude of roughly unity. The flow direction is horizontal and the sample is viewed

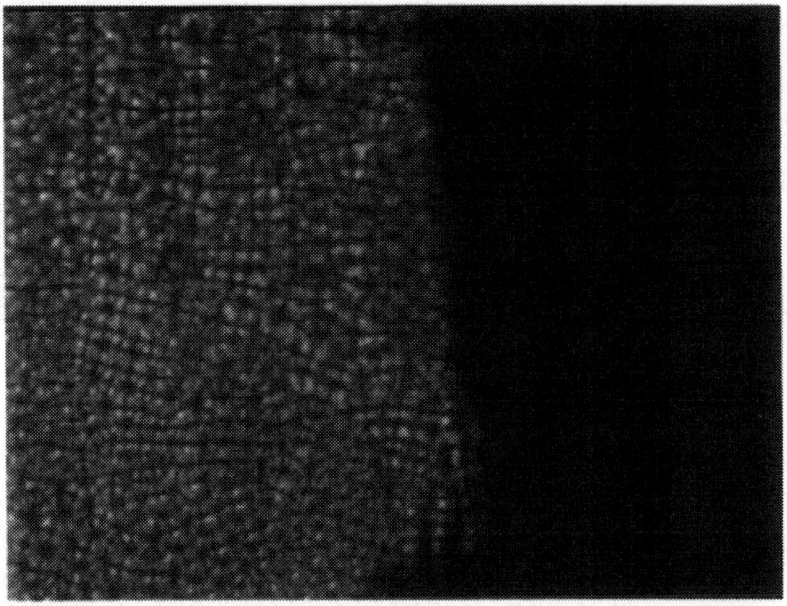

Figure 6. Images of 8CB in the smectic-A state at roughly 27.5 degrees C undergoing oscillatory shearing at a frequency of 10 s^{-1} and an amplitude of around unity. The gap is 300 μm. In (a), the the lower plate is about to move to begin its stroke to the right, while in (b) it has completed its stroke and is about to return to the left.

along a direction parallel to the velocity gradient. The sample is viewed between crossed polarizers using a X20 objective; the field of view is about 500 μm wide. The two images Figures 6a and 6b were both taken roughly when the instantaneous velocity was zero, that is, at the two extremes of the oscillatory cycle. Hence Figure 6a corresponds to the beginning of a forward "stroke" of the moving plate to the right, while 6b corresponds to the end of that stroke. If the strain γ is represented as $\gamma = \gamma_0 sin(\omega t)$, then at beginning of the stroke the phase ωt is a multiple of 2π, while at the end of the stroke the phase has increased by π. Both images 6a and 6b show the grid pattern of a parabolic focal conic texture. However, in 6a, the right half of the image is a black homeotropic "monodomain", while in 6b, this black monodomain is been converted by the shearing strain during the stroke into a focal conic grid of lower contrast than that of the left of 6b. This lower, "weaker", grid pattern disappears during the next half cycle (the "left-moving" or "reverse" stroke), which brings the image back to that of Fig. 6a. Thus, during each full cycle, the homeotropic region is converted to a weak grid pattern during the "forward" stroke, and then back to a homeotropic pattern during the "backward" stroke.

The sample texture therefore is little changed during the course of a full cycle. However, there is some change! The fraction of the sample occupied by the grid pattern is slightly reduced from one cycle to the next. This is shown in Figure 7a-b, which is a pair of images both taken at roughly the same phase in the strain cycle, near the beginning of the rightward stroke. Figure 7a-b shows that the grid pattern of a parabolic focal conic texture is slowly eaten up the black homeotropic monodomain, as time progresses. Thus, when the "weak" grid texture created during the forward stroke disappears during the backward stroke, it takes with it a small bit of the "strong" grid that had existed before the down stroke began.

The reduction in areal fraction occupied by the grid pattern can be monitored by measuring the intensity of light "I" transmitted through the sample without as a function of time during the oscillatory shearing. Before the oscillatory shearing begins, "I" is low because of the high density of defects that scatter light. As large-amplitude oscillatory shearing progresses, "I" increases, because of the gradual disappearance of focal conic defects that scatter light. Defining I_{fin} to be the "final" transmitted light intensity achieved after 20 minutes of oscillatory shearing, the complementary quantity, $I_{fin} - I$, normalized by I_{fin}, is plotted versus time during oscillatory shearing in Figure 8. $I_f - I$ is a therefore a measure of the intensity of light scattered by the defects. Also plotted in Figure 8 is the normalized time-dependent apparent complex modulus G^* measured during the alignment process of 8CB in a cone-and-plate rheometer [35]. The comparison in Figure 8 is qualitative; the optical and the rheological experiments were

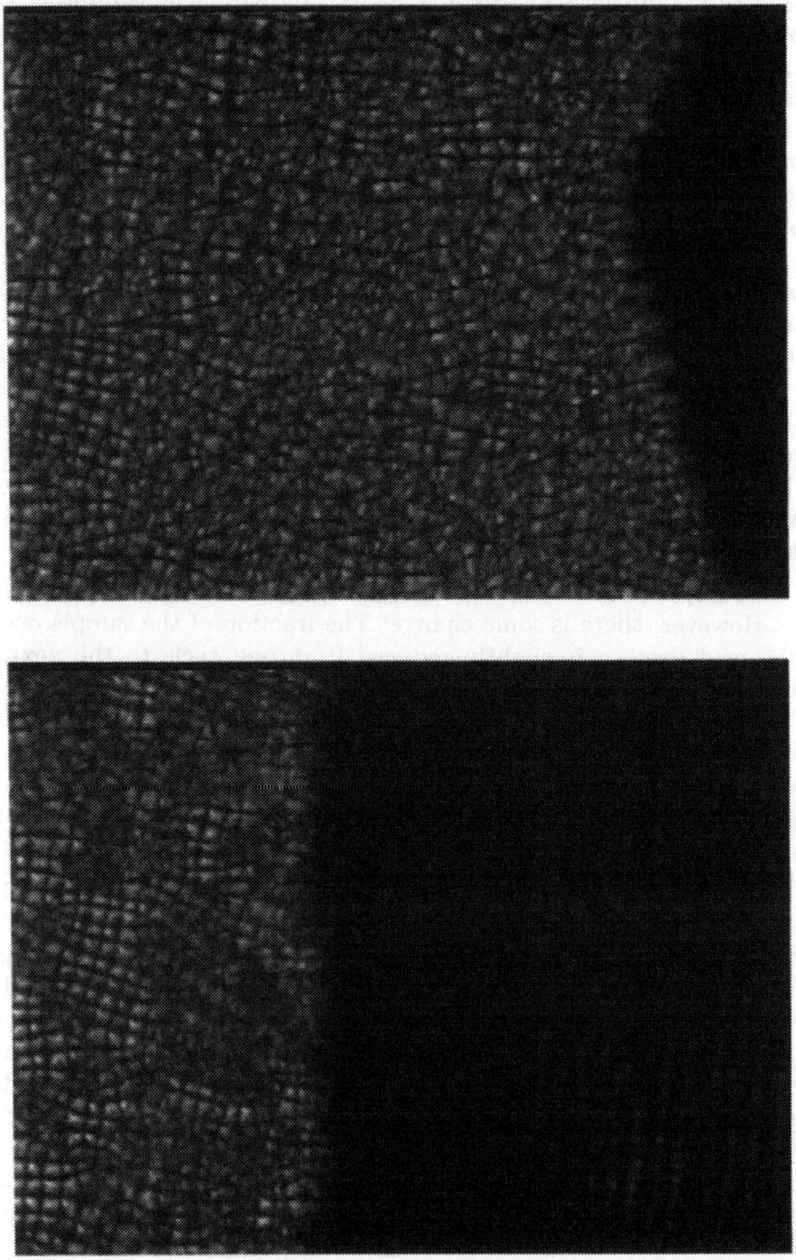

Figure 7. **The same as in Figure 6, except that both images were taken at roughly the same phase angle in the oscillatory shearing cycle, but separated in time by 10 secs, corresponding to around 15 cycles of shearing)**

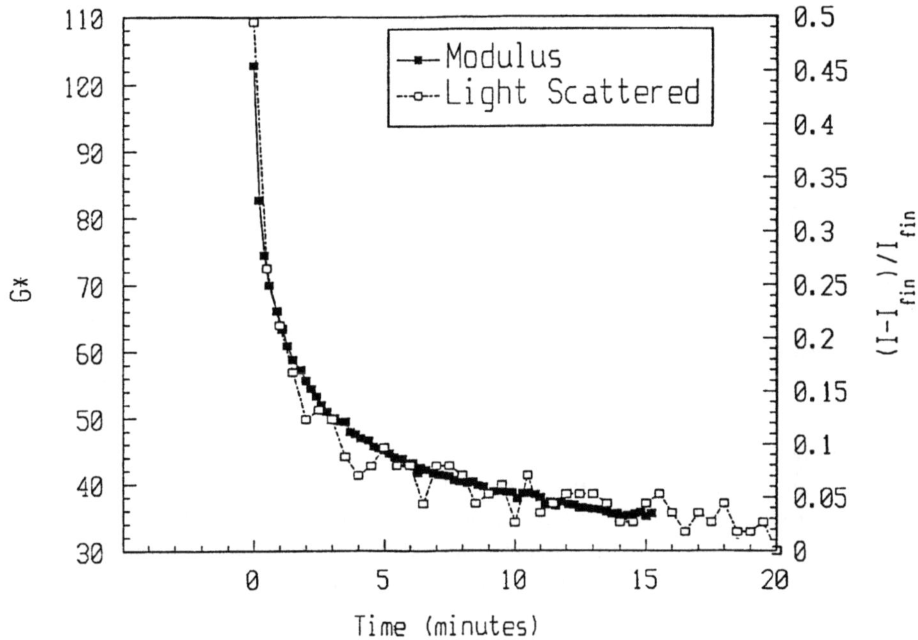

Figure 8. Normalized light scattered from smectic 8CB compared to the apparent complex modulus G^* in dyn/cm^2 versus time during large-amplitude oscillatory shearing at a frequency of 10 s^{-1} and a strain amplitude of roughly unity. The rheological data were obtained in a cone-and-plate rheometer, while the optical data were obtained in with the sample between parallel glass disks.

carried out in different devices with possibly different boundary conditions. The rheometer used to measure G^* in Figure 8 was a cone-and-plate device with steel fixtures of radius 25 mm and a cone angle of 0.04 radians, while the optical measurements were carried out in a parallel disk device with glass plates separated by 0.3 mm. Nevertheless, there is in Figure 8 a striking coincidence in the time dependence of the alignment process, probed either optically or mechanically. This would seem to confirm that at the (rather low) frequencies probed in these experiments, the modulus of a defect-ridden smectic is dominated by contributions from the defects.

We now consider briefly the case of oscillatory shear imposed on a "thick" 1000 micron-thick sample. This sample was initially filled with a dense "cobweb"-like network that seemed to respond largely *elastically* to the deformation. That is, the motion of the lines more or less tracked the instantaneous macroscopic strain; the maximum positive and negative deflections of the disclination lines largely coincided with the peak positive and negative displacements of the oscillating plate.

Nevertheless, the motion of the disclination lines was highly non-uniform. Some lines stretched in a direction parallel to the shearing displacement while other lines stretched in a direction orthogonal to it. This highly *non-affine* deformation of the disclination network presumably arises from the strong dependence of the modulus of the smectic on layer orientation. Since for some orientations of the layers the modulus is close to that of a liquid, while for others it is is close to that of a solid, the imposed strain tends to *localize* itself in regions of the sample whose modulus is closest to that of a liquid.

As in the case of the "thin" 300 μm thick sample, little change in the defect pattern could be perceived over the coarse of a single cycle of oscillatory strain. Only after many cycles could it be seen that the defect texture had coarsened to an observable extent. Thus, the mechanism by which coarsening occurs in either thick or thin 8CB samples is rather weak; it produces large changes in orientation only after many cycles. A gradual alignment also occurs for lamellar block copolymers that is at least superficially similar to that of small-molecule smectics [35, 37, 40].

6. Summary and Unsolved Problems

The behavior of small-molecule nematic liquids in shearing flows is complex, but can to a large extent be understood using the Leslie-Ericksen theory. In a shearing flow the Leslie-Ericksen theory predicts large differences between the behavior of flow-aligning and that of tumbling nematics. The most interesting and complex behavior is exhibited by tumbling nematics. In sudden start-up of fast shearing, the director in an initially well aligned tumbling nematic rotates through multiple periods of π in the shearing plane. Associated with these rotations are multiple oscillations of stress that slowly damp in a torsional shearing apparatus out over a few hundred units of strain. In a plate-and-plate device (and probably also in a cone-and-plate apparatus), the damping of the oscillations is associated with radial nonuniformity in the director orientation. The nonuniformity can be monitored by the motion of twist walls, across which the director orientation jumps. If the shearing direction is reversed before many disclinations appear, the damping of the stress oscillations can be reversed, as the twist walls are expelled from the sample. For a gradually increasing shear rate, a different progression occurs in tumbling nematics. At low Ericksen number, Er $<$ 10 or so, with homeotropic (i.e., perpendicular) boundary conditions at the solid surfaces, the director in the bulk of the sample rotates towards the flow alignment angle, reaching a steady-state angle set by the balance of viscous and Frank elastic stresses. Eventually, at a critical Ericksen number, the director begins to rotate out of the shearing plane, and at still higher Er, there is a second instability to the formation of roll cells oriented parallel to flow. Eventually, disclination lines form. At high shear rate, the density of disclination lines seems to follow a simple scaling law; however, the detailed mechanisms by which disclinations form are largely unknown.

By contrast to nematics, the flow properties of smectics are much less understood. In almost no cases can a monodomain smectic withstand a large shearing strain without forming defects. Such defects, once formed, rapidly proliferate. Unlike nematic defects, smectic defects generate a solid-like rheological response, and make large contributions to the rheological properties of the smectic, especially to the elastic properties at low frequencies. Large amplitude oscillatory shearing can drive out many of these defects, but the mechanism by which this occurs is so far largely unknown.

References

1. Frank, F.C. (1958) *Disc. Faraday Soc.* **25**, 19.
2. de Gennes, P.G. and Prost, J. (1993) *The Physics of Liquid Crystals*, Oxford Univ. Press, New York.
3. Leslie, F.M. *Arch. Rat. Mech. Anal.* **28** (1968), 265.
4. Skarp, K., Lagerwall, S.T., and Stebler, B. (1980) *Mol. Cryst. Liq. Cryst.* **60**, 215.
5. Kroger, M. and Sellers, H.S. (1995) *J. Chem. Phys.* **103**, 807.
6. Archer, L.A. and Larson, R.G. (1995) *J. Chem. Phys.* **103**, 3108.
7. Kuzuu, N. and Doi, M. (1984) *J. Phys. Soc. Jpn.* **53**, 1031.
8. Gu, D.-F., Jamieson, A.M., and Wang, S.Q. (1993) *J. Rheol.* **37**, 985.
9. Gu, D.-F. and Jamieson, A.M. (1994) *J. Rheol.* **38**, 555.
10. Moldenaers, P., Yanase, H., and Mewis, J. (1990) in *Liquid-Crystalline Polymers*, Ed. Weiss, R.A., Ober, C.K., A.C.S. Symp. Series 435, Amer. Chem. Soc., Washington, D.C.
11. Burghardt, W.R. and Fuller, G.G. (1991) *Macromolecules* **24**, 2546.
12. Mather, P.T., (1996) work in progress.
13. Carlsson, T. (1984) *Mol. Cryst. Liq. Cryst.* **104**, 307.
14. Zuñiga I. and Leslie, F.M. (1989) *Liq. Cryst.* **5**, 725.
15. Pieranski, P., Guyon, E., and Pikin, S.A. (1976) *J. Phys. (Paris)* **37,C1**, 3.
16. Han, W.H. and Rey, A.D. (1993) *J. Non-Newt. Fluid Mech.* **48**, 181.
17. Mather, P.T., Pearson, D.S., and Larson, R.G. (1996), *Liq. Cryst.*, **20**, 539.
18. Larson, R.G. and Mead, D.W. (1993) *Liq. Cryst.* **13**, 151.
19. Manneville, P. and Dubois-Violette, E. (1976) *J. Phys. (Paris)* **37**, 285.
20. Larson, R.G. (1993) *J. Rheol.* **37**, 175.
21. Dubois-Violette, E., Durand, G., Manneville, P., and Pieranski, P. (1978) *Solid State Phys., Suppl.* **14**, 147.
22. Srinivasarao, M. and Berry, G.C. (1991) *J. Rheol.* **35**, 379.
23. Gähwiller, C. (1972) *Phys. Rev. Lett.* **28**, 1554.
24. Manneville, P. (1981) *Mol. Cryst. Liq. Cryst.* **70**, 223.
25. Cladis P. and Torza, S. (1976) *Coll. Interf. Sci.* **4**, 487. Oxford Univ Press, New York.

26. Horn, R.G. and Kléman, M. (1973) *Ann. Phys.* **3**, 229.
27. Rosenblatt, Ch.S., Pindak, R., Clark, N.A., and Meyer, R.B. (1977) *J. Phys. (Paris)* **38**, 1105.
28. Ostwald, P. and Allain, M. (1985) *J. Phys. (Paris)* **46**, 831.
29. Sethna, J.P. and Kléman, M. (1982) *Phys. Rev. A* **26**, 3037.
30. Lavrentovich, O.D. (1986) *Sov. Phys. JETP* **64s**, 984.
31. Marignan, J., Malet, G., and Parodi, O. (1978) *Ann. Phys.* **3**, 221.
32. Duke, R.W. and Chapoy, L.L. (1976) *Rheol. Acta* **15**, **548**.
33. Tamamushi, B. (1974) *Rheol. Acta* **13**, 247.
34. Panizza, P., Archambault, P., and Roux, D. (1995) *J. Phys. II France* **5**, 303.
35. Larson R.G., Winey, K.I., Patel, S.S., Watanabe, H., and Bruinsma, R. (1993) *Rheol. Acta* **32**, 245.
36. Safinya, C.R., Sirota, E.B., Bruinsma, R.F., Jeppesen, C., Plano, R.J., and Wenzel, L.J. (1993) *Science* **261**, 588.
37. Koppi, K.A., Tirrell, M., Bates, R.S., Almdal, K., and Colby, R.H. (1992) *J. Phys. II, France* **2**, 1941.
38. Goulian, M. and Milner, S.T. (1995) *Phys. Rev. Lett.* **74**, 1775.
39. Mather, P.T., Pearson, D.S., and Larson, R.G. (1996), *Liq. Cryst.*, **20**, 527.
40. Kannan, R.M. and Kornfield, J.A. (1994) *Macromolecules* **27**, 1177.

STRUCTURAL PROPERTIES OF WORMLIKE MICELLES

F. LEQUEUX, S.J. CANDAU
Laboratoire d'Ultrasons et de Dynamique des Fluides Complexes
U.R.A. au CNRS n°851, Université Louis Pasteur
4, rue Blaise Pascal, 67070 Strasbourg Cedex, France

1. Introduction

Surfactants in aqueous solutions can associate into various structures : spherical micelles, vesicles, lamellar phases, sponge phases, etc... They can also self-assemble into very long flexible cylinders : in that case, the aggregates are called wormlike micelles, threadlike micelles, rodlike micelles, or giant micelles. Actually, only a few surfactants, with appropriate counterions self-assemble into very long cylindrical aggregates. The most popular are CTAB (in presence of NaSal or KBr), $CPCLO_3$, and lecithin/bile salt mixture. The diameter of the cylinders is about 5nm, the persistence length about 20 nm and the total length can reach several micrometers (cf. Fig. 1).

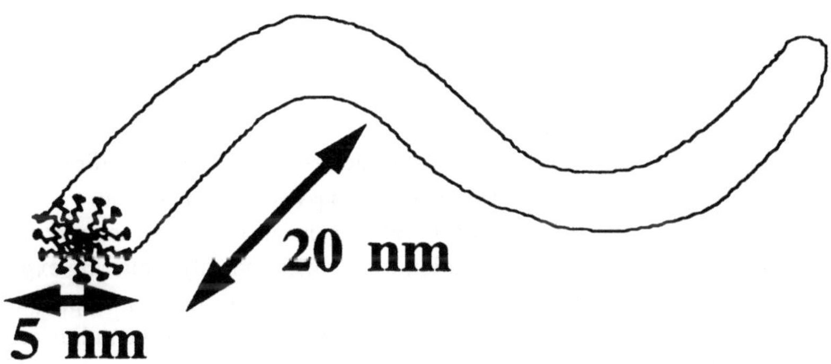

Figure 1. Worm-like micelle.

T. McLeish (ed), Theoretical Challenges in the Dynamics of Complex Fluids, 181-190.

The solutions of worm-like micelles are then reminiscent of polymer solutions, and exhibit a dilute regime, a semi-dilute viscoelastic regime, and a polyelectrolyte regime (in salt free solutions of ionic surfactants).

However, there are some crucial differences between wormlike micelles and polymer solutions : 1) The length distribution of the wormlike micelles is in general exponential and the average size is a function of the surfactant concentration [1][2]. 2) Scission and recombination of the micelles occur continuously due to thermal fluctuations [1][2]. 3) The cylinders can form loops [3] and interconnections [3][4].

2. Micellar growth

The equilibrium statistics of micelles are controlled by the competition between the mixing entropy that tends to promote the formation of small micelles and the scission energy E_{break} that conversely favours micellar growth. For neutral surfactants, E_{break} can be identified with E_c, the energy associated with the surfactant packing near the cylinder end-caps.

The length distribution $C(L)$ of neutral equilibrium polymers has been calculated using a mean field approach and is given by [1][2] :

$$c(L) \propto \exp[-L/\bar{L}] \tag{1a}$$

$$\bar{L} \simeq \phi^{1/2} \exp(E_c/2k_BT) \tag{1b}$$

where ϕ is the surfactant volume fraction.

Equations (1) are also appropriate to the description of dilute, inflexible micellar rods, for which they reduce to the results of a mass-action calculation.

Direct experimental verifications of the $\phi^{1/2}$ dependence of $\bar{L}$ are not easy to perform as the classical methods of characterization of polymers (mainly scattering experiments) require an extrapolation to zero polymer concentration. In this respect the most convincing results were obtained by comparing measurements of the radius of gyration and of the hydrodynamic radius in the regime where the micelles are small rods [5].

For charged micelles there are two opposite contributions to the energy of scission, E_{break}. These are the repulsive energy of the surface charges that favors breaking of micelles and the end cap energy that favors a growth of micelles. The micellar growth is also modified by the increased entropy of the counter ions near the end caps. This leads to a complicated behavior that has been described by Mackintosh et al. [6] and that is illustrated in Fig. 2 showing the variation of $\bar{L}$ with surfactant volume fraction for neutral and charged micelles.

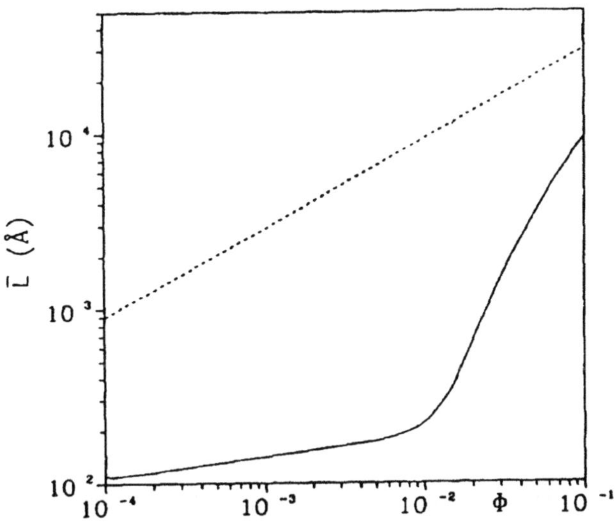

Figure 2. Micellar length *vs*. surfactant volume fraction for $E_c = 20\ k_BT$. The upper line is for neutral micelles and the lower line is for micelles with one effective charge per Bjerrum length and a radius of 20 Å (from ref.[6]).

The most striking feature of the growth law is the existence of a dilute regime in which the micelles are nearly monodisperse and adopt the shape of a small stiff cylinder. The average micelle length $\bar{L}$ varies slowly with concentration. At higher concentrations, $\bar{L}$ is smaller than for neutral micelles but the growth is more rapid. In this regime the length distribution is very broad as in the case of neutral micelles. Then in a third regime the growth is moderate to eventually lead to micelles of same size as in the neutral case.

The verification of the above behaviour by means of static experiments is very difficult to achieve because of the effect of electrostatic interactions. On the other hand, rheological experiments show clearly the abrupt change of length at the cross-over between dilute and semi-dilute regime [7].

3. Micellar flexibility

The first unambiguous evidence of micellar flexibility was given by Porte et al. in a study combining dynamic light scattering and magnetic birefringence experiments on a CPyBr system [8].

In the magnetic field, stiff cylindrical micelles reorient to produce an optical response proportional to the length of the micelles. In the case of worm-like micelles, parts of any given micelle, separated by contour length much larger than the persistence length ℓ_p, align independently in the magnetic field roughly as if they were unconnected small rods of length ℓ_p. Therefore, upon inducing a micellar growth by addition of salt from stiff micelles of length less than ℓ_p to worm-like micelles, one observes a cross-over between a regime of increasing magnetic birefringence to a saturation plateau. The cross-over occurs as $\bar{L} \simeq \ell_p$. The latter, measured by dynamic light scattering was found to be of the order of 200 Å.

Information on the micellar shape is also provided by scattering experiments. Thus a study combining light scattering and small angle scattering data allowed to measure the form factor of solutions of cetylpyridinium chlorate over a very wide range of scattering wavevector q [9]. In the high q range, the data fit the form factor of a rigid cylinder with radius $r_c = (20 \pm 1)$Å (the value quoted being consistent with the length of the surfactant molecule). In a lower q range, corresponding to spatial scales larger than l_p the form factor is characteristic of a flexible coil with excluded volume. The cross-over between the two regimes provides an estimate of the persistence length which is found to be $\ell_p = (170 \pm 50)$Å, in good agreement with the previous finding.

The worm-like micelles can also be directly observed by cryomicroscopy, but one has to be aware of possible artefacts. In particular the films used for transmission microscopy are very thin, and depletion effects are clearly present. It doesn't seem possible to obtain reliable estimates of the persistence length and of the average micellar length.

4. Connected wormlike micelles

Upon decreasing the surface per polar head for a given hydrophobic tail size, one encounters first spherical micelles, then wormlike micelles that combine both cylindrical (the mean body) and spherical (the end-caps) curvatures. Upon decreasing further the surface per polar head, the end cap energy increases and the micelles get longer and longer. The curvature of the micelles tends to be reduced until a transition to a lamellar or a vesicular phase takes place (see phase diagram, next paragraph). However, one can envision another scheme in which the extremities, positively curved, are replaced by connections (negatively curved) (Fig. 3).

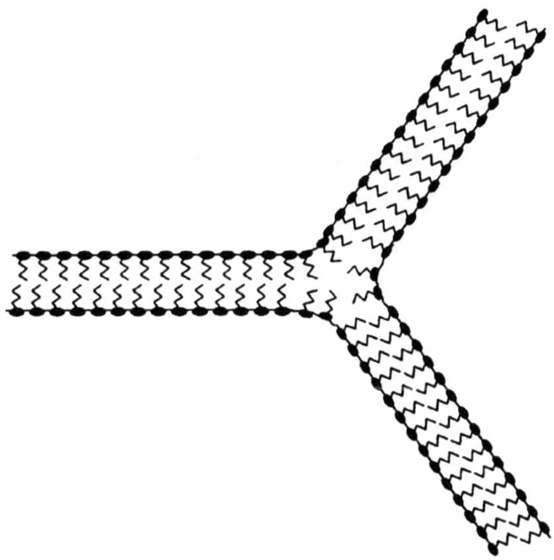

Figure 3. Three-fold branched micelle.

The existence of intermicellar connections accounts quite well for some behaviours observed in micellar solutions in presence of high salt contents [10]. The possibility of connections has been first proposed by G. Porte and co-workers about 10 years ago [11]. But, their effect on the rheological properties of worm-like systems has been modeled [12] and observed [13] only 2 years ago (see chapter by Cates - this volume). The static picture of a connected network is difficult to model and requires sophisticated theoretical tools [14]. The presence of connections is generally expected in systems with a high salt concentration (in order to screen the repulsion between polar heads).

Direct observation of branched micelles is rather difficult. The systems of connected worm-like micelles seem to be rather sensitive to a thermal shock. In addition, it is difficult to discriminate between the overlapping and the crosslinking of two micelles. However, recently several groups have shown clearly the presence of three-fold connections [3][4].

5. Phase diagram

At a given temperature, the structure of the self-assemblies of surfactants is strongly dependent on surfactant and salt concentrations. Fig. 4 represents a schematic diagram surfactant concentration - salt concentration that merges from the ensemble of static and

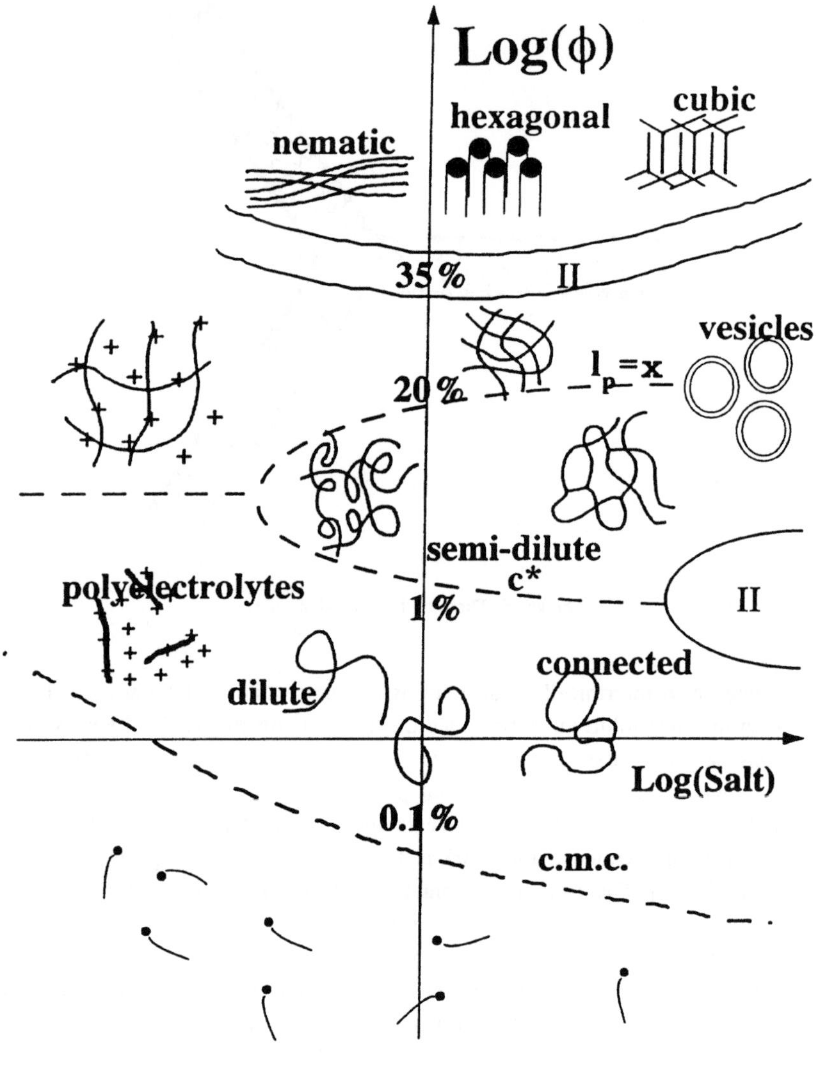

Figure 4. Schematic diagram : surfactant volume fraction ϕ versus salt concentration.

dynamical studies, performed during mainly the ten last years. This diagram must be considered as indicative. Not all the structures represented show up for a given system and at a given temperature. Also the different domains are more or less extended depending on the system.

Upon scanning the diagram by varying the surfactant concentration at given salt content, we encounter a series of transitions or crossover lines. We discuss below the static properties of the systems obtained in the different domains delimited by these lines.

5.1. DILUTE REGIME

- At very low concentration, the surfactant is molecularly dispersed, up to the critical micellar concentration (cmc).
- Above the critical micellar concentration (cmc), micelles form, spherical or ellipsoidal in salt free systems, then turning into a worm-like shape upon increasing the salt concentration to eventually become branched at very high salt content.

5.2. SEMI-DILUTE REGIME

This domain extends in between the crossover line $c = c^*$ between dilute and semi-dilute regimes and the line where the correlation length, that can be visualized on the main distance between two contact points, becomes equal to the persistence length ℓ_p. If the electrostatic interactions are screened, the persistence length is the natural one and is of the order of 200 Å. The presence of electrical charges stiffens the micelles, and at low salt contents the crossover dilute/semi-dilute coincides with the line $\xi = \ell_p$ where now ℓ_p is increased due to the electrostatic contribution. In that case the crossover is very sharp because of the enhanced growth for $c > c^*$ (cf. Fig. 2). Upon increasing the salt concentration, c^* decreases as a result of the micellar growth, as discussed in paragraph 2.

At salt concentrations large enough that the electrostatic interactions are screened, the systems behave as semi-dilute solutions of neutral polymeric chains [2][15]. In particular, it was found that the scaling law of classical polymers describes the behaviour of the static correlation length and of the hydrodynamic correlation length (ξ - $\xi_H \sim \phi^{\sim -0.77}$), that of the scattered intensity at zero $q(I(o) \sim \phi^{\sim -0.3})$ and of the shear modulus G $\sim \phi^{\sim 2.3}$). Examples are given in Fig. 5. The dynamics of these systems has been described theoretically by M. Cates (see his chapter - this volume).

At higher salt concentration, the micelles form connections. It must be stressed that this doesn't happen through a true crossover. There are no detectable differences between the structural properties of the linear and connected micelles. Conversely, the dynamics properties are very sensitive to the structure, and in fact, the existence of

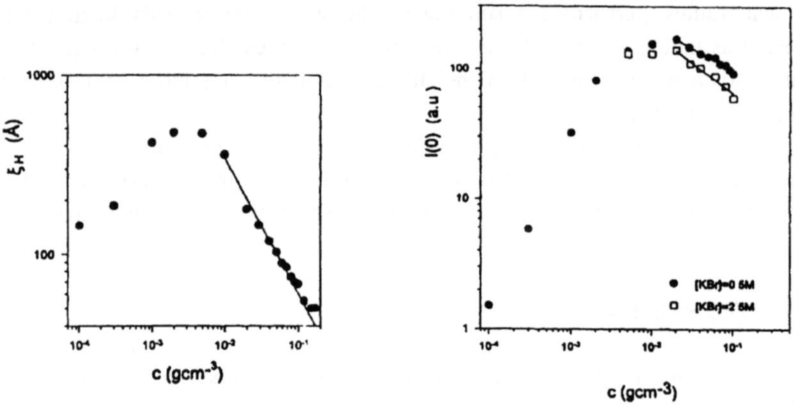

Figure 5. Effect of surfactant-concentration (CTAB) on
a) the dynamic (ξ_H) correlation lengths ;
b) the scattered intensity extrapolated at zero-q.
[KBr] = 0.5 M, T = 33.2°C (from ref.[17]).
For $c<c^*$, ξ_H characterizes the average micellar size and increase
upon increasing surfactant concentration.
For $c>c^*$ one observes the scaling behaviour of classical polymers in good solvent.
The straight line is the best linear fit and has a slope -0.76.
The full line drawn through the scattered intensity data represents the best fit of the expression derived from the renormalization group theory for classical polymers and taking into account the micellar growth.

connections has been revealed through rheological studies [13].

At very high salt concentration, there is a phase separation with a critical point [18]. At the moment very little is known about the structure of the systems in the vicinity of the critical point and within the two phases formed upon demixing.

In this domain, the temperature is an important parameter. Under some conditions one can obtain lamellar phases or vesicles. In the latter case, there is a domain of coexistence between vesicles and cylindrical micelles.

5.3. CONCENTRATED REGIME

This regime is located above the line corresponding to $\xi = \ell_p$. On the left side of the phase diagram, the behaviour is that of classical polyelectrolytes. In particular, SANS experiments reveal the presence of a peak at a wavevector q^* that varies like $\phi^{1/2}$ and a shoulder at $q = 2q^*$ [19]. In this regime a surprising result has been reported concerning the plateau modulus that was found to be considerably reduced by the electrostatic interactions [20][7]. There is not at the moment clear interpretation of this effect. It was proposed that the decrease of the plateau modulus was due to a pre-nematic effect involving collective fluctuations of ordered micelles. But an increase of the entanglement length, or contour length between two consecutive entanglements, would also account for this effect.

As the salt concentration is increased, the electrostatic interactions are screened. In that case, and more or less independently of the salt content, the micelles are locally rigid. One observes a peak in the structure factor whose position q^* varies as $\phi^{1/2}$ [21]. Also a lesser increase of the plateau modulus with surfactant concentration was observed [21][22] upon approaching the isotropic nematic transition, which was interpreted as the signature of a pre-nematic effect like in polyelectrolyte systems.

At higher concentrations, after a phase separation one observes a nematic or an hexagonal phase whose location depends on the salt content [23]. In some cases, a cubic phase can be observed above the connected phase.

References

1. Cates, M.E. (1987) Reptation of living polymers : dynamics of entangled polymers in the presence of reversible chain-scission reactions, *Macromolecules* **20**, 2289–2296.
2. Cates, M.E. and Candau, S.J. (1990) Statics and dynamics of worm-like surfactant micelles, *J. Phys. Condens. Matter* **2**, 6869-6892.
3. Knoblich, A., Masumoto, M., Murata, K. and Fujiyoshi, Y. (1995) Cryogenic transmission electron-microscopic studies of micellar structure correlated with solution viscosity on perfluorooctyl sulfonates and their mixtures with a nonionic surfactant, *Langmuir* **11**, 2361-2366.
4. Danino, D., Talmon, Y., Levy, H., Beinert, G. and Zana, R. (1995) Branched thread-like micelles in an aqueous-solution of a trimeric surfactant, *Science* **269**, 1420-1421.
5. Missel, P.J., Mazer, N.A., Benedek, G.B., Young, C.Y. and Carey, M.C. (1980) Thermodynamic analysis of the growth of sodium dodecyl sulfate micelles, *J. Phys. Chem.* **84**, 1044-1057.
6. MacKintosh, F.C., Safran, S.A. and Pincus P.A. (1990) Self-assembly of linear aggregates : the effect of electrostatics on growth, *Europhys. Lett.* **12**, 697-702.
7. Kern, F., Lequeux, F., Zana, R. and Candau, S.J. (1994) Dynamical properties of salt-free viscoelastic micellar solutions, *Langmuir* **10**, 1714-1723.
8. Porte, G., Appell, J. and Poggi, Y. (1980) Experimental investigations on the flexibility of elongated cetylpyridinium bromide micelles, *J. Phys. Chem.* **84**, 3105-3110.
9. Appell, J. and Marignan, J. (1991) Structure of giant micelles : a small angle neutron scattering study, *J. Phys. II France* **1**, 1447-1454.

10. Lequeux, F., Candau, S.J. (1994) Dynamical properties of worm-like micelles - Deviations from the classical picture, *ACS Symposium Series* **578**, 51-62.
11. Porte, G., Gomati, R., El Haitamy, O., Appell, J. and Marignan, J. (1986) Morphological transformations of the primary surfactant structures in brine-rich mixtures of ternary systems (surfactant/alcohol/brine), *J. Phys. Chem.* **90**, 5746-5751.
12. Lequeux, F. (1992) Reptation of connected wormlike micelles, *Europhys. Lett.* **19**, 675-681.
13. Khatory, A., Lequeux, F., Kern, F. and Candau, S.J. (1993) Linear and non-linear viscoelasticity of semi-dilute solutions of wormlike micelles at high salt content, *Langmuir* **9**, 1456-1464.
14. Elleuch, K., Lequeux, F. and Pfeuty, P. (1995) Crosslink effects on equilibrium polymers, *J. Phys. I France* **5**, 465-474.
15. Candau, S.J., Hirsch, E. and Zana, R. (1984) New aspects of the behaviour of alkyltrimethylammonium bromide micelles : light scattering and viscometric studies, *J. Physique* **45**, 1263-1270.
16. Schurtenberger, P. and Cavaco, C. (1993) On the application of polymer renormalization group theory to polymer-like microemulsions, *J. Phys. II France* **3**,1279-1288.
17. Buhler, E., Munch, J.P. and Candau S.J. (1995) Dynamical properties of wormlike micelles - A light scattering study, *J. Phys. II France* 5, 765-787.
18. Appell, J. and Porte, G. (1983) Cloud points in ionic surfactant solutions, *J. Phys. Lett. France* **44**, L689-L695.
19. Schmitt, V., Schosseler, F. and Lequeux, F. (1995) Structure of salt-free wormlike micelles : signature by SANS at rest and under shear, *Europhys. Lett.* **30**, 31-36.
20. Schmitt, V. and Lequeux, F. (1995) SANS spectra and elastic plateau modulus in a charged wormlike micelles solution : effect of salt, *J. Phys. II France* **5**, 193-197.
21. Schmitt, V., Lequeux, F., Pousse, A. and Roux, D. (1994) Flow behavior and shear induced transition near an isotropic/nematic transition in equilibrium polymers, *Langmuir* **10**, 955-961.
22. Shikata, T. and Pearson, D.S. (1994) Phase transitions in entangled networks of wormlike micelles, *Langmuir* **10**, 4027-4030.
23. Van der Schoot, P. (1996) The hexagonal phase of wormlike micelles, *J. Chem. Phys.* **104**, 1130−1139.

MECHANICAL PROPERTIES OF BILIQUID FOAMS

F. LEQUEUX & P. BOLTENHAGEN
Laboratoire d'Ultrasons et de Dynamique des Fluides Complexes
U.R.A. au CNRS n°851, Université Louis Pasteur
4, rue Blaise Pascal, 67070 Strasbourg Cedex, France

1. Introduction

An emulsion is a dispersion of droplets of one fluid, mixed into an immiscible phase of another fluid, and stabilised by surfactants. Let us consider now an emulsion of simple oil droplets into a water phase. If the oil concentration is below the close packing one (oil volume fraction of 64% for monodisperse oil droplets), this solution is dilute. But it is easy to concentrate the oil, by centrifugation for instance, and thus to get semi-dilute samples (above 64%) and concentrated samples (about 98% of oil). These concentrated samples consist in arrays of nearly polyhedral droplets of oil separated by a thin water film. They are usually called biliquid foams, as they are similar to classical foams, with the air phase replaced by an oil phase. Actually, they can be very stable if they contain adequate surfactant and if they are nearly monodisperse. The monodispersity kills the ripening since the chemical potential of oil depends on the radius of the droplet.

Actually these biliquid foams exhibit a complex mechanical behavior : at first view there are just fluids with a yield stress : they are nearly elastic when submitted to a small stress, while they can flow when submitted to large stresses. But we will see how this picture is schematic, and shares some similarities with the cubic phase discussed in this volume by C. Toprakcioglu.

We will first explain how the interfacial tension generates the elasticity of the systems. Moreover, these systems also exhibit a loss modulus and we will explain why usual linear approachs fail to explain the observed loss modulus. Actually, the packing of the cells is in general amorphous. We will see how the disorder due to the random close packing affects strongly the linear mechanical properties. We will then present the classical picture of this yield stress fluid, in terms of droplets rearrangements induced by flow. But this mean field approach is in disagreement with creep experiments. We will try finally to give a rough picture of the effect of disorder on the flow properties, describing the flow in terms of the dynamics of glassy systems.

T McLeish (ed), Theoretical Challenges in the Dynamics of Complex Fluids, 191-201

2. Linear Elasticity

We will now review the present approach for linear mechanical properties of these biliquid foams.

2.1. PRINCEN'S MODEL FOR FOAM ELASTICITY [1]

Let us first consider a foam with nearly 100% oil. When this foam is submitted to a small strain, the cells are deformed.

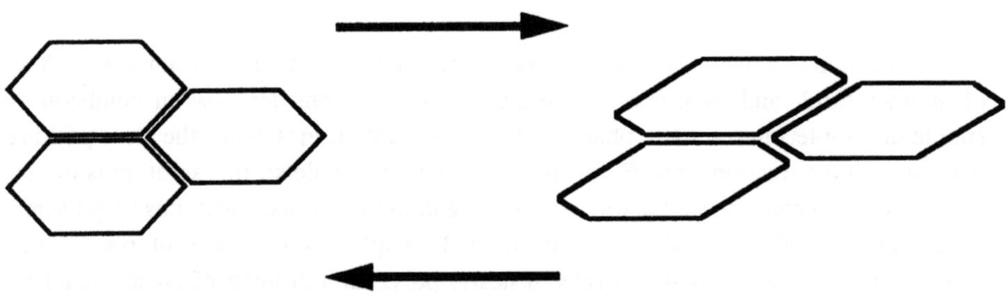

Figure 1. - Schematic 2d cells submitted to a simple shear strain : the total surface area of the interfacial films increase and this confers elasticity to the fluid.

This leads to two interesting features (fig. 1):
- the orientational distribution of the films, which is isotropic at rest, becomes anisotropic.
- the total surface area S of the films increases like the square of the strain.

One can in fact easily show, from geometrical considerations, that:

$$S=S_0(1+n.\gamma^2) \tag{1}$$

where S_0 is the initial surface, γ the strain, and n a numerical factor which depends on the details of the cell shape. Hence the Free Energy F, which is simply the interfacial tension times the film area, is written per unit volume :

$$F=\Gamma(1+n.\gamma^2)\frac{4\pi R^2}{\frac{4}{3}\pi R^3} \tag{2}$$

where Γ is the interfacial tension and R the droplets radius (when they are spherical). The stress is then the derivative of the Free Energy with respect to the strain :

$$\sigma=\frac{\partial F}{\partial\gamma}\propto\frac{\Gamma}{R}\gamma \tag{3}$$

This relation shows that the system is a Hookean solid with an elastic modulus equal to the interfacial tension divided by the droplet radius, up to a geometrical factor. Actually, this numerical factor depends on the structure and has been computed for various 2d and 3d ordered foams [2]. In order to model some linear viscoelastic properties of foams, we will now present an effective medium approximation.

2.2. 3D EFFECTIVE MEDIUM APPROACH

We will first introduce Palierne's model [3] of the viscoelasticity of semidilute emulsions. Then, we will explain, using Hemar's self-consistent model [4], how we can get a general relation between the film's mechanical properties and the foam macroscopic viscoelasticity.

2.2.1. *J.F. Palierne's model for semi-dilute viscoelastic emulsions*

Palierne's model [3] consists in the calculation of the macroscopic viscoelasticity G of an emulsion of viscoelastic droplets (volume fraction Φ, radius R and viscoelastic modulus of the droplets G_{inc}) dispersed in a viscoelastic matrix of modulus G_{mat} with general interfacial properties. The interfacial properties must be described with an interfacial tension Γ, but also with a complex dilational viscoelastic modulus β' corresponding to the elastic and loss response of the interface when the surface area oscillates, and with a shear viscoelastic modulus β'' corresponding to a shear of the interface at constant surface area [5]. The calculation also includes the mechanical interactions between the droplets. The derived macroscopic viscoelastic modulus G is written:

$$G = G_{\text{mat}} \cdot \left[\frac{1 + \frac{3}{2}\Phi \frac{E}{D}}{1 - \Phi \frac{E}{D}} \right] \tag{4}$$

where:

$$\begin{aligned} E = {} & 2(G_{\text{inc}} - G_{\text{mat}})(19G_{\text{inc}} + 16G_{\text{mat}}) + 48\beta'\Gamma/R^2 \\ & +32\beta'' (\Gamma + \beta')/R^2 + 8\Gamma(5G_{\text{inc}} + 2G_{\text{mat}})/R \\ & +2\beta' (23G_{\text{inc}} - 16G_{\text{mat}})/R + 4\beta'' (13G_{\text{inc}} + 8G_{\text{mat}})/R \end{aligned} \tag{5}$$

and

$$\begin{aligned} D = {} & (2G_{\text{inc}} + 3G_{\text{mat}})(19G_{\text{inc}} + 16G_{\text{mat}}) + 48\beta'\Gamma/R^2 \\ & +32\beta'' (\Gamma + \beta')/R^2 + 40\Gamma(G_{\text{inc}} + G_{\text{mat}})/R \\ & +2\beta' (23G_{\text{inc}} + 32G_{\text{mat}})/R + 4\beta'' (13G_{\text{inc}} + 12G_{\text{mat}})/R \end{aligned} \tag{6}$$

Actually, this is a generalisation of the Einstein relation for the viscosity of suspension, that one may easily recover, for $G_{inc} \to \infty$. This model is now commonly used in the field of polymeric emulsion, but it can be also used in self-consistent way to describe foam viscoelasticity.

2.2.2 *Y. Hemar's model for 3d foams*

A foam can be viewed as an array of droplets, each one being surrounded by about 12 neighbours. Hemar's approximation [4] consists in replacing the neighbouring droplets by a continuous matrix. Lets take some droplets X, occupying a volume fraction Φ, and replace their surrounding by a viscoelastic matrix of modulus G_{mat}. We can then use the relations (4,5,6) to obtain the macroscopic modulus G of this matrix containing the droplets. Then, within the effective medium approximation, the modulus of the matrix alone is set to be the same as the modulus of the matrix containing the droplets. This leads to the self-consistent relation :

$$G = G_{mat} \tag{7}$$

which gives, whatever the volume fraction Φ of the oil droplets, the relation between the interfacial properties, the droplets viscoelasticity and the macroscopic modulus:

$$E = 0 \tag{8}$$

First, in the case of a foam with only interfacial tension, equations (5) and (8) reduces to a Princen relation :

$$G = \frac{\Gamma}{2R} \tag{9}$$

where the numerical coefficient is very close to the one obtained with tetrakaidecaedres cells [2], when $1/R$ is replaced by the interfacial area per volume unit. Let us now consider the effect of the two interfacial viscosities.

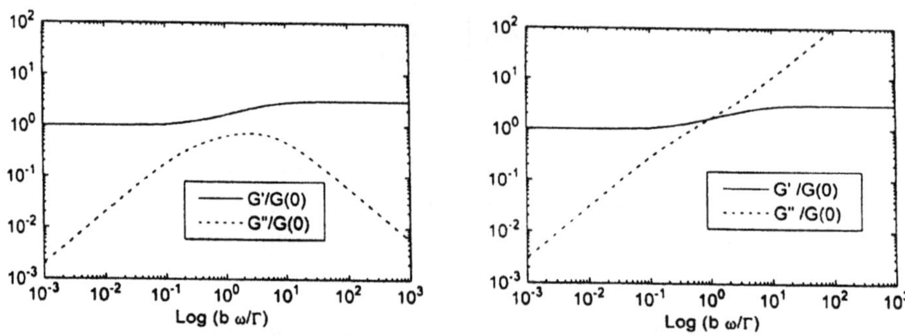

Figure 2. Macroscopic elastic and loss modulus for a foam with films with a newtonian interfacial dilational viscosity b (a), or with a newtonian interfacial shear viscosity b (b).

On figure 2a are plotted the macroscopic elastic and loss moduli, as a function of frequency, in the case of an interface with only an interfacial tension Γ and a dilational interfacial viscosity b such that $\beta'=i\omega b$. In that case, we can observe that the elastic modulus is only multiplied by a factor 3 in the whole range of frequency, and that the loss modulus exhibits a maximum ! This is a very surprising result : it means actually that it is possible to shear a foam without modifying both the surface area of the film and the volume of the cells (In all the models, the fluids are assumed to be incompressible). Then, if we submit the sample to an oscillatory strain, at low frequency the surface area is always minimum and this gives relation (9), but at high frequency, with a dilational shear viscosity, the surface area tends to be kept constant. Hence, the elastic modulus is increased, but the loss modulus goes to zero.This is not at all the case if we consider a system with no dilational interfacial viscosity. For instance, (see fig. 2b), in the case of an interfacial tension and an interfacial shear modulus, we get a simple behavior : the loss interfacial shear modulus is then nearly additive to the macroscopic loss modulus, up to a geometrical factor depending smoothly on the frequency.

2.3. LINEAR LOSS MODULUS ?

Actually, these results cannot account for the measured loss modulus [6] of biliquid foams. In fact, Cates et al [7] have demonstrated that all the dissipation phenomena that could occur in biliquid foams are very fast with typical time of about 10^{-6} s. However, the actual measurements show a minimum or a plateau of the loss modulus between 10^{-3} Hz and 10^{2} Hz [6]. This is totally inconsistent and shows that there are others mechanisms in foams that create dissipations, as we will see later on. But first, we will see that even the relation between the static elastic modulus and the interfacial tension must include the effect of position disorder, especially when the foams are wet, i.e. when the volume fraction of oil is between 64% and 95%.

2.4. EFFECT OF DISORDER ON LINEAR ELASTICITY

Actually, detailed calculations of the elastic modulus of crystalline foams [8], taking into account the volume fraction of water, and thus, the Plateau border (i.e. the interstices between the oil droplets), do not fit at all with the experimental data. As explained in [6], the measured elastic modulus is always smaller than the calculated one. This is due to a dramatic effect of position disorder. The stress propagation is completely modified in a disordered media : it has been shown that the stress in fact localises on some paths in granular media [9]. Then, only some of the droplets are subject to the stress, and as a result, the measured elastic modulus is smaller than the one estimated in ordered material. We will see now that the disorder has even more dramatic effects on the non-linear mechanical properties of foams.

3. Flow and non linear mechanical properties

One of the successes of the Princen model [1] was to give one of the most simple image of a fluid with a yield stress. However, if his picture is roughly applicable, a detailed investigation shows that the mechanical behavior is very different from the meanfield one introduced by Princen. In that case, also, the effect of the disorder leads to a complex and not fully understood behavior.

3.1. PRINCEN'S MODEL FOR THE YIELD STRESS

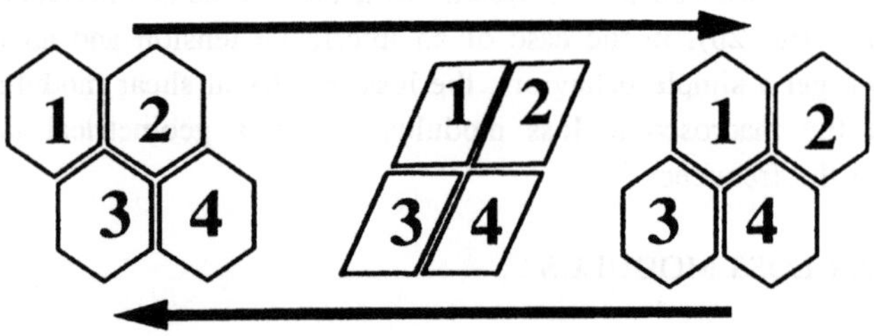

Figure 3. Schematic effect of the strain on a periodic foam. The length of interfaces between the cells 1 and 4 vanishes. This is mechanically unstable and a new film appears spontaneously between the cells 2 and 4. This process provides and yield stress for the fluid.

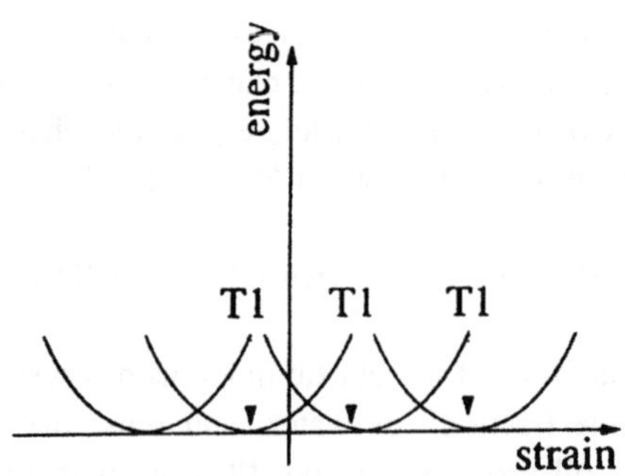

Figure 4. Energy representations of the *T1* rearrangements. The dash lines correspond to irreversible motions.

Let us consider a 2d foam constituted by a periodic array of hexagonal cells. Each cell has a constant area (corresponding to the volume in 3d), and the edges have a line tension (corresponding to the 3d interfacial tension). Let us now strain this foam, while minimising the energy , or the total edges length [1]. As shown in fig. 3, the lengths of some edges drop to zero. When the length of one edge vanishes, the system becomes unstable and a new edge appears instantaneously. This process is usually called *T1*. By this mean, the system, while creeping, regenerates and remains always close to the initial hexagonal array. We can then plot roughly the energy versus the strain, in fig.4 . As a result of this naive picture, if the foam is submitted to a stress smaller than the stress σ_y necessary to reach the T1 process, then the system will not flow and the response will be elastic. But, if the stress is larger than σ_y, then the stress will be sufficient to make the system cross all the possible *T1*, and the foam will indefinitely flow. So this periodic foam model, the Princen model actually, describes a fluid with a yield stress, which is purely elastic below the yield stress σ_y, and which simply flows above the yield stress. In this respect it exhibits a similarity to surfactant periodic cubic phases discussed in this volume.

One can immediately see that a distribution of the heights of the energy barriers in fig.4, will completely modify this simple mechanical behavior.

3.2. CREEP AND VISCOELASTIC MEASUREMENTS

Quite recently, we have performed some very accurate creep measurements [10]. We have prepared a foam with oil droplets of .5 μm diameter, at an oil concentration of about 90%. We have placed carefully the sample into a rheometer, in a cone/plane cell with rough surfaces to avoid wall slipping. We have indeed verified that the deformation is affine. We have then submitted the sample to a constant stress for some time. While the stress is loaded, the sample deforms instantaneously (faster than 10^{-1}s in fact). We call the instantaneous strain γ_i. Then the sample flows extremely slowly with a viscosity of about 10^6 Pa.s.. After some time, we unload the stress and the sample partially recovers by a strain of γ_r. This is indeed the recovery strain.

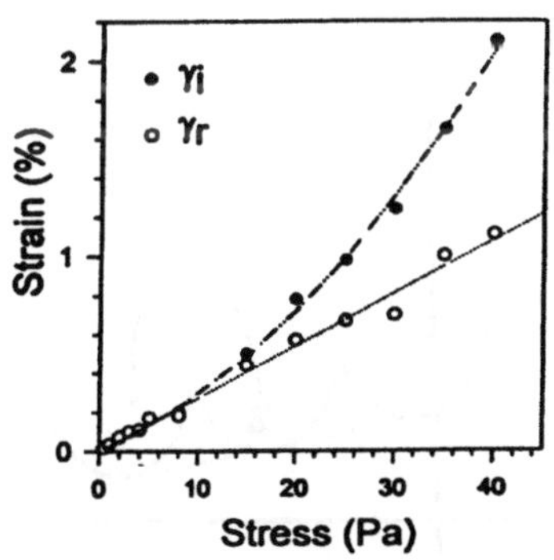

Figure 5. Instantaneous and recovery strains in creep measurements, as functions of the shear stress. One sees that the response is non-linear above a strain of 2.10^{-3} /.5μm oil droplets, oil volume fractions 90% .

On fig 5, we have plotted the instantaneous and the recovery strains as functions of the applied stress. We see clearly that although the recovery is proportional to the strain up to a few per cent of deformation, the instantaneous strain is not proportional to the stress above a strain equal to 2.10^{-3}. This means that the viscoelastic response is not linear above a strain of a few 10^{-3} ! Moreover, the system flows, it possesses a finite Newtonian viscosity, and this also means that some spontaneous rearrangements occur even at rest. In order to try to give a clear picture of these mechanisms, we will first recall how a simple liquid flows.

3.3. EYRING'S APPROACH OF LIQUID VISCOSITY

A liquid is constituted of molecules which are temporarily trapped into the cages formed by their neighbours [10]. The cage picture, even if it is mostly qualitative, gives a good idea of what really occurs in a liquid. At rest, the motions of the molecules can be described as random hopping from one cage to a neighbouring one, and these jumps are thermally activated. What happens, now, if a liquid is submitted to shear stress ? The liquid flows, which means that the average displacement of the molecules is no longer zero, but corresponds to the velocity field in the flow. This comes from the fact that a hop in one direction is more easy than in the opposite direction. Actually, under the action of stress, the cages lose their symmetry, and the random walk become biased, as illustrated in fig.6. The combination of all these directed walks creates the flow. Moreover, it is easy to show that in that case, the viscosity is Newtonian. To summarize, a Newtonian flow is induced by random thermal hoping of the objects between cages, biased by the application of a stress. How can this apply to our foam? A rough estimation of the energy barrier in the T1 model gives 10^7 kT, so it seems completely unbelievable that these barriers could be thermally crossed. But this is also inconsistent with the fact that the system flows, with a Newtonian viscosity. We will try to get insight in this picture.

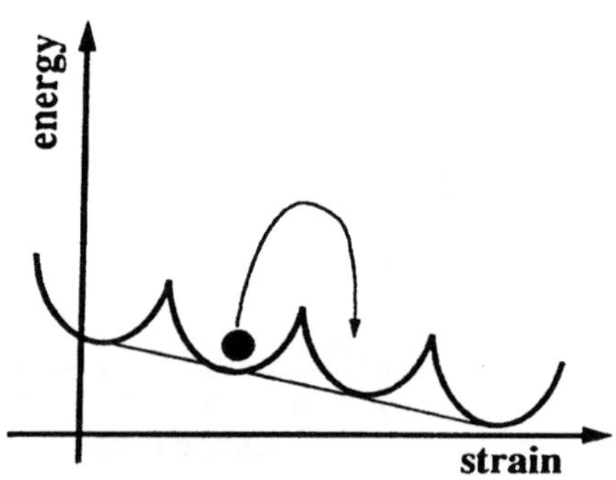

Figure 6. Naive picture of the viscosity : molecules hop from cages to cages. The stress biases the random hoping and this induces the flow.

3.4. FOAMS AND GLASS DYNAMICS

It is well known that glasses flow, even if it takes centuries for a window glass to deform noticeably under gravity. Recently, a very crude but quite successful model for glass transition has been proposed [12]. The principle of the model is the following : let us consider a glass of soft spheres as an array of particles, each one being trapped in an energy well. Let choose randomly the depth of each well. If one particle is thermally hopping from its well to a new one, it modifies the configurations around it. This can be modelled by considering that the jump of one particle modifies randomly the depth of the other particles wells. Writing just the evolution of population in the energy wells, both by thermally induced hoping, and by modification of the depth of each well induced by the jump of the other particles, provides a very good picture of the glass transition.

Our foam is disordered. This means that, a priori, the energy barrier necessary for a rearrangement similar to a T1 process, should be distributed from zero to a value of the order of 10^7 kT. So the droplets that are trapped by very low energy barriers could thermally rearrange. These rearrangements will modify the stress field far around them - because elastic fields have long range decay. These modifications could decrease new barriers up to thermally crossable values, new hops could then occur and the process play again. This could explain the fact that the foam actually flows.

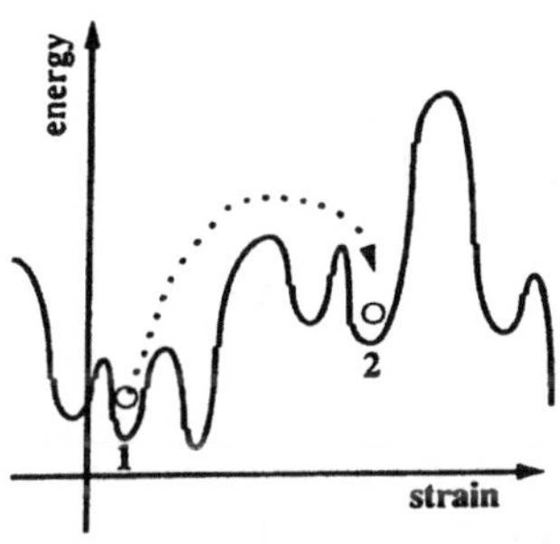

Figure 7. Each droplet is trapped in a random energy field. Loading the stress can induce hopping of the droplets, which is not the case when the stress is unload. Displacement from 1 to 2 corresponds them to the excess of instantaneous strain, compared to recovery strain in figure 6.

More precisely, each droplet can be viewed as trapped in a random energy field. In fig.7, we have roughly represented such a situation. Applying a stress to the system corresponds to a tilt of the whole picture, as in fig.6. So the application of the stress can induce a displacement from position 1 to position 2. But when the stress is unloaded, the particle remains trapped in position 2. This explains qualitatively the non-linear behavior : the recovery is linear as it is due only to the medium elasticity, while the instantaneous strain is not linear, exhibiting some irreversible motions in the system.

3.5. MECHANICAL REGIMES

It seems, from the fact that the system flows,that thermal energy could induce some very rare rearrangements.The mechanical energy per droplet provided is of the order of the thermal one, *kT*, for a strain γ_c such that:

$$G \gamma_c^2 R^3 \simeq kT \tag{10}$$

where *G* is the measured elastic modulus. This provides a crossover value for the strain, below which most of the events are similar to those occurring at equilibrium, and above which the phenomena should be very different from those at equilibrium. Actually the critical strain γ_c is about 10^{-3} . It probably corresponds to the beginning of the non-linear mechanical response.

However, there is another crossover for the strain corresponding to an apparent yield stress. A crude measure of the shear rate/shear stress behavior indicates that there is a yield stress, below which the system flows extremely slowly, and above which the system flows rapidly, while actually fracturing. Going back to the *T1* picture, we can see that if the stress is large enough for the system to cross most of the barriers, then the system will flow very quickly.This new crossover *i.e.* the apparent yield stress, corresponds typically to the mean value of the energy barrier, and thus to a strain of the order of unity. Hence the apparent yield stress should be equal to the elastic modulus. Actually the experimental yield stress is not equal to the elastic modulus, but is ten times smaller. This is a simple effect of disorder.

On the other hand, and as explained previously, for a stress above the apparent yield stress, the system has to wait, after each rearrangement, that somewhere fluctuations and relaxations allow for a new rearrangement.

As a conclusion, we have probably three mechanical regimes :

1) for very small stresses, the mechanical energy is small compared to the thermal one, the response is probably linear, and the system flows very slowly.

2) for an intermediate value of the stress, the mechanical energy is larger than the thermal one, but does not allow the system to cross all its energy barriers. The response becomes non-linear, the instantaneous strain is non-linear, but the recovery strain is still nearly linear. Actually, in this regime, one observes that the loss modulus is clearly non linear, but the elastic modulus seems roughly linear.

3) for a large value of the stress, nearly all the barriers could be crossed, the T1 picture of Princen is quite good except that we don't have flow but fracture. The experimental apparent yield stress is about one tenth of the elastic modulus, while the periodic models predict a ratio of about 1. This is another effect of the disorder.

4. Challenges and Comments

In fact, most of the problems, in the mechanics of foams, remain open. How does the stress propagate in this structural material and affect the value of the elastic modulus.

The loss modulus, which is observed, is not at all explained by the surfactant and the interface microhydrodynamics. Probably, rearrangements must be responsible for a part of the loss modulus, but the high frequency behavior (see also Ramaswamy, this book) remains a challenge.

We have identified at least three regimes, but what happens in each regime has to be carefully investigated.

If a rearrangement occurs somewhere, how can it modify the stress and does it modify the water and the surfactant concentrations on long distances ? Moreover, how does this affect the system and induce new rearrangements ?

References

[1] Princen H.M.(1982) Rheology of Foams and Highly Concentrated Emulsions I *J. Coll. Interf. Sci.* **91** 160-174

[2] Reinelt D.A. and Kraynik A.M. (1993) Large elastic deformation of three-dimensional foams and highly concentrated emulsions *J. Coll. Interf. Sci.* **159** 460- 470

[3] Palierne J.F. (1990) Linear Rheology of Viscoelastic Emulsions with Interfacial Tension *Rheologica Acta* **29** 204-214

[4] Hemar Y., Hocquart R., and Lequeux F.(1995) Effect of Interfacial Rheology on Foams Viscoelasticity, an Effective-Medium Approach *J. Phys. II France* **5** 1567-1576

[5] Edwards D.E., Brenner H. and Wasan D.T., (1991) *Interfacial Transport Processes and Rheology*, Butterworth-Heinemann

[6] Mason T.G., Bibette J. and Weitz D.A. (1995) Elasticity of compressed emulsion *Phys. Rev. Lett.* **75** 2051-2054

[7] Buzza D.M.A., Lu C.-Y.D., and Cates M.E. (1995) Linear Shear Rheology of Incompressible Foams *J. Phys. II France* **5** , 37-52

[8] Buzza D.M.A. and Cates M.E. (1994) Uniaxal Elastic Modulus of Concentrated Emulsions *Langmuir* **10** 4503-4508

[9] See in Charmet J.C., Roux S. and Guyon E. (1990) *Disorder and Fracture* Plenum Press, New York

[10] Boltenhagen P., Bibette J. and F. Lequeux *in preparation*

[11] Hansen J.R. and MacDonald I.R. (1986) *Theory of simple liquids*, Academic Press, London

[12] Bouchaud J.P., Comtet A. and Monthus C. (1995) On a dynamical model of glasses *J. Phys I France* **5** 1521-1526

EFFECT OF FLOW ON LYOTROPIC PHASES

Didier ROUX
Centre de Recherche Paul Pascal, CNRS
Avenue Schweitzer
33600 Pessac, France

1. Introduction

Surfactants in solution are known to self-organize in many different structures. Among which, one can distinguish between finite size micelles, uni-dimensional linear micelles leading to polymer like systems [1] and bi-dimensional membranes or films.

This article will mainly deal with two-dimensional objects obtained in mixing surfactant with water and eventually an hydrophobic solvent. In the introduction we will focus on the equilibrium properties of such phases (structure and stability), introducing most of the information needed for the understanding of non-equilibrium behaviour. We will then describe in detail what we have observed when shearing lyotropic phases.

1.1 EQUILIBRIUM PROPERTIES OF MEMBRANE PHASES: GENERALITIES

When surfactant molecules aggregate into two-dimensional structures, they can make either surfactant films separating hydrophobic regions of space (oil) from water or bilayers embedded in water. The first case refers to microemulsions [2, 3] when the second cases refer to lamellar, sponge or vesicle phases. We will mainly describe the membrane properties rather than the film ones, while most of what we say is valid for both cases, some simplifications occur for bilayers that makes the description more universal.

1.2 ELASTIC PROPERTIES OF MEMBRANES

Let us consider the deformation of a flat bilayer made of amphiphilic molecules. Canham [4] and Helfrich [5] have proposed to describe the energetic cost of this deformation in the framework of thin film elastic theory [6]. Within this model the elastic cost of a deformation for small displacement from the flat reference state is:

$$E = \int \frac{1}{2}\kappa\left[\left(\frac{1}{R_1}+\frac{1}{R_2}\right)^2 + \overline{\kappa}\frac{1}{R_1}\frac{1}{R_2}\right] dS \tag{1}$$

The first part of the integral refers to the mean curvature ($1/R_1+1/R_2 = C$) and κ is the elastic bending constant. The second part refers to the Gaussian curvature ($H = 1/R_1 R_2$) and $\overline{\kappa}$ is the Gaussian bending constant. Both κ and $\overline{\kappa}$ have units of energy. While the mean curvature is a function of the exact shape that the surface takes, the Gaussian curvature is a topological invariant as demonstrated by Gauss and Bonnet [7].

T McLeish (ed), Theoretical Challenges in the Dynamics of Complex Fluids, 203-233.

This means that the elastic contribution from the Gaussian curvature is only a function of the topology of the surface, namely:

$$E_g = \overline{\kappa}\int \frac{dS}{R_1R_2} = 4\pi\,\overline{\kappa}\,(n_s - n_h) \tag{2}$$

where n_s is the number of independant surfaces and n_h the number of handles, a handle being defined as a passage between two pieces of surface (see figure 1).

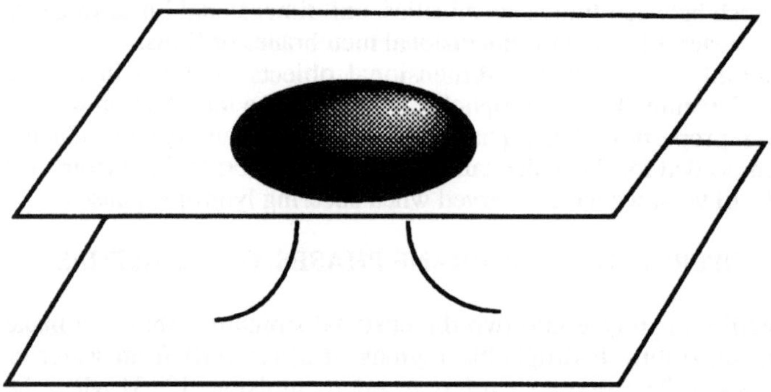

Figure 1: A handle connects two pieces of surface.

1.3 STRUCTURE OF MEMBRANE PHASES

When surfactants are mixed with water, and if they have a prefered organization for bilayers, there are several ways for these bilayers to fill up space homogeneously, avoiding either seams or edges. One of the most obvious organizations is the lamellar phase where the membranes are packed in a one-dimentionnal array (see figure 2a). This lamellar phase is a liquid crystalline phase of smectic A type [8]. Another type of organization corresponding to a liquid isotropic phase is the vesicle phase composed of separated little "bags" (see figure 2b). While vesicles have been described long ago as metastable phases which can be prepared by ultrasonication of a lamellar phase with excess water, it is only recently that these phases have been shown unambiguously to exist at thermal equilibrium [9,10,11]. A third type of organization has been more recently proposed, it consists of membranes randomly connected in space (see figure 2c), this phase has been called the sponge phase and corresponds most often to a phase labeled L3 in phase diagrams [12,13].

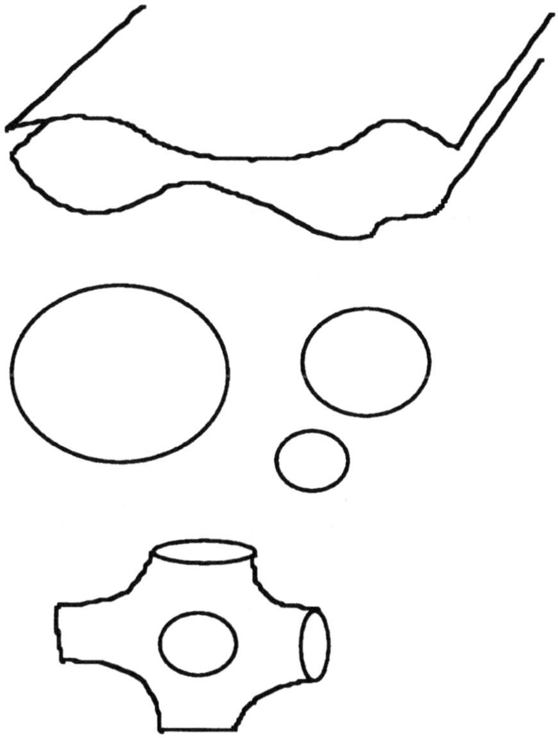

Figure 2. Different ways of organizing membranes in space, 2a lamellar phase, 2b vesicle phase, 2c sponge phase.

1.4 STABILITY OF MEMBRANE PHASES

In the limit of flexible membranes (both κ and $\overline{\kappa}$ are not too large compared to $k_B T$) and rather dilute phases (low surfactant volume fraction), the stability of the different phases can be efficiently described as a competition between elasticity and entropy. This is true for the three types of organization we have briefly mentioned and in the absence of long range electrostatic interactions.

1.4.1 *Lamellar phase.*

The lamellar phase is characterised by its smectic distance d corresponding to the repeating distance between membranes. This distance is fixed by the amount of solvent added to a given quantity of surfactant until a limiting distance is reached after which the system undergoes a phase transition. In certain cases, even in the absence of long range electrostatic interactions, very dilute lamellar phases can be obtained [13-16]: the d-spacing reaching in certain cases several thousand Angstroms.

The stability of these dilute lamellar phases can only be explained by the existence of a long range repulsive interaction: the undulation force. This interaction, entropic in origin, has been first described theoretically by Helfrich [17]. It is due to the limitation of the thermally excited undulations by the existence of neighbouring membranes (at a distance d). This interaction has been quantitatively measured by Safinya et al. [14, 18]. The interaction potential per unit area can be expressed as:

$$V(d) = \mathrm{coef}\,\frac{(k_B T)^2}{\kappa}\,\frac{1}{d^2} \tag{3}$$

the amplitude is inversely proportional to the bending constant κ and it varies as the inverse of the square of the distance between membranes.

The ensemble of interacting membranes constituing the lamellar phase is described by its elastic energy [8]:

$$E = \frac{1}{2}\,\overline{B}\left(\nabla_z u\right)^2 + \frac{1}{2}\,K\left(\nabla_\perp^2 u\right)^2 \tag{4}$$

u is the relative distance between two membranes, $\overline{B}$ and K are respectively the compressional elastic constant at constant chemical potential (for more explanations see [19]) and K the bending (splay) constant of the smectic phase. Using the Helfrich expression for the interactions between layers, it is possible to calculate the elastic constants of the smectic phase [20,21]:

$$\overline{B} = \mathrm{coef.}\,\frac{(k_B T)^2}{\kappa}\,\frac{1}{d^3} \tag{5}$$

$$K = \mathrm{coef.}\,\frac{\kappa}{d} \tag{6}$$

One of the most interesting properties of these scaling laws is that taking into account that d can vary continuously from 50 A to 5000 Å, K varies typically by two orders of magnitude when $\overline{B}$ varies by six orders of magnitude.

1.4.2 *Sponge phase*

If, instead of a constrained membrane such the one we have considered in the lamellar organization, one consider a "free" membrane, it is possible to define a persistence length ξ_K [22]. As for linear polymers, below this length the membrane can be considered as flat in average while above it is crumpled. However, in contrast to the polymer case, this length varies exponentialy with the bending constant κ

$$\xi_K = e^{\frac{\kappa}{k_B T}} \tag{7}$$

When diluting a lamellar phase and assuming $\overline{\kappa}$ to be zero, one can reach a situation where d becomes of the order of ξ_K. In this situation and because there is no special cost assiocated with the change in topology ($\overline{\kappa}$=0), one expects a phase transition from the lamellar phase to the sponge phase [13]. The sponge phase can be either described as a phase of random surfaces or as a molten cubic phase. It is an isotropic liquid phase, however due to the particularity that this surface divides space into two equivalent volumes it has a remarkable property of symmetry [23].

When $\overline{\kappa}$ is considered different from zero it may play a very important role in controling the topology of the phase [24]. Typically $\overline{\kappa}$ can be considered as a chemical potential associated with the topology [25]. depending upon its value it favours either the sponge phase or the vesicle phase.

1.4.3 *Vesicle phase.*
Upon even more dilution or making $\overline{\kappa}$ more and more negative, one expects the sponge phase to be destabilized toward a vesicle phase. The statistical mechanics of the vesicle phase can be described again as an equilibrium between the mixing entropy and bending elasticity [26]. This treatment leads to the distribution of sizes of the vesicle population. Some more accurate estimations of the entropy (taking into account the undulation entropy) have been proposed [27,11] and are still the subject of a theoretical debate [28].

1.5 CONCLUSIONS

The universal description of the structures and the stability of membrane phases is directly related to the competition between elasticity and entropy. The fact that most of the other types of interactions and molecular details can be integrated in this phenomenological description simplifies and generalized the theory. In terms of experimental behaviour, the signature of this universality is easily demonstrated by the similarity of many phase diagrams in the dilute regime [21]. In fact, as poposed by Leibler and Huse, one can draw a kind of universal phase diagram (see figure 3) where most of the physical ideas described above are used to draw the respective stability regions of the different phases.

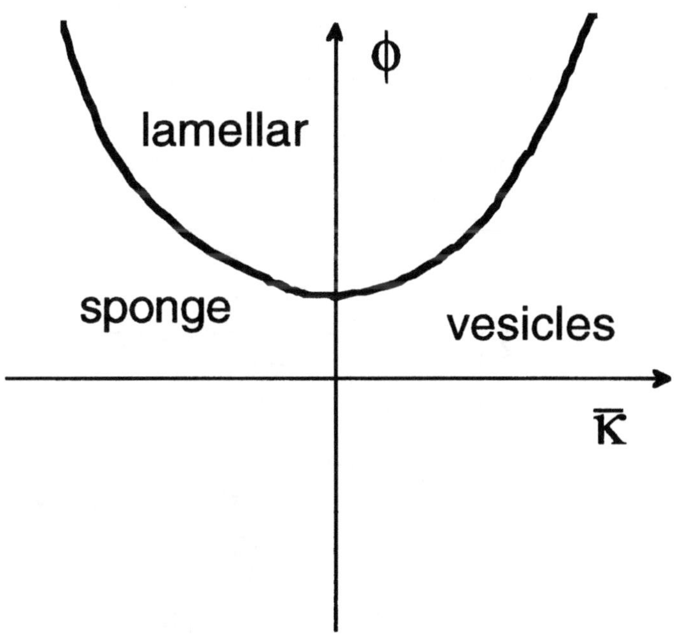

Figure 3. Schematic phase diagram of a system of membranes. ϕ is the membrane volume fraction, $\overline{\kappa}$ is the gaussian bending constant and κ is assumed to be of order 4-8 k_BT.

In conclusion, equilibrium properties of dilute membrane phases are relatively well understood. However non equilibrium properties are much less known. Among the most interesting properties to study is the effect of flow and the relation between structure and rheology.

2. Effect of Flow on Membrane Phases

Since membrane phases are constituted of surfaces arranged in space, one expects a strong coupling between the flow and the structure of these phases. In order to study the effect of flow on the structure, we have developed a series of shear cells. Most of the cells have been developed to study the structure using scattering techniques. We can perform x-ray and light scattering in house but also neutron scattering (at the Laboratoire Léon Brillouin in Saclay) using specially designed Couette cells. We can also make microscopic observations using a plate/plate cell and measure dielectric properties in the three directions of space. We briefly describe the different cells we built then we will mention some of the transitions that can be observed under shear.

2.1 SHEAR CELLS

Because the systems we study are made of two to five constituents, they are very sensitive to evaporation. Consequently, all the cells we designed are sealed (at least evaporation is minimized). Most of the cells are Couette cells (two cylinders: one rotating the rotor, the other one not moving, the stator).

2.1.1 *Light and neutron scattering cells*

This was the first cell built [29]. The rotor is made either of glass or transparent plastics. The main part is an oven that can be thermostated (using a water bath). On the four sides of the oven large windows allow either a direct observation with eyes or a laser beam to go through the cell. The stator is also the cover of the oven, it is made of the same material as the rotor. The controlled temperature water run also into the stator to assure a better temperature control. A system of mirrors allows the laser beam to penetrate from above inside the stator and to go, one time only, through the gap between the rotor and the stator (see figure 4).

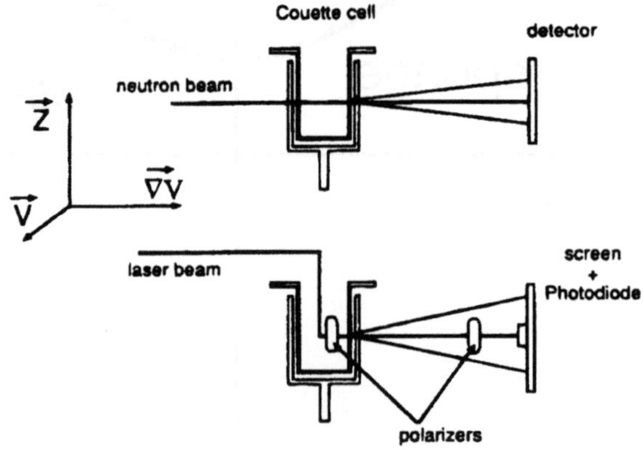

Figure 4. Light scattering cell

The neutron scattering cell is based on the same principles, the only difference is that quartz is used for both the stator and the rotor. The windows of the oven are also made of quartz.

2.1.2 *X-ray cell*
The x-ray cell is also built on the same design. It is however more compact and is made of Lexan an x-ray transparent polymer [30]. The rotor is reinforced by an aluminium core leaving just a small passage for the x-ray beam in the center. In the core the water of the thermostated bath can circulate (see figure 5). This cell allows observation in the velocity/vorticity direction (the x-ray beam going through the center of the cell). It is also possible to send the beam tangentially and to observe in the shear/vorticity direction.

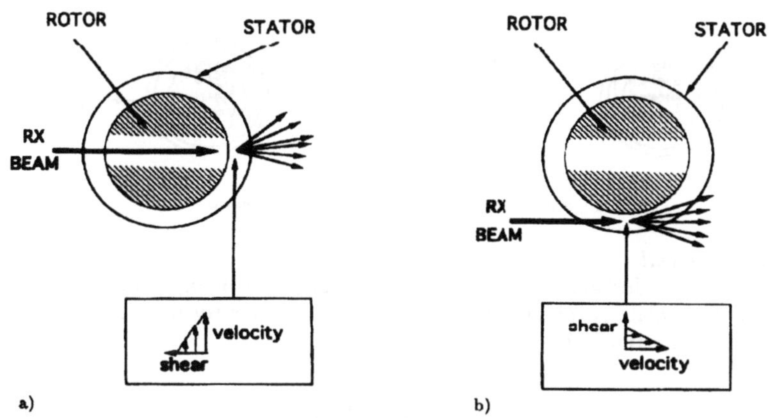

Figure 5. X-ray cell

2.1.3 *Cells for dielectric measurements*
In order to be able to measure dielectric properties in the three main directions (shear, velocity and vorticity) we built a basic cell with several options as seen in figure 6. Depending on the location of the electrodes we can access to the three different directions. The dielectric properties can be detected using an impedencemeter (Hewlwet Packard HP1100).

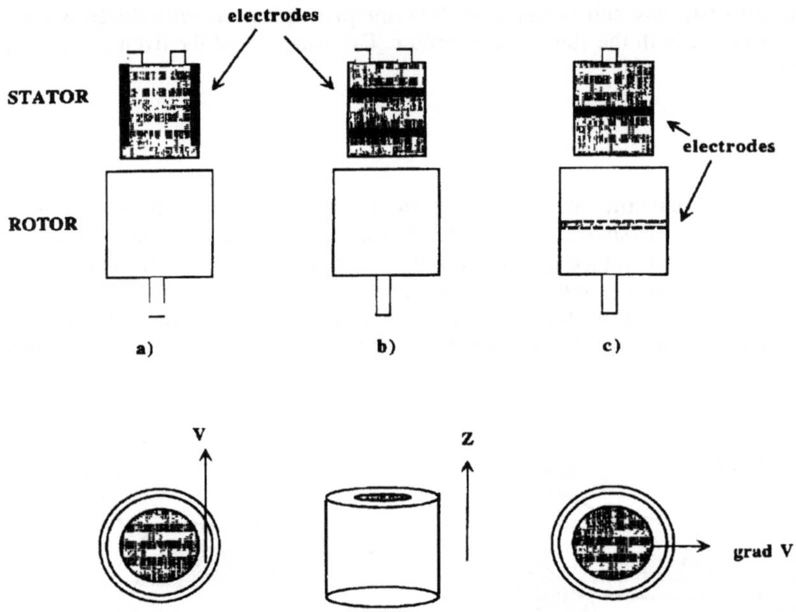

Figure 6 Shear cells for dielectric measurements

2.1.4 *Plate/plate cell*
This cell has been designed for microscopic observation using a regular polarizing microscope. One plate is rotating, the shear rate varies linearly from the center to the edge of the cell (see figure 7).

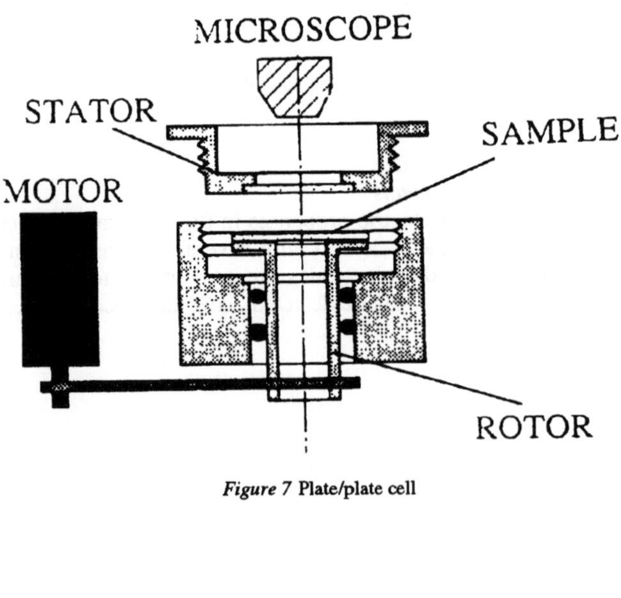

Figure 7 Plate/plate cell

2.2 PHASE TRANSITIONS UNDER SHEAR

When flow is applied to a lyotropic phase linear deformation of the structure will happen. This can be classically observed using flow-birefringence measurements. In certain cases, more dramatic effects can be observed, which are clearly non-linear. It is not yet clear how to treat these observations. However, in the following, we will try to distinguish between what we call a phase transition under shear and shear transitions. A phase transition under shear can be considered as the stabilization of a phase which exists somewhere in the phase diagram using shear starting from another phase.

Typically starting from an isotropic phase, it seems that with shear one can go through a phase transition and stabilize either a nematic or a smectic phase. This seems to us different from shear transitions which can lead to structure that are metastable or are not observed at equilibrium. We will mainly deal with the latter case, however before going into more details on this subject let us very quickly review some of the phase transitions under shear that have been observed.

2.2.1 *Isotropic/Nematic transition under shear*

This topic is described in more details in the article by Cates [1]. However since it is a nice illustration of a phase transition produced by shear, let us briefly describe the experimental facts.

Typical phase diagrams of living micelle systems exhibit for relatively high concentration of surfactant a nematic phase [32,33]. This nematic phase is understood to be of the calamitic type, i.e. made of rod like particles. However, since they are flexible and not rigid, they have to be considered as polymers. At higher concentration in surfactant there may exist in certain cases a hexagonal phase.

When the isotropic phase (typically above c*) is sheared a clear transition is observed by different techniques. Using rheology, a plateau appears at a well defined shear rate $\dot{\gamma}_c$ in the stress/shear curve when using a shear rate control rheometer [32,33]. When a stress control rheometer is used a jump in shear rate can eventually be observed [32]. Using neutron scattering, below the critical shear rate $\dot{\gamma}_c$ the neutron scattering pattern remains isotropic and a continuous anisotropy appears above $\dot{\gamma}_c$ (see figure 8).

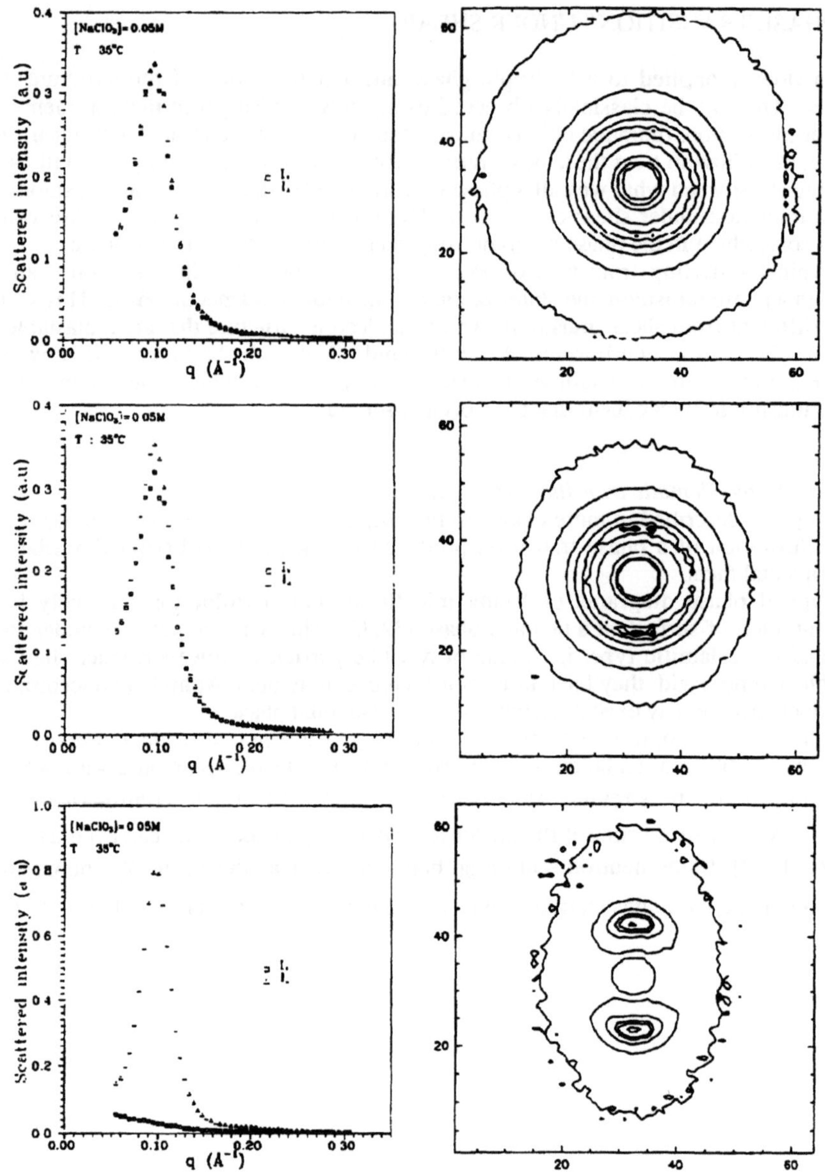

Figure 8. Appearance of the anisotropy using neutron scattering and shearing a cylindrical micellar system. Below $\dot{\gamma}_c$ (8a), above $\dot{\gamma}_c$ (8b and 8c).

If the anisotropy is plotted as a function of the shear rate, it clearly starts at finite shear rate (see figure 9). The extrapolated value corresponds very precisely to the appearance of the plateau in the stress/shear rate curve.

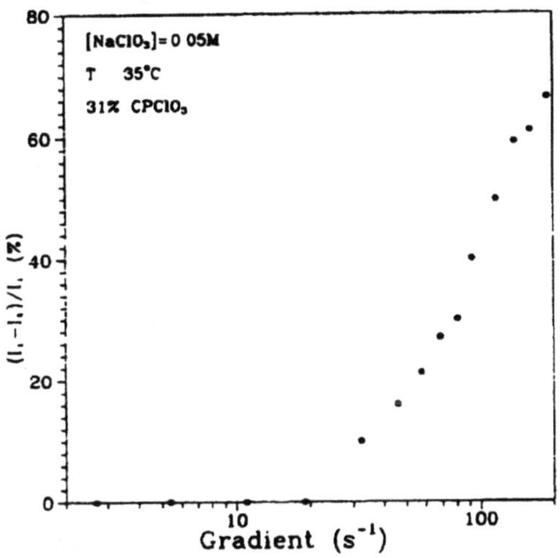

Figure 9. The anisotropy measured by neutron scattering as a function of the shear rate [32].

2.2.2 *Sponge to lamellar transition*

Another transition belonging to the same type is the sponge-to-lamellar transition. Again using both neutron scattering and birefringence measurements, it was possible to show that non linear anisotropy appears at a finite shear rate [34]. Figure 10 shows the neutron scattering pattern at different shear rates. Figure 11 shows the evolution of this anisotropy as a function of the shear rate.

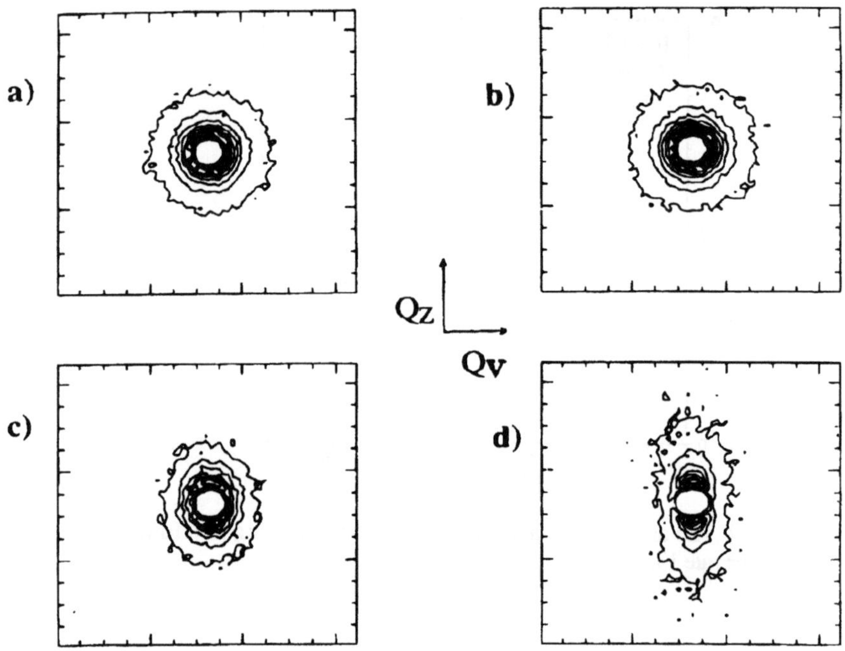

Figure 10 Neutron scattering pattern of a sponge phase at different shear rate. a) 0s-1, b) 18 s-1, c 720 s-1 and d) 2880 s-1

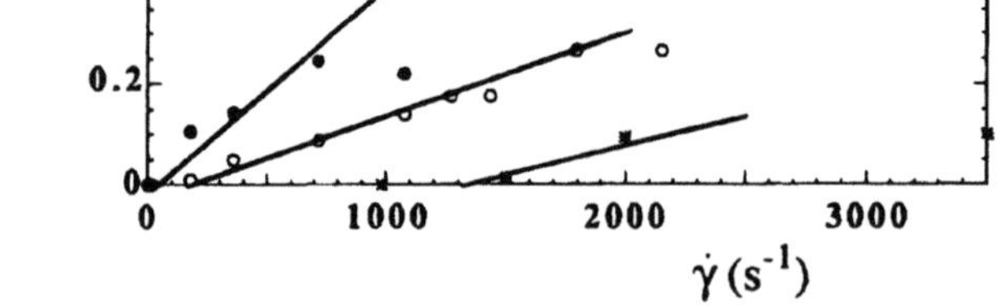

Figure 11 Anisotropy as a function of the shear rate for different concentrations of a sponge phase.

Analysis of the evolution of the critical shear rate at which this nonlinear behaviour appears seems to indicate that it might be related to the caracteristic frequency of the membrane fusion [34].

3. Lamellar phase under shear: shear diagram

We just saw that certain isotropic phases exhibit a non linear behaviour above a well defined shear rate. This behaviour can be interpreted as a phase transition driven by shear. However, even if theoretical models exists in both cases (isotropic to nematic [1] or isotropic to smectic [35]), it is not clear whether the actual description is sufficient to understand this behaviour [1].

We will now turn to a rather different type of transition. Indeed, as we will see, when a lamellar phase is sheared many transitions can be observed. However, these transitions are not phase transitions since they do not really affects the microscopic nature of the phase. They can be more exactly refered as "textural transitions" or even better "orientational transitions" in the way that they affect the long range organization of the phase through the nature and the organization of textural defects (dislocations, disclinations or point defects). They are, however, well defined out-of-equilibrium transitions (instabilities) and can be characterized using most of the techniques that we have developed.

3.1 SHEAR DIAGRAM OF THE SDS SYSTEM

The first system studied leading to a well define shear diagram is a classical lamellar phase, extensively studied at equilibrium, made of water, sodium dodecyl sulphate (SDS), pentanol and dodecane [13b]. The lamellar phase is composed of a water layer surrounded by two surfactant films each bilayer being separated by a dodecane layer. In adding dodecane the lamellar spacing can change continuously from 50 to 400 Å. The lamellar phase is stabilized by undulation interaction [14, 18]. As already explained the effect of the swelling is mainly to change the smectic elastic constants $\overline{\mathbf{B}}$ and K.

Under shear this system presents two clear transitions separating three different steady states of orientation (see figure 12). At very low shear rate (below typically $1s^{-1}$) the lamellar phase is mainly oriented with the director in the shear direction but defects (probably dislocations) are present in both the velocity and the vorticity direction (state 1). Above a well defined critical shear rate, a completely different orientation is observed: an instability develops leading to a state of close packed multilayered vesicles (state 2). As we will see the caracteristic size of these vesicles is a function of the shear rate. Increasing the shear rate even more, one reaches a region where the multilayered vesicle state coexists (when the shear rate is kept constant) with a rather well oriented state (state 3). As long as the shear rate increases the amount of the shear cell filled up with the vesicles state decreases to be replaced by the new oriented state. Again, above a well caracterized limit, the shear cell is now filled up with this oriented state with no remaining vesicles. Above this last limit no more changes are observed. In the coexisting region both the vesicle state and the oriented state are organized in stripes regularly spaced in the vorticity direction [29].

3.2 LIGHT SCATTERING OBSERVATION

One of the most spectacular signatures of the transitions is obtained using small angle light scattering. Indeed, a laser beam sent through the sample during shear leads the following observations. In state 1 and 3 one can just observe small angle scattering around the laser beam. However all along state 2 (the vesicles state) a well defined ring of scattering can be observed on a screen placed at some distance from the cell (typically from 10 to 40 cm). Figure 13 shows the evolution of the ring of scattering when the shear rate is increased.

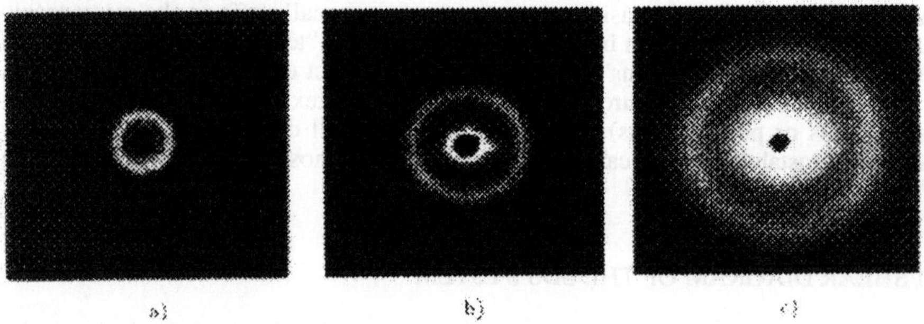

Figure 13. Light scattering pattern observed in the vesicle state under shear at different shear rates. The shear rate increases from a to c.

3.3 X-RAY SCATTERING

As already explained we can use x-ray scattering to understand the orientations of the lamellar phase under shear [36]. Indeed, since we have access to all the directions the intensity of the Bragg peaks (coming from the lamellar spacing) tell us directly the amount of membranes oriented in a given direction. To illustrate this feature we have represented on figure 14 the "pure orientation" of a smectic state together with the expected x-ray pattern observed in the two geometries: the velocity/vortex direction and the shear/vorticity direction depending whether the x-ray beam is sent through the center of the cell or tangentially.

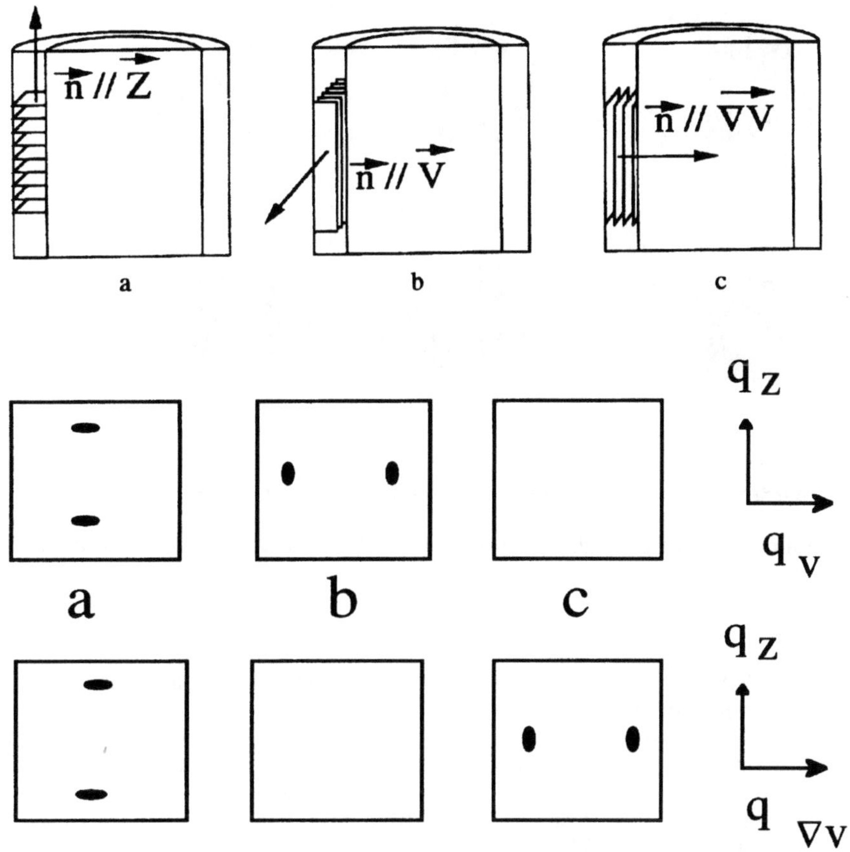

Figure 14. The pure orientation of a smectic state in the shear cell and the expected x-ray patterns observed in the velocity/vorticity direction (x-ray beam sent through the center of the Couette cell) and in the shear/vorticity direction (x-ray beam sent tangentially to the Couette cell).

Figure 15 shows the x-ray data obtained in the three different states described above. Since the vorticity direction is in common, it is easy to reconstruct the different proportions of membranes (layers) pointing out in the different directions.

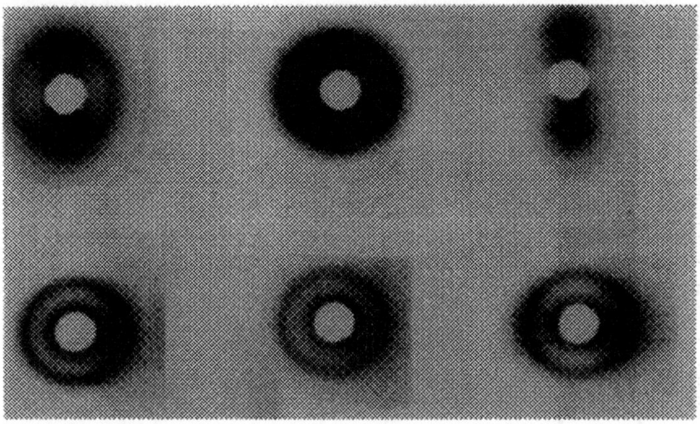

Figure 15. Two dimensional x-ray patterns observed in the two geometries accessible.

The following table gives the proportions of each orientation calculated from the x-ray pattern.

TABLE 1. percentage of membranes in each direction in the three different states.

state	vortex (a)	shear (b)	velocity (c)
I	30%	10%	60%
II	33%	33%	33%
III	20%	0%	80%

3.4 MICROSCOPIC OBSERVATIONS

Direct space observations can be obtained using the plate/plate cell under a microscope (using polarized ligth). In state 2 the texture looks regular with a well characterised length of modulation of the orientations (see figure 16). This characteristic length is the same observed using light scattering.

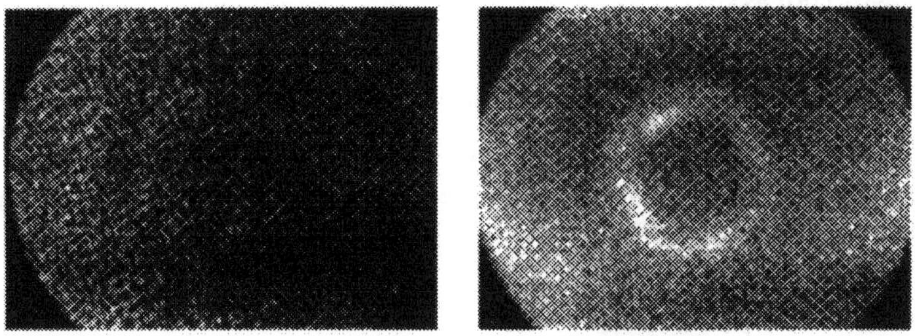

Figure 16. Typical texture observed under polarizing microscope. This texture is observed both under shear and at rest when stopping the shear quickly.

3.5 CONDUCTIVITY MEASUREMENTS

We can also observed the different states of orientation using conductivity [37]. Figure 17 shows the result of measurements in the three principal directions as a function of the shear rate. Clearly, state 1 and 3 are anisotropic while state 2 looks rather isotropic. The transitions can be observed. At first sight, the conductivity seems to be related to the amount of membranes in the different directions (see table 1).

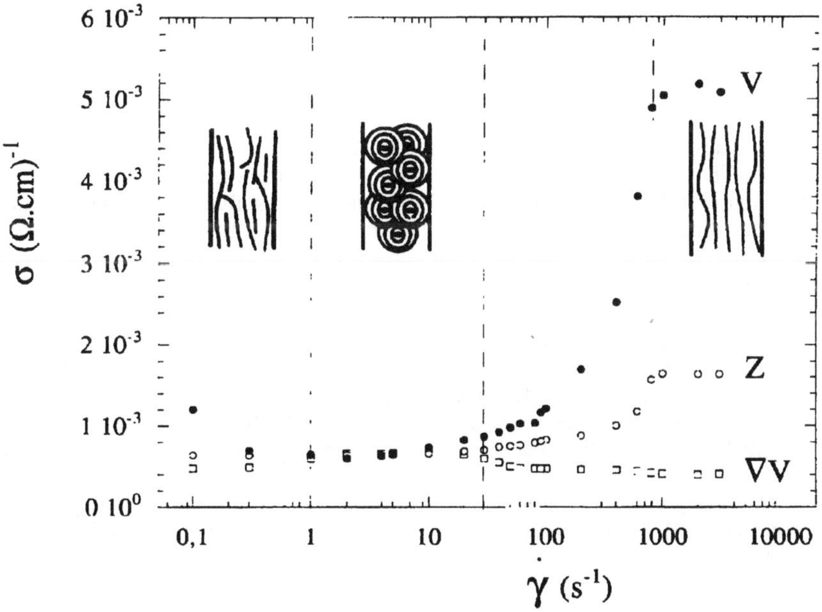

Figure 17. Conductivity measurements under shear in the different directions as a function of the shear rate for a sample containing 60% dodecane.

3.6 CONCLUSIONS

Using different techniques of observation of the structure under shear we have characterized a shear diagram of a lamellar phase. All the techniques used show a well defined behaviour caracterized mainly by a lamellar to vesicles instability happening above a well defined critical shear rate. The multilayered vesicles formed under shear have the structure of onions and are packed in space. Figure 18 presents a typical arrangement of such a structure.

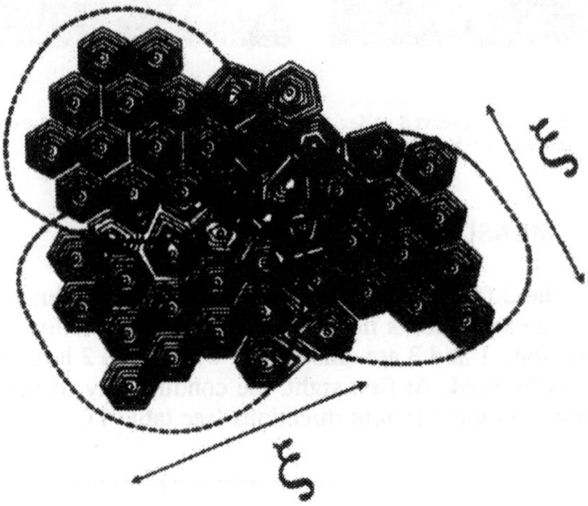

Figure 18. A representation of a cut of the multilayered vesicles state (onion state) obtained in shearing a lamellar phase.

The shear diagram presents regions of steady states separated by dynamic transitions. The fact that we are dealing with steady states means that whatever the experimental path followed, after waiting long enough it is always the state mentioned on the shear diagram that is found. Typical times to reach the steady states are very dependent on the shear rate and the path chosen but can vary from a few minutes to several hours.

One particularity of the onion state is that if instead of changing the shear rate very slowly, the shear rate is stopped quickly, the onion state can be quenched and the structure formed under shear can persist for several hours to several months (depending upon the system studied). In certain cases, it is even possible to add water (or another solvent) to be able to dilute the multilayered vesicles and a colloidal suspension of these vesicles can be obtained. It is a new method for preparation of well controlled vesicles [38].

Recently using a different system, we were able to obtain freeze fracture pictures of the onion state [39]. Figure 19 shows typical freeze fracture photos obtained.

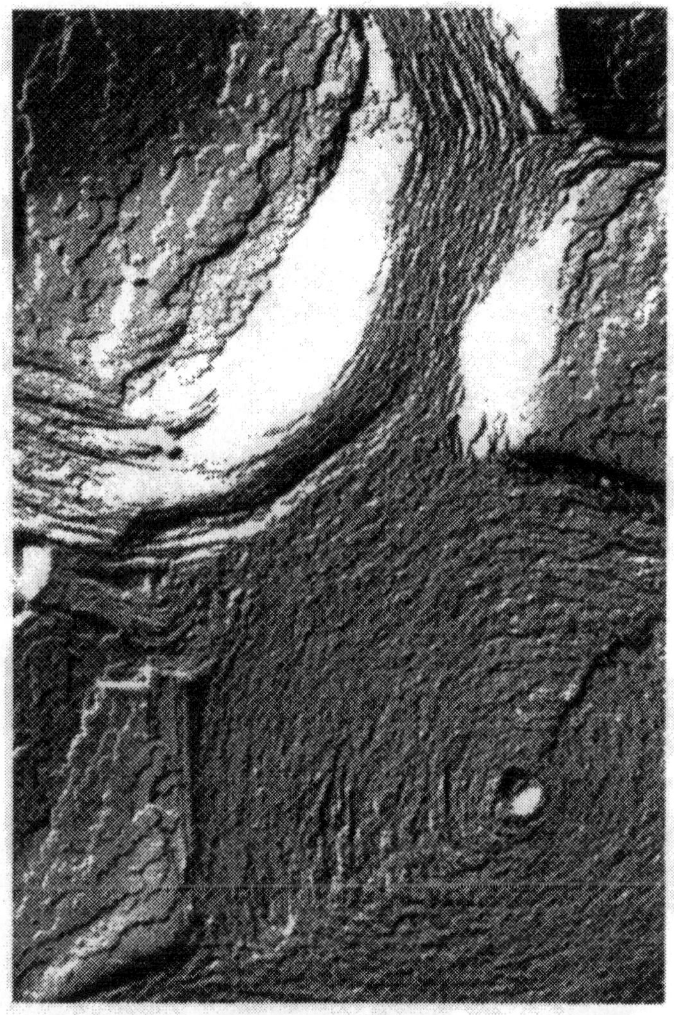

Figure 19. Freeze fracture photos of the onion state. These pictures were obtained by T. Gulik [39].

4. Properties of Lamellar Phase under Shear.

4.1 EVOLUTION OF THE ONION SIZE

We have seen that the onion size was changing in the steady state as a function of the shear rate. We have studied different systems and have always found the same scaling law for the onion diameter D namely:

$$D \propto \frac{1}{\sqrt{\dot{\gamma}}} \tag{8}$$

Figure 20 shows the experimental behaviour found for onion phase prepared from the AOT/NaCl/water system [38].

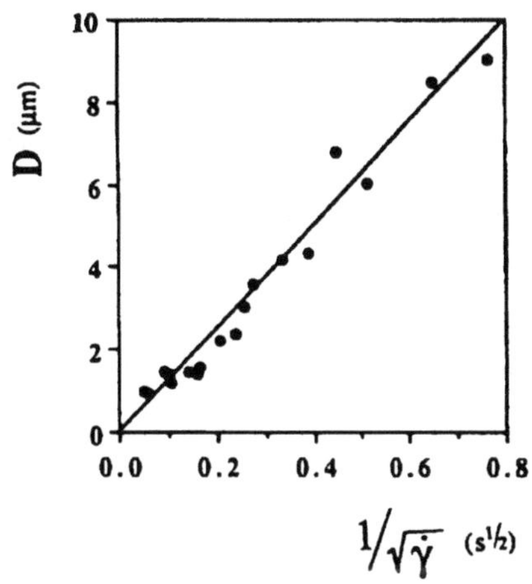

Figure 20. Variation of the experimental diameter of onions as a function of the shear rate.

There are several ways to derive this scaling but up to now there is no way which seems completely satisfactory. The easiest way to understand the origin of this power law is a generalization of an argument given by Taylor [40,41]. The basic idea is that the shear stress which has the effect of making the onion smaller is balanced by the elastic stress which favours bigger onions. The size results from this competition.

$$S_{viscous} = \eta \, \dot{\gamma} \tag{9}$$

$$S_{elastic} = 4\,\pi\,(\kappa + 2\,\bar{\kappa})\frac{R}{d}\frac{1}{R^3} \tag{10}$$

leading to :

$$R = \sqrt{\frac{4\pi(\kappa + 2\bar{\kappa})}{\eta\dot{\gamma}d}} \tag{11}$$

It is implicitly understood that the onion elasticity is only due to curvature energy, an assumption which is not completely obvious.

Another way to derive the scaling law is to consider that under shear one can reach the instability limit of one of the hydrodynamic modes of the lyotropic smectic phase [43]. The most interesting mode is the one which exists at low frequency and depending upon the q-vector it reads [19,42]:

$$\omega = \frac{\kappa}{\eta d} q_z^2, \text{ when q is in-plane (undulation mode)}$$

and

$$\omega = \mu \overline{B} q^2$$

when q is out-of-plane and in the limit of small wave vectors (baroclinic mode).

with μ being a dissipation coefficient ($\mu = d^2/\eta$) and $\overline{B}$ the compressibility at constant chemical potential (see equation 5).

Using the onion size for the unstable wave vector ($q = 1/R$) and the shear rate as the characteristic frequency of the instability ($\omega = \dot{\gamma}$), one gets:

$$R = \sqrt{\frac{\kappa}{\eta d \dot{\gamma}}}, \text{ if the undulation instability is used or:}$$

$$R = \sqrt{\frac{(k_B T)^2}{\kappa \eta d \dot{\gamma}}}, \text{ if the baroclinic instability is used.}$$

in all cases we get the right scaling but with coefficients that are different functions of the elastic coefficients.

More recently, we have proposed a microscopic model illustrating the way the onions can change size [44]. Two different limiting cases are examined depending whether permeation is allowed or not leading to different scaling laws. When permeation is allowed we recover the square root scaling with again a coefficient which is function of the elastic properties. The mechanism of the change in size can briefly be described as follows. When under external pressure an onion will loose material (water and surfactant) from the center, where the stress is maximum, to the outside. The rate of leakage will be limited by permeation. It is assumed that during this process the number of layers can rearrange freely. We are studying in more details the transient kinetics of the change of size in order to check this mechanism [45].

4.2 RHEOLOGICAL BEHAVIOUR

Since we have now described in details the experimental behaviour of a lyotropic lamellar phase under flow, we will now present the rheological behaviour. We will limit ourselves to the steady behaviour, the linear study of the viscoelasticity is presented by Cates in his contribution [46].

First let us notice that if we want to study the steady state behaviour each experimental point has to reach this steady state. This may take a long time (from 10-20 min to several hours). Consequently, it is not possible to use the automating run of most rheometers and the stress versus shear rate has to be constructed point per point waiting enough time for equilibrium.

Figure 21 and 22 present such an experimental curve obtained for the SDS system. The experiments have been done on a stress controlled rheometer [47].

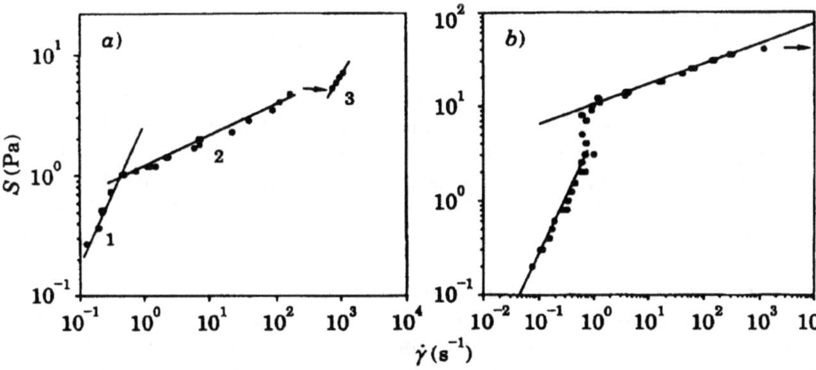

Figure 21. Stress versus shear rate curve of a SDS lamellar phase. The numbers refer to the different steady states described in section 2. a) is a sample containing 69% of oil, b) contains 50%.

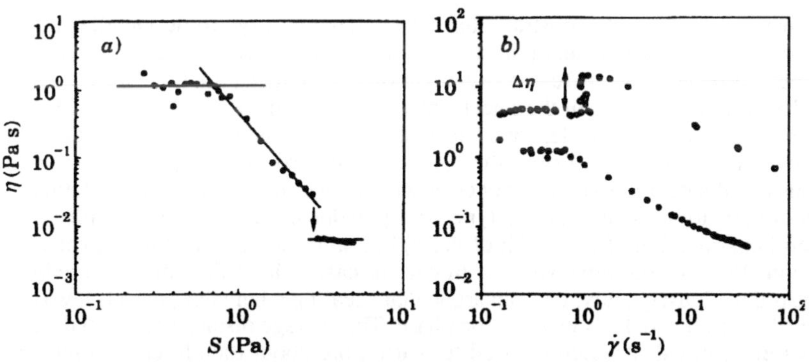

Figure 22. Viscosity versus stress curves of a SDS lamellar phase. The curve corresponds to 71% of oil (a) and 69% and 50% respectively in b.

Clearly, each steady state observed with scattering techniques has a clear rheological signature. State 1 (partially oriented system) is very viscous but has a Newtonian behaviour. The viscosity is typically of the order of a few thousands times the solvent viscosity. When the limit of the first instability is reached (from oriented to onion state) two different behaviours can be observed. At high dilution, the stress versus shear rate curve exhibits a break in the slope going from 1 to 0.2 (in logarithm scale). This is the signature of a change from Newtonian to shear thinning as indicated on figure 22a. The shear thinning regime is characteristic of the onion state. For more concentrated samples before reaching the onion state and the shear thinning regime with the same power law as the dilute samples, there is first a well defined shear thickening regime. In fact, the stress jumps from one value to a higher one with no change in the shear rate. The transition from oriented to onions is then well described by a jump in the viscosity (see figure 22b). At higher shear rate when the instability between the onion state and the better oriented state is reached, we observe a jump in shear rate. Indeed, while increasing the stress above a well defined value, the shear rate increases until it stabilizes at its new value. The third state is again newtonian but with a viscosity which is now only a few times the viscosity of the solvent.

One should notice the differences in using a stress controlled rheometer and a shear controlled one. For as regards the first transition, as we have described, using a stress control rheometer leads to the following observation for concentrated samples: as we reach the transition, the stress is increased with no effect on the shear rate until we reach another (higher) limiting value for the stress where onions are formed and we start to describe the shear thinning onion state. If a shear controlled rheometer is used, once the limiting shear rate is reached a jump in the stress can be observed. Indeed, the stress goes up (needing some time corresponding to a transient behaviour) until it reachs the high value.

Concerning the second transition, the behaviour is exactly symmetric. Using a stress controlled rheometer allows us to see a jump in shear rate as soon as the transition stress is reached. Using a shear controlled rheometer will show a plateau with inside the shear cell a coexistence between onions and oriented regions as seen with the direct space observations [29].

4.3 HAND-WAVING EXPLANATIONS

All the experimental facts demonstrate that when shearing a lyotropic lamellar phase one can observe out-of-equilibrium transitions between steady states. These states are all lamellar but correspond to different organizations of defects. In fact, they can be considered as different textures of the smectic phase. Among the better characterized state is the onion state that might be described as a lattice of crossed disclination lines. This lattice is not perfectly organized since the light scattering exhibit a ring of scattering, it corresponds to a disordered state. It is not clear whether we should describe the nature of the onion organization as of a glassy type or a powder of small monocristallites. A more careful examination of the width of the light scattering peak will help to answer this question. We will see however that in certain cases, the onion state can exhibit long range order. Consequently, one can argue that shearing a lamellar phase is a way to produce well controlled defects in a smectic phase. The average distance between defects being precisely defined, it might be used to learn more about the effect of defects in liquid crystalline phases.

We would like to explain the reasons for these different transitions to exist. However, there is no general way of adressing this question. We can only give some hand waving argument to help to understand the experimental observations.

As a start for looking for explanation we can use the work of Oswald et al. [48,49] done on thermotropic systems. Working on an aligned smectic phase, these authors have described what they called a plastic flow corresponding to the first state we observed. In this regime the defects are assumed to move with the flow. This displacement is possible by permeation and the high viscosity observed is due to the dissipation taking place in this permeation process. They also have shown that above a critical velocity the permeation process is no longer able to go fast enough to unable the dislocations to follow the flow. Then, a normal stress develops leading to an undulation instability. At this stage the authors conclude to a loss of the orientation of the smectic phase.

We believe that this limiting velocity corresponds to the shear rate at which the oriented/vesicles transition takes place. Indeed, once the undulation instability develops the system is very unlikely to be able to flow because of the very high density of defects such as parabolic focal conics (see article by R. Larson, this volume). The system has to find another type of organization to be able to flow - it is where the onion state appears which can be described as a lattice of disclinations. The comparison between the lattice size of the undulation instability and the onion size seems to be able to explaine qualitatively the two regimes observed (continuous or discontinous transition) [47]. The origin of the second transition (between the onion state to the high shear oriented state) is less clear. We can only argue that at very high shear rate there are no defects in the flow direction and this is probably due to the fact that dislocations have no time to develop and the system prefer to compress the layers in order to adapt small changes of thickness rather that developing dislocations as it will do at rest.

5. Comparison with a thermotropic system 8CB.

The oriented/vesicles transition has been observed in many different lyotropic systems. Since thermotropic liquid crystals under flow have been more studied and there was no mention of this type of behaviour, we wanted to see how general was our approach. For that we studied one of the most known thermotropic liquid crystal molecule: the 8CB [50]. This one component material can be considered as a rod like molecule giving a isotropic/nematic (I/N) phase transition at 40°C, a nematic to smectic A transitions (N/A) at 33°C and a smectic A to crystalline transition at 29°C.

This system has been recently extensively studied under flow by Safinya *et al.* in a series of very nice papers [30, 36, 51, 52]. While they were mainly interested in the effect of flow on smectic fluctuations in the nematic phase, they however did some x-ray measurements in the smectic phase. They show that at high shear rate a well oriented phase was prefered but with the normal of the layers in the vortex direction rather than in the shear direction as for lyotropic systems. They also have observed that at lower shear rates, there was a mixture of orientations rather than a pure one. Since we wanted to test the validity of our aproach in terms of transitions under flow, we decided to perform rheological measurements in the smectic state to see how the system was going from a rather disordered state at low shear rate to a well oriented one at high shear rate.

Figure 23 presents a stress versus shear rate curve obtained at 31°C (two degrees below the N/A transition). This curve is caracteristic of a transition between a shear thinning state (called I) and a newtonian one (II). In the first part of the curve corresponding to state I, the stress goes as square root of the shear rate (fig. 23b). The hysteresis during the transition is obtained if we don't wait for a very long time on the way up and down during the rheological measurements. If we wait long enough to reach the steady state, the transition happens at the same place on the way up and down.

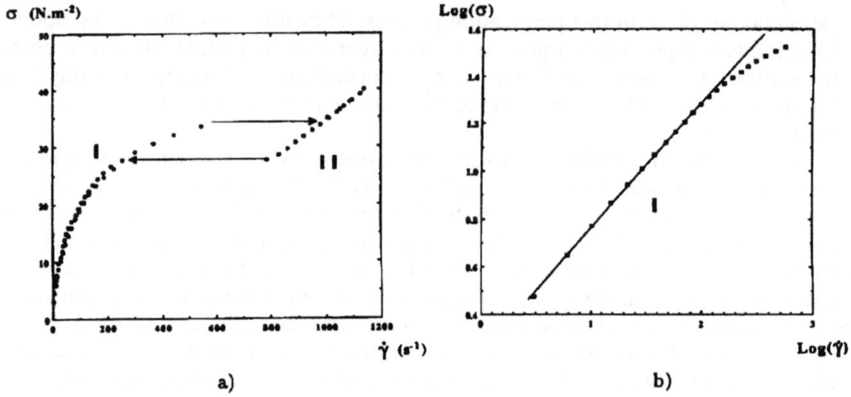

Figure 23. Stress versus shear rate curve obtained on a stress controlled rheometer for the 8CB system at 31°C.

In order to characterize the two states observed we performed x-ray using the home made x-ray shear cell. Figure 24 shows the results of the x-ray pattern obtained in both geometries (centered or tangential) for the two states I and II below and above the transition. As already observed by Safinya, state II corresponds to a well aligned sample with the normal of the layers in the vortex direction (figure 2a). State I can be caracterized by no layers at all in the direction of the flow but a rather homogeneous pattern in the two other direction. This is compatible with multilayered cylinders packed in the flow direction as shown on figure 24a. However, we were unable as for the lyotropic system to do light scatering because of the very high turbidity of samples in state I. Consequently, we have no information on the width of the cylinders.

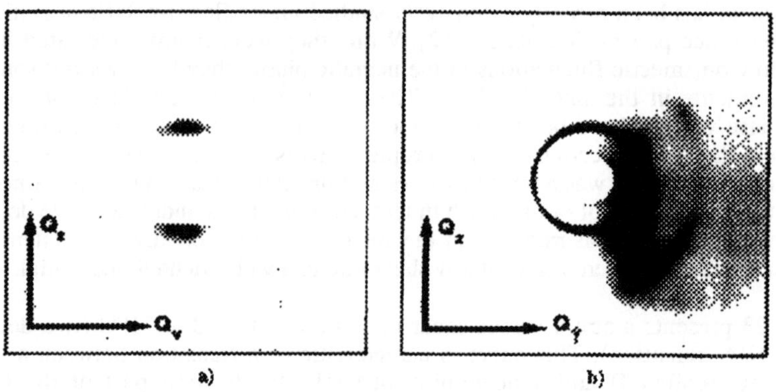

Figure 24(a) X-ray pattern obtained in the two geometries (centered and tangential to the couette cell). This corresponds to stateI (low shear).

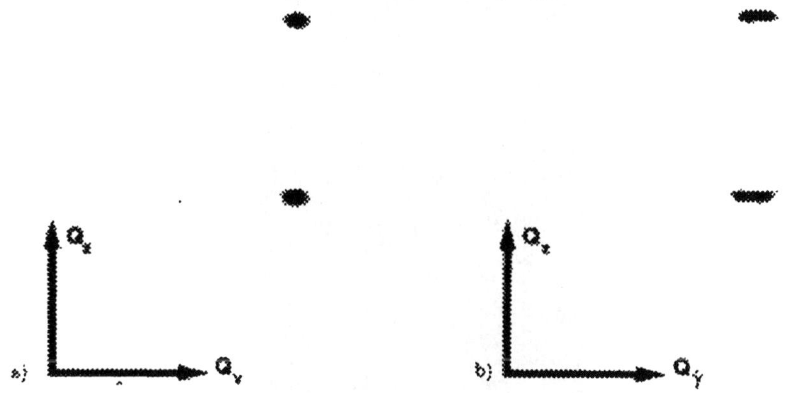

Figure 24.(b) X-ray pattern obtained in the two geometries (centered and tangential to the couette cell).

Interesting behaviour is found when this transition is studied as a function of temperature [50]. It seems that the discontinous transition between state I and II disappears just before 32°C (one degree below the N/A transition). Scaling laws recalling critical phenomena can be extracted from the temperature variations.

6. More Exotic Transitions.

Studying many diffrent lyotropic systems, we were able to observe other types of transitions under shear. We will now briefly describe transitions involving onion states. These transitions have been studied with another lamellar system : SDS/octanol/NaCl/water. However, premilinary experiments have shown that these transitions can also be seen with the SDS/dodecane or AOT/water systems.

When shearing a sample containing around 20% of membrane (9% of SDS and 11% of octanol in weight %) in 80% of brine (20g/l of NaCl in water), we first observe at low shear ($\dot{\gamma} = 1\ s^{-1}$) the formation of the onions that can be easily caracterized by light scattering with the appearance of an isotropic ring of scattering (see figure 25a).

Increasing the shear rate one observes a quite different behaviour than the one described for the SDS/dodecane system [53]. Indeed, instead of having a transition at high shear rate toward a partially oriented state, the ring develops a modulation in its intensity that increases with the shear rate until the pattern stabilizes into six dots of scattering the two vertical ones being less intense (see figure 25 b and c).

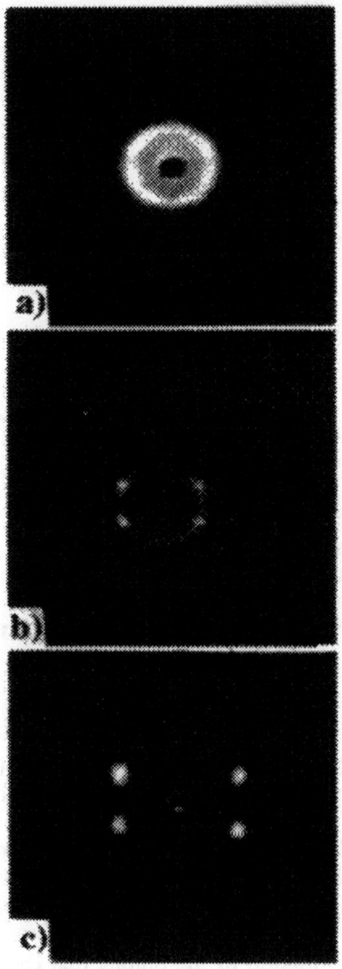

Figure 25. Evolution of the light scattering pattern as the shear rate increases. a) 1s-1, b) 10 s-1, c 1000 s-1.

It is rather clear that the system undergoes a disorder/order transition under shear.

The neutron scattering pattern is also quite interesting to follow along the transition. Figure 26 shows the evolution of the neutron scattering pattern as the shear rate increases. Again, a very strong modulation of the scattering intensity appears progressively.

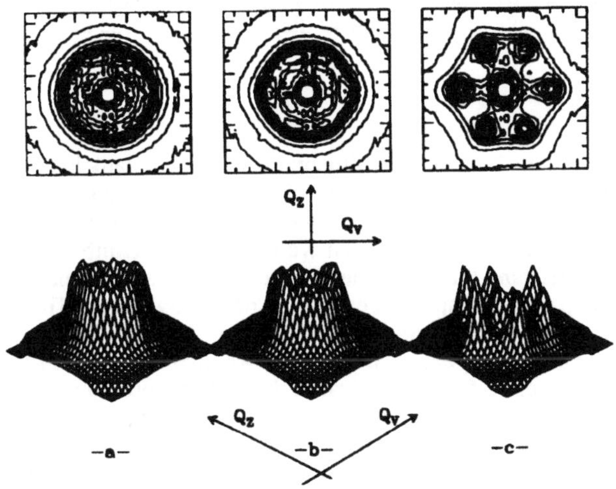

Figure 26. Evolution of the neutron scattering pattern as the shear rate increases. a) 1.3 s-1; b) 10 s-1, c) 780 s-1.

It is easy to reconstruct the long range organization of onions leading to such patterns. In fact, in order to make the neutron scattering and the light scattering compatible, the onions are organized in planes (in the velocity/vorticity directions). Inside these planes the onions are packed in an hexagonal lattice and the planes are sliding over one each others in the shear direction keeping the orientation of the crystalline array. Figure 27 shows a two dimensional representation of this onion organization.

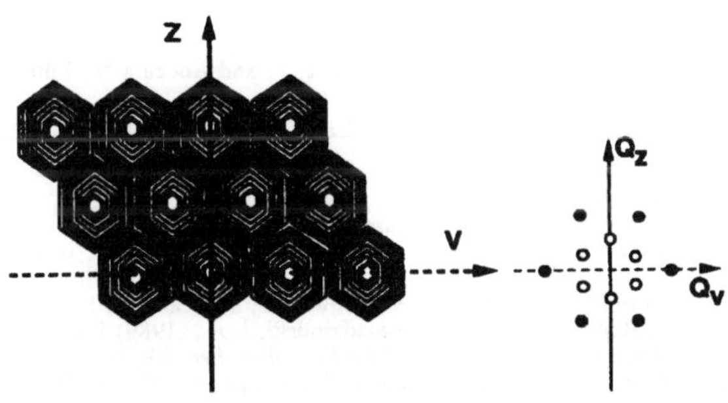

Figure 27. Two dimensional representation of the onion organization leading to the light and neutron scattering patterns observed.

This transition has many features in common with what is called the shear ordering transition for colloidal systems [54, 55].

Studying this system more systematically, we recently observed an even more fascinating behaviour [56]. For a slightly different membrane concentration, above the layering transition that has been described just above, we observed a jump in the onion size. Indeed, increasing even more the shear rate, a new population of onions appears that are much bigger than the first one (10 to 70 mm instead of 1-4 mm). The two populations coexist first then the second population remains alone. Changing temperature, we were able to map the shear diagram for all these transitions [56].

7. Conclusions

It is somehow surprising how the experimental behaviour of lamellar phase under shear is rich and diversified. We believe that the best way to describe this behaviour is to use non-equilibrium physics and to talk about steady states, transient states and out-of-equilibrium transitions. The first step that we were able to climb was to show that the relevant way to understand this behaviour was to build up a shear diagram localizing in the experimental parameter space the regions of steady states. These diagrams are equivalent for their use to the phase diagrams obtained at equilibrium except that they refer to out-of-equilibrium phases. Depending whether shear or stress is controlled different behaviour can be found. Typically, depending upon the transition observed, coexisting states can occupy most of the area of a shear diagram or can be completely omitted. This is experimentally important to realize that fact since it is always easier to study "pure" states (only one type of orientation) rather than a mixed state.

Because these transitions are out-of-equilibrium ones, one may also expect transitions between steady states and time dependent states (oscillating with time or even chaotic). We do have some indications that these transitions exist but some experimental difficulties have yet to be overcome.

We believe also to have demonstrated that the rheological consequences of all these transitions can be understood in terms of change in the structure (at some scale) of the complex fluid phase. The use of many different techniques such as the scattering ones but also dielectric properties should help to connect the structure of the fluids under shear and rheology.

8. References

1. Cates, M.,Lequeux F. and Candau, S. *this volume*
2. See Physics of Amphiphilic Layers, Langevin, D., Meunier, J., and Boccara N., Editors Springer-Verlag Berlin and Heidelberg 1987
3. See Gelbart, W., Ben-Shaul, A. and Roux, D., Micelles, Membranes, Microemulsions and Monolayers Springer-Verlag, New York 1994
4. Canham P. B. : *J. Theor. Biol.* (1970) **26**, 61
5. Helfrich, W., *Z. Naturforsch*, (1973) **28a**, 693
6. Landau L. D. and Lifshitz *Theory of elasticity* (Pergammon, New York, 1970)
7. *Statistical mechanics of membranes and surfaces*, Eds Nelson D. R., Piran T. and Weinberg S. World Scientific (1989)
8. P. G. de Gennes, *The physics of liquid crystals*, G. M. Brown editor (Academic Press, New York, 1975)
9. Kaler, E. W., Murthy, K. A., Rodriguez, B. E. and Zasadzindski, J. A., (1989) *Science* **245** , 1371
10. Cantu, L., Corti, M., Musolino, M; and Salina P., (1990) *Europhys. Lett.* **13** , 561
11. Hervé, P., Roux, D., Bellocq, A.M., Nallet, F. and Gulik-Krzywicki, T. (1993) *J. Phys. II France* **3**, 1255
12. Porte, G., Marignan, J., Bassereau, P., May, R. (1988) *J. Phys. (Fr)* **49**, 511
13. Cates, M. E., Roux, D. Andelman, D., Milner, S. and Safran, S. (1988) *Europhys; lett.* **5**, 733 (Erratum **7**, 94)
13. Roux, D. and Bellocq, A. M. (1985) in *Physics of amphiphiles* Degiorgio and Corti M. Editors (North-Holland, Amsterdam)
14. Safinya, Roux, D., Smith, G. S., Sinha, S. K., Dimon, P. Clark, N. and Bellocq, A.M. (1986) *Phys Rev. Lett.* **57**, 2718
15. Larche, F., Appell, J., porte, G, Bassereau, P. and Marignan, J., (1985) *Phys. Rev. Lett.* **56**, 1700
16. Strey, R., Schomacker, R., Roux, D., Nallet, F. and Olsson, U. (1990) *J. Chem. Soc. Faraday Trans.* **86**, 2253
17. Helfrich, W. (1978) *Z. Naturforsch* **33a**, 305
18. Roux, D. and Safinya, C. R. (1988) *J. de Phys. (Fr)* **49**, 307
19. Nallet, F., Roux, D. and Prost (1989) *J. de phys. France* **50**, 3147
20. Lubensky, T. C., Prost, J. and Ramaswamy, S. (1990) *J. de Physique France* **51**, 933
21. Roux, D., Nallet, F. and Safinya, C. in [3]

22. de Gennes, P. G. and et Taupin, C. (1982) *J. Phys. Chem.* **86,** 2294
23. Roux, D., Coulon, C. and Cates, M. E. (1992) *J. Phys. Chem.* **96**, 4174
24. Porte, G., Appell, J., Bassereau, P., Marignan, J. (1989) *J. Phys. (Fr)* **50**, 1335
25. Huse, D., Leibler, S. (1988) *J. Phys. (Fr)* **49**, 605
26. Israelachvili, J. (1985) in *Physics of amphiphiles* Degiorgio and Corti M. Editors (North-Holland, Amsterdam)
27. Simons, B. D. and and Cates, M. E. (1992) *J. Phys. II France* **2**, 1439
28. Morse, D. C. and Milner, S. T. (1994) *Europh. Lett.* **26** , 565
29. Diat, O., Roux, D. and Nallet, F. (1993) *J. Phys. II France* **3**, 1427
30. Safinya, C. R., Sirota, E. B, Plano, R. J. and Bruinsma R. F. (1990) *J. Phys. Condens. Matter* **2**, SA 365
32. Schmitt, V., Lequeux., F., Pousse, A. and Roux, D., (1994) *Langmuir* **10**, 955
33. Berret, J.F., Roux, D.C., Porte, G. and Lindner, P. *Europh. Letters* (1994) **7**, 521
34. Diat,O. and Roux, D., (1995) *Langmuir* **11**, 1392
35. Cates, M.E. and Milner, S. T., (1989), *Phys. Rev. Lett.,* **62,** 1856
36. Safinya C. R., Sirota E. B. Bruisma R. F., Jeppesen, C. Plano, R. J. and Wenzel, L. J. *Science* (1993)
37. Soubiran, L., Roux, D. and Coulon, C. (1995) *Europhys. Lett.*
38. Diat, O. and Roux, D. (1993), *J. Phys. II France* **3**, 9
39. Gulik, T., Dedieu, Roux, D., Degert, C. and Laversanne, R. (1996) *Langmuir*
40. Taylor, G. I., *Proc. R. Soc. A* (1932) **138,** 41
41. Taylor, G. I. (1934), *Proc. R. Soc. A* **146**, 501
42. Brochard, F. and de Gennes, P. G., (1975), *Pramana Suppl.* **1**, 1
43. Cates, M.,E., private communication
44. Prost, J., Leibler, L. and Roux, D. unpublished
45. Panizza, P., Colin, A., Roux, D. to be published
46. Panizza, P., Roux, D., Vuillaume, V., Cates, M. E. and Lu, D. *Langmuir* (1996) **12**, 248 , Cates, M.,E., in this issue
47. Roux, D., Nallet, F. and Diat, O., (1993), *Europhysics Letters*, **24**, 53
48. Oswald, P. and Kléman, M. (1983)*J. Phys. (Paris) Lett.* **43** L-411
49. Oswald, P. and Ben-Abraham, S.,I. (1982) *J. Phys. (Paris)* **43**, 1193
50. Panizza, P., Archambaut, P. and Roux, D. (1995), *J. phys. (France)*, **5**, 303
51. Safinya, C. et al. *Phys. Rev. Lett.* (1991) **66**, 1986
52. Bruinsma R. and Safinya C. (1991) *Phys. Rev.* **A43**, 5377
53. Diat, O., Roux, D. and Nallet, F. *Phys. Rev. E*, (1995), **51**, 3296
54. Ackerson, B. J. and Clark, N. A., *Phys. Rev. A* (1984) **30**, 906
55. Tomita, M. and Van de Ven, T. G. M., *J. colloid Interface Sci.* (1984) **99**, 374
56. Sierro, P. and Roux, D. submitted to Phys. Rev. Lett.

SURFACTANT CUBIC PHASES

C. TOPRAKCIOGLU
Department of Physics
University of Patras
Patras 26 500, Greece

Surfactant molecules by virtue of their amphiphilic nature, are capable of stabilizing interfaces between two immiscible liquids such as oil and water thereby dividing aqueous and paraffinic phases into finely dispersed microdomains. The geometric features of these surfactant sheets that form the interface between the immiscible bulk components depend on the composition and temperature. A rich phase behavior is often exhibited by such ternary systems and the resulting diversity of structures ranges from liquid crystalline to random ones. Lamellar and hexagonal liquid crystalline phases have long been known to exist both in binary and ternary systems, but it was the pioneering work of Luzzati and co-workers [1-3] that firmly established the existence of cubic liquid crystalline phases in binary lipid-water mixtures. More recently, Scriven [4] suggested that the microstructure of some ternary phases could be based on triply periodic minimal surfaces of cubic symmetry, and thus paved the way to non-classical descriptions of these systems [5-10]. A number of studies have since demonstrated that the cubic phases formed by ternary systems are well-described by such triply periodic minimal surface structures [11-16]. Furthermore, it was clearly shown that in the phase diagrams of certain amphiphilic systems phase transitions involving transformations between cubic structures of different symmetries and topologies occur within regions previously assigned as being one-phase [11-14,17]. Very recently, rheological studies of ternary cubic phases [17,18] revealed some of the fascinating features of the dynamical behavior of these systems.

We begin our discussion by reviewing some basic geometric ideas required to describe the interfacial topology and energetics of amphiphilic layers, and proceed to a brief, elementary description of triply periodic minimal surfaces. We then present and discuss experimental results on the phase behavior, structure and rheology of ternary cubic phases.

1. Minimal Surfaces and Surfactant Interfacial Topology

1.1. MATHEMATICAL DESCRIPTION OF MINIMAL SURFACES

A point P on a surface S in R^3 is characterized by a mean curvature H

$$H = \frac{1}{2}(\kappa_1 + \kappa_2) \tag{1}$$

and a Gaussian curvature G

$$G = \kappa_1 \kappa_2 \tag{2}$$

T. McLeish (ed.), Theoretical Challenges in the Dynamics of Complex Fluids, 235-256 where κ_1, κ_2 are the principal curvatures, i.e. the curvatures associated with the principal directions at that point. A well-known theorem (the Gauss-Bonnet theorem) links the Gaussian curvature to the Euler characteristic χ, a topological invariant of the surface

$$\int_S G dA = 2\pi\chi \tag{3}$$

with

$$\chi = 2(1-g) \tag{4}$$

where g is the genus of the surface [19,20]. A *minimal surface* (MS) is most simply defined as a surface on which $H = 0$ everywhere. A plane represents the trivial case of such a surface, but more generally, a minimal surface consists of balanced saddle points (i.e. $\kappa_1 = -\kappa_2$). This implies that for a MS $G \leq 0$. One of the simplest non-trivial examples is the minimal surface:

$$z = F(x, y) = x(x^2 - 3y^2) \tag{5}$$

called the "monkey saddle" by Hilbert [21,22] since it has the shape of a saddle (see *Figure 1*), with three depressions (two for the monkey's legs and one for its tail !).

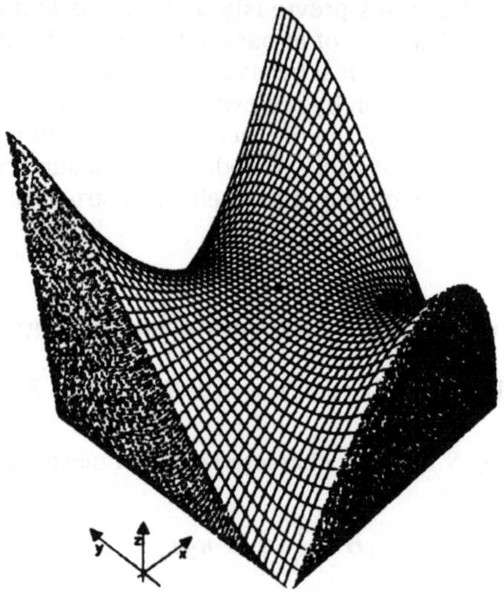

Figure 1. Monkey saddle surface (see eqn. 5). The dot indicates the origin. (Adapted from ref. 22)

It follows from the Laplace-Young equation that a MS must satisfy the differential equation [22,23,24]

$$(1+F_y^2)F_{xx} - 2F_xF_yF_{xy} + (1+F_x)^2 F_{yy} = 0 \tag{6}$$

where F_x, F_{xy} etc. are partial derivatives with respect to the variables x and y.

In general, describing a minimal surface in the Monge notation, i.e. as $z = F(x,y)$, can be difficult or sometimes impossible, so we have to use a method that will enable us to calculate the coordinates (x,y,z) of the surface. In 1866 Weierstrass published a set of equations which allows such a calculation [25,26]. The Weierstrass method proceeds in two stages. First through a Gauss (normal) map from the minimal surface to the unit sphere, followed by stereographic projection onto the complex plane. If the doubly transformed surface S'' is described by a complex function $R(\omega)$, then a point P'' with coordinates $(\omega_\alpha, \omega_\beta)$ on S'' corresponds to the point P on S with coordinates (x,y,z) in real space given by:

$$x = \mathrm{Re}\int_{\omega_0}^{\omega_1} (1-\omega^2)R(\omega)d\omega \tag{7}$$

$$y = \mathrm{Re}\int_{\omega_0}^{\omega_1} i(1+\omega^2)R(\omega)d\omega \tag{8}$$

$$z = \mathrm{Re}\int_{\omega_0}^{\omega_1} 2\omega R(\omega)d\omega \tag{9}$$

where $\omega = \omega_\alpha + i\omega_\beta$.

For the Schwarz-P surface (see *Figure 2*), for example [27],

$$R(\omega) = \left(\omega^8 - 14\omega^4 + 1\right)^{-1/2} \tag{10}$$

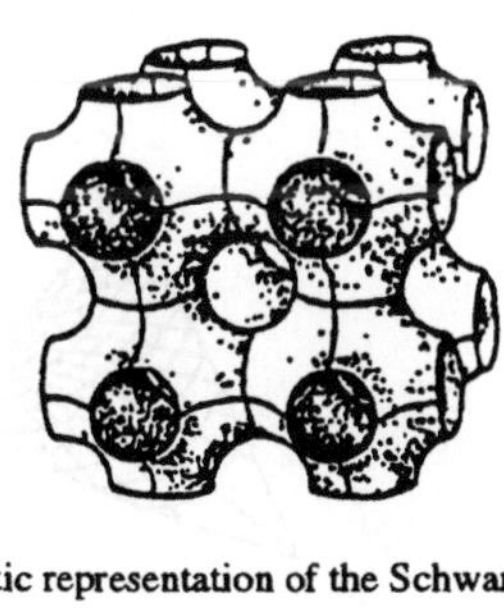

Figure 2. Schematic representation of the Schwarz-P triply-periodic minimal surface

with the integration domain shown in *Figure 3*. The Weierstrass method guarantees the surface obtained is a minimal surface, for the vector $(x(u,v), y(u,v), z(u,v))$ satisfies Laplace's equation with respect to the parameters u and v [26,28].

An interesting transformation discovered by Bonnet [29,30] allows one to transform one minimal surface to another while preserving distances along the surface- i.e. the Bonnet transformation is isometric. Mathematically, the transformation amounts to multiplying the integrand of the Weierstrass equation by $e^{i\theta}$ where θ is called the Bonnet angle. Two surfaces related by the Bonnet transformation are called *associate* surfaces; they are called *adjoint* surfaces if $\theta = 90^o$.

The study of minimal surfaces was inspired by the work of Plateau [31] who demonstrated that a MS can be produced by dipping a closed wire frame into a soap solution. Weierstrass and others were attracted to the Plateau problem, but it was Schwarz [34] who from the mid-1860's if not earlier, visualized his solution surfaces to the Plateau problem as fundamental domains (patches) for the construction of *triply periodic* minimal surfaces free of self-intersection. It turned out that such surfaces can indeed be created by symmetry operations (e.g. rotations, translations etc.) performed on a fundamental minimal surface patch. Schwarz went on to discover the *F* (also known as *D*) and *P* surfaces, among others (see *Figure 3*). The search for triply periodic minimal surfaces (TPMS) - also called infinite periodic minimal surfaces (IPMS) - has continued, and one of the most notable discoveries has been that of the Gyroid (G) surface by Schoen [33] (see *Figure 3*). TPMS subdivide space into two distinct, usually congruent subspaces (there are exceptions, as in the case of the I-WP surface where the two subspaces are of unequal volumes).

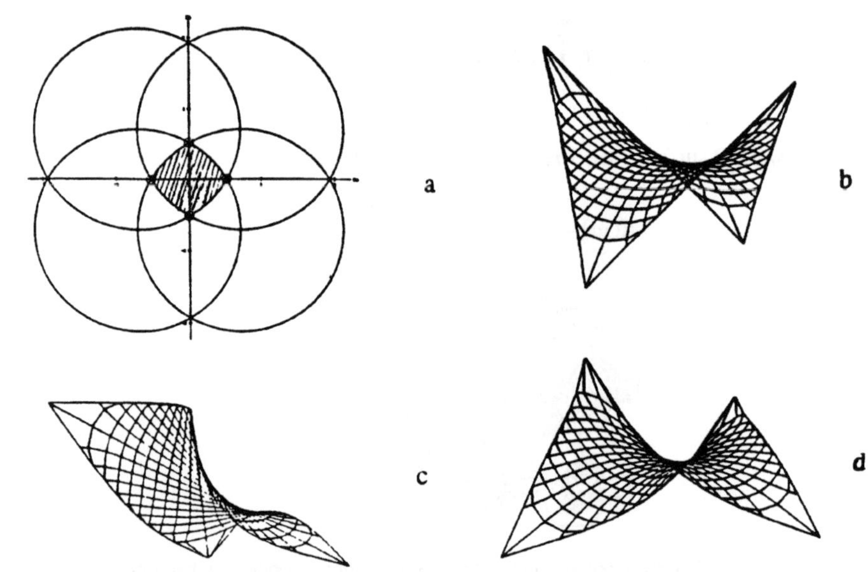

Figure 3. a) Integration domain for the D, P, and G minimal surfaces (see eqns. 7-10). b) Fundamental surface patch of the D-surface (θ=0°), c) Fundamental patch of the P surface (θ=90°). d) Fundamental patch of the G surface (θ=38.015°). A triply periodic minimal surface can be constructed in each case by symmetry operations involving the fundamental element of each surface (see Figure 2). (Adapted from ref. 26)

A useful way of visualising the structure corresponding to a particular TPMS is via its "skeletal net" which describes the two intertwined "labyrinths" that are separated by the minimal surface. This is illustrated in *Figure 4* where it can be seen that the Schwarz-P surface is based upon two interlaced simple (primitive) cubic networks while the Schwarz-D surface is described by two intertwined diamond networks. The labyrinths of the *G* surface are a little more difficult to visualize and it is worth noting they are enantiomorphic! When both intertwined labyrinths contain the same "material" in each case, the *P*, *D*, and *G* surfaces have the space groups Im3m, Pn3m, and Ia3d respectively,. It is important to point out that the *P*, *D*, and *G* surfaces are associate (i.e. related via the Bonnet transformation). Moreover, the *P* and *D* surfaces are adjoint.

More recently, triply periodic surfaces of constant *non-zero* mean curvature everywhere have been computed and much progress has been made towards proving their mathematical existence [5,34,35,36]. These surfaces can be thought of as generalizations of the minimal surfaces.

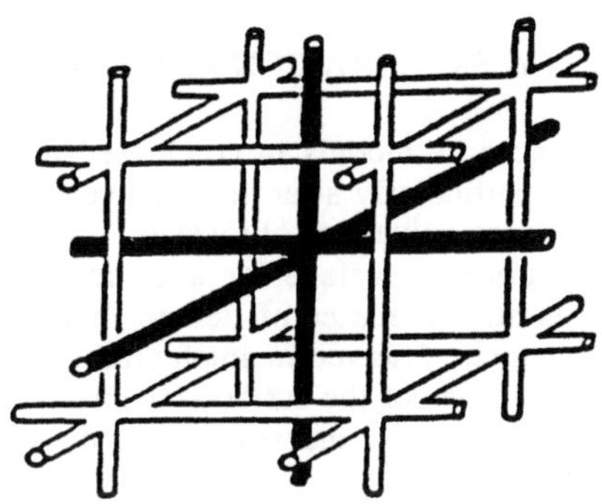

a

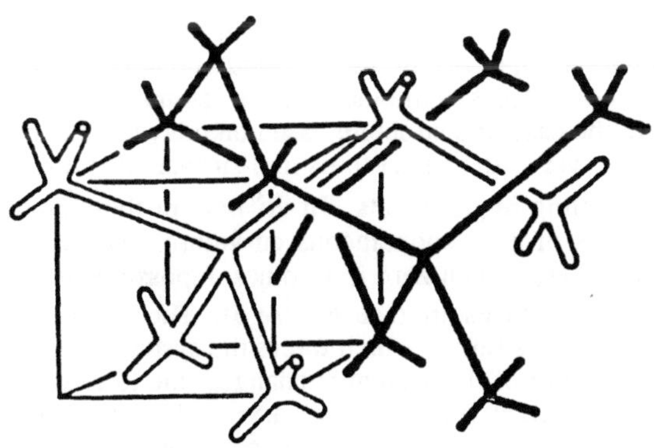

b

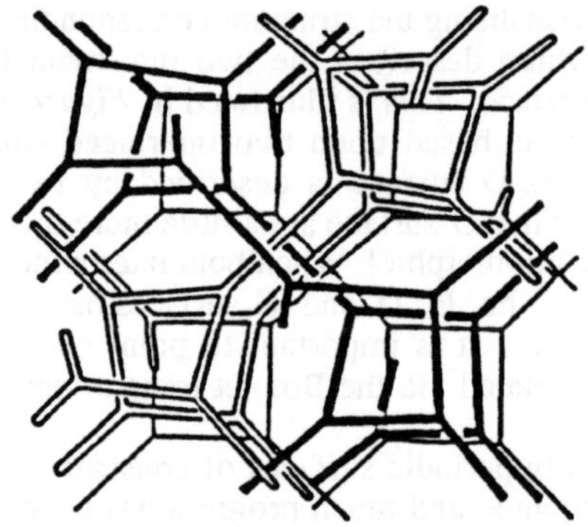

Figure 4. Skeletal nets of the P, D, and G triply periodic minimal surfaces.
a) The interpenetrating Schwarz-P labyrinths with simple cubic symmetry
b) The Schwarz-D labyrinths with double diamond symmetry
c) The enantiomorphic intertwined labyrinths of Schoen's Gyroid (G)-surface

1.2. CURVATURE ELASTICITY, FRUSTRATION AND PHASE TRANSITIONS

The importance of minimal surfaces as models in condensed matter structure has rapidly increased in recent years, particularly after the discovery of numerous cubic liquid crystalline phases through the investigations of Luzzati [1-3] and others. Exactly why nature selects a particular minimal surface for a given system (at a certain composition and temperature) is still a matter of debate, but it is already well-established that the variety of structures observed in amphiphilic systems can often be understood by focusing on the properties of the surfactants interfaces. In particular, transitions between different structures can be understood in terms of the minimization of the curvature elastic free energy.

A geometric insight into the constraints faced by surfactants sheets in forming self-assembled structures in three-dimensional space can be gained by consideration of a concept developed by Charvolin and Sadoc [6]. The forces acting upon a system of amphiphilic layers are of various origins, and are not yet fully understood. It should be clear, however, that whatever their origin, their *normal* components tend to maintain constant spacing between interfaces while their *lateral* components tend to determine interfacial curvature. In the case of a structure based on bilayers, it is obvious that due to the symmetry of the bilayer with respect to its mid-surface, it is not always possible to satisfy these two conflicting requirements. In R^3 both of these conditions are met only in lamellar structures of zero spontaneous curvature. When the spontaneous curvature of the interfaces, H_o, is non-zero it becomes impossible to accomodate both constant spacing and constant curvature (see *Figure 5*). The result, as Charvolin and Sadoc point out, is a state of frustration which forces the system to adopt structures of compromise. As the curvature of the system is changed due to changes in molecular interactions within the surfactant film, e.g. by varying the temperature, transitions between different structures take place (see *Figure 6*). Thus, cubic phases are often found sandwitched between lamellar and hexagonal phases.

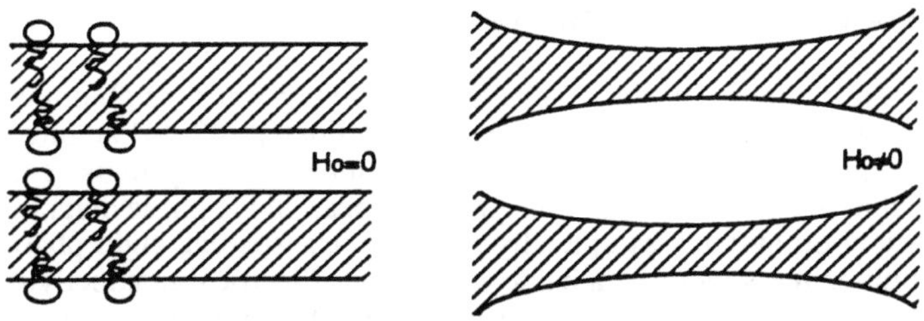

Figure 5. Schematic representation of a) lamellar structures showing zero spontaneous curvature (i.e. $H_0=0$) and constant spacing between interfaces. b) Incompatibility of non-zero curvature and constant interfacial separation leading to frustration.

At lowest order in a curvature expansion, the elastic free energy of the surfactant interface per unit area may be expressed as [37]

$$F = 2K(H - H_o)^2 + \overline{K}G \tag{11}$$

where H_o is the spontaneous curvature, H and G are the mean and Gaussian curvatures respectively, K is the bending modulus (or bending rigidity), and $\overline{K}$ is the elastic modulus of Gaussian curvature (saddle-splay constant) of the interface. To obtain the phase diagram of a system, the Helfrich free energy (eqn. 11) can be minimized under the constraints of constant total interfacial area (conservation of surfactant molecules at the interface) and constant total enclosed volume(conservation of internal constituent volume). The phase diagram implied by the Helfrich free energy thus involves an understanding of the roles of H_o, K and $\overline{K}$. Numerous treatments based on such an approach have been reported [e.g. 38-41]. Without going into the details of these calculations we may quickly point out that generally $\overline{K}$>0 favours saddle-shaped surfaces (i.e. $G < 0$) irrespective of H_o, while it can be shown that if $H_o = 0$ then the lamellar structure with zero mean and Gaussian curvatures represents the ground state for $\overline{K}$<0. If $K > kT$ and $H_o \neq 0$ constant mean curvature structures (e.g. spheres, cylinders etc.) may be formed, including structures based on triply periodic minimal surfaces.

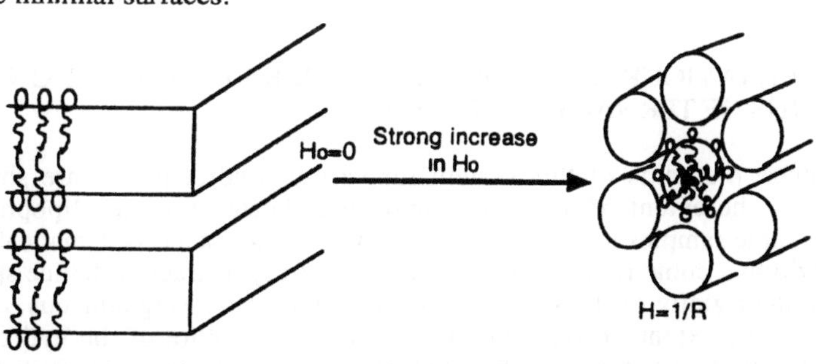

Figure 6. Increase in curvature causes frustration and forces the lamellar system to adopt a new (hexagonal) structure.

2. Experimental Results and Discussion

There is a vast literature of phase behaviour involving water/amphiphile/oil systems thanks to the work of many investigators, notably Fontell [42-44] among others, who mapped the phase diagrams of many binary and ternary systems, and identified numerous cubic regions. Ternary cubic phases appear in the phase diagrams of certain water/surfactant/oil systems as optically isotropic, transparent, rigid phases whose x-ray diffraction patterns index to a cubic symmetry. Electrical conductivity and NMR measurements in some of these phases suggest that they are continuous both in water and oil. The evidence, therefore, points in the direction of bicontinuous structures of cubic symmetry.

The type of structure we shall invoke to describe such phases is illustrated schematically in *Figure 7* and consists of a bilayer decorating a triply periodic minimal surface. As will be shown later, analysis of the x-ray diffraction patterns of the cubic phases makes it possible to determine not only the *symmetry*, but also the *topology* of their structure and thus identify the particular minimal surface on which the structure is based. We shall consider here a representative surfactant system that has been studied extensively: The ternary system water/didodecyldimethylammonium bromide (DDAB)/hydrocarbon [e.g. see 11-14].

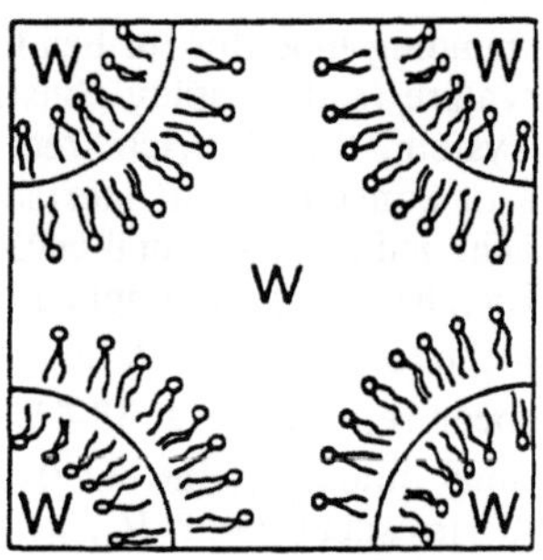

Figure 7. Schematic diagram of the cross-section of the Schwarz-P unit cell showing a bilayer structure decorating the minimal surface. W denotes the aqueous regions of the system. The oil is contained within the bilayer, i.e. in the regions between the tails of the two opposing surfactant monolayers. Note that the two distinct interpenetrating aqueous labyrinths are separated by the surfactant bilayer

2.1. PHASE BEHAVIOUR AND SMALL-ANGLE X-RAY DIFFRACTION PATTERNS OF TERNARY CUBIC PHASES

The partial phase diagrams of this system for various hydrocarbons are shown in *Figure 8* where the extent of the cubic region is indicated for each lipophile. On examining the cubic samples it is found that their properties change systematically with composition within a cubic region. In particular, the x-ray diffraction lattice spacings increase while the rigidity of the samples decreases with increasing aqueous content. The variation in the extent of the cubic domain with the nature of the hydrocarbon solvent has been studied extensively by Maddaford [14] and can be attributed to the degree of solvent penetration into the surfactant tail region. With decreasing aqueous volume fraction, the curvature of the surfactant interface must become greater towards the aqueous phase, (i.e. more negative, by convention). With smaller, highly penetrating molecules, the extent of sustainable negative curvature is greater, and this leads to a large cubic region, as in the case of toluene, spanning a wide compositional range in the phase diagram. The least negative curvature tolerated by the system within a cubic structure sets the boundary of the cubic phases at the high aqueous content side of the diagram (Note that the curvature referred to here is that of the interface where the surfactant *head groups* are located. Consequently, while the surface bisecting the bilayer has $H = 0$, the water/oil interface may have non-zero curvature).

The small angle x-ray (and neutron) powder diffraction patterns of the cubic phases show Bragg peaks which index to a cubic structure whose space group depends on the particular composition of the ternary mixture. Furthermore, some parts of the cubic region are found to exhibit two cubic phases in coexistence. Clearly, transitions between different cubic structures occur within what was once thought to be a region of uniform structure. A typical result can be seen in *Figure 9* where six Bragg reflections are clearly visible and index to a diamond structure of space group Pn3m related to the Schwarz-D surface.

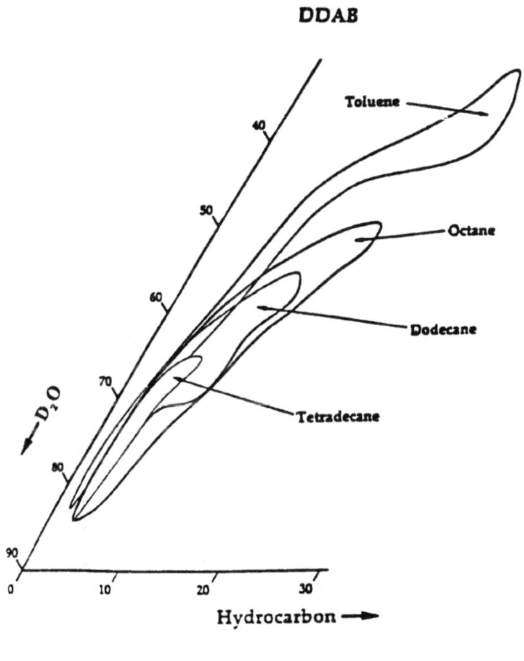

Figure 8. Partial phase diagram at 25°C showing the cubic region in four ternary systems consisting of DDAB/D_2O/toluene, DDAB/D_2O/octane, DDAB/D_2O/dodecane, andDDAB/D_2O/tetradecane.

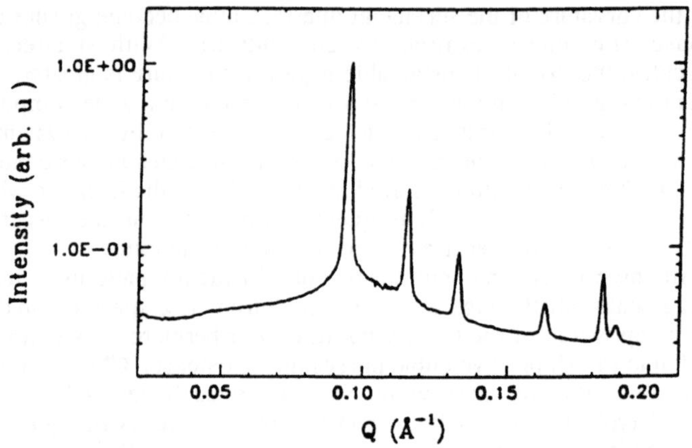

Figure 9. Small-angle X-ray diffraction pattern from a sample in the double diamond region of the system DDAB/D_2O/dodecane with composition (weight %) 41.74/47.56/10.71 respectively. The Bragg reflections $\sqrt{2},\sqrt{3},\sqrt{4},\sqrt{6},\sqrt{8},\sqrt{9}$ are consistent with the space group Pn3m.

A large number of diffraction experiments have been carried out by several groups on the DDAB system with various hydrocarbons [5,11,12-14] and the results reveal the existence of at least three structures based on the *G*, *D* and *P* triply periodic minimal surfaces with space groups Ia3d, Pn3m and Im3m respectively. With increasing water volume fraction the transition sequence *G*—> *D*—> *P* seems to be general in these systems. At high aqueous content there are at least two more cubic phases of as yet unknown structure. As an example, we may consider the results shown in Table 1. where on increasing the water content the lattice constant is seen to increase and eventually a *D*—> *P* transition is observed. The increase in lattice parameter as a function of aqueous volume fraction for cubic phases with different hydrocarbons but with Im3m symmetry is shown in *Figure 10*.

It turns out that this variation of the lattice constant with composition contains information that can be related to the symmetry and topology of the structure. The approach we shall present here is due to Hyde [8] and is based on the construction of two parallel surfaces on either side of the minimal surface and each displaced a distance $\ell/2$ from it (i.e. this is the model of a bilayer decorating a minimal surface). Clearly, there are two types of surface to consider, namely the water/oil interface where the surfactant head groups are situated, and the *minimal surface* which forms the mid-surface of the bilayer. The areas per unit cell of these two surfaces are related by [8]:

$$\frac{A}{2} = S(0)\left(1 + \langle G\rangle_{\substack{unit\\cell}} \ell^2\right) \tag{12}$$

TABLE 1. Variation of lattice parameter and $D \rightarrow P$ transition in the system DDAB/D_2O/octane.

Composition DDAB/D_2O/-Octane	Space Group	Lattice Parameter (A)
48.21/36.44/15.34	Pn3m	83
47.73/37.22/15.04	Pn3m	86
44.82/40.95/14.23	Pn3m	92
43.84/46.12/10.03	Pn3m	98
40.10/51.70/8.20	Pn3m	98
38.18/54.00/7.82	Pn3m	99
36.95/57.35/5.70	Pn3m	107
36.22/56.36/7.42	Im3m	132
35.23/54.84/9.93	Im3m	137
34.94/57.90/7.16	Im3m	140
32.98/60.26/6.75	Im3m	153
29.80/64.10/6.10	Im3m	167
29.29/64.71/6.00	Im3m	170
27.79/66.52/5.69	Im3m	184
26.44/68.15/5.41	Im3m	187
24.09/70.97/4.93	Im3m	209
22.13/73.33/4.53	Im3m	229

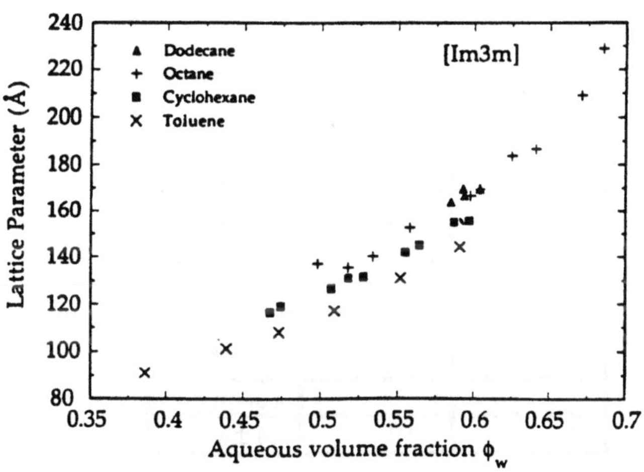

Figure 10. Variation of lattice parameter with aqueous volume fraction in various DDAB/D_2O/hydrocarbon systems in the body-centered cubic (bcc) structural region of the cubic phases.

where A is the total surfactant head group area per unit cell, and S is the surface area at the centre of the bilayer (i.e. S is the surface area of the minimal surface per unit cell). We use the Gauss-Bonnet theorem [8,19] and introduce the dimensionless area (i.e. the normalized surface to volume ratio per unit cell $\frac{S}{V^{2/3}}$) $S^* = S\alpha^{-2}$ where α is the lattice constant, to obtain

$$S^* = \frac{A}{2\alpha^2} - \frac{2\pi\chi\ell^2}{\alpha^2} \tag{13}$$

Thus a plot of $\frac{A}{2\alpha^2}$ against $\frac{2\pi\ell^2}{\alpha^2}$ will generate a straight line whose slope gives χ and whose intercept yields S^*(for a given composition, α and ℓ are determined experimentally by x-ray diffraction [11,14]). Since both S^* and χ are known for various minimal surfaces such plots can be usefully employed to identify the minimal surface associated with a particular cubic phase thus providing a complete characterization of its structure. Examples of such plots can be seen in *Figures 11 and 12,* while Table 2 shows the S^* and χ values extracted from the data and their agreement with the theoretical values. More recently it has been possible to grow single crystals of cubic phases [47] the diffraction patterns of which can provide the ultimate crystallographic characterization of these fascinating structures.

TABLE 2. Comparison of S^* and χ values with the theoretical ones.

Pn3m (D-surface)

	Theory	Experiment
χ	-2	-2.05 ± 0.26
S^*	1.9189...	1.97 ± 0.03

Im3m (P-surface)

	Theory	Experiment
χ	-4	-4.26 ± 1
S^*	2.3455...	2.30 ± 0.07

We may summarize the phase behaviour and structural studies by noting that the transitions between cubic symmetries appear to be driven primarily by interfacial curvature. The aqueous volume fraction is mainly responsible for setting the mean curvature of the interface, while the interaction of the oil with the surfactant tails can affect the spontaneous curvature. For a given composition, the system adopts that structure which best optimizes the (generally) divergent values of the actual and preferred curvatures at the expense of alternative structures. Calculations show, however, that the free energy difference between various minimal surface structures is very small [e.g. see 15,41]. So a delicate energy balance appears to be responsible for the transitions observed between different cubic structures. At this point it may be appropriate to ask why nature seems to show such a preference for the associate surfaces P, D, and G which appear to be the most commonly observed out of all possible minimal surface candidates. A plausible answer to this question has been proposed by Helfrich [46]. He points out that ideally the Gaussian curvature on the surface should be everywhere such that the curvature elastic free energy is minimized. In practice, however, there are flat points in a TPMS and G cannot be uniform on any minimal surface. The next best thing is minimum relative variance of $\langle G \rangle$. It appears that among the large number of TPMS the surfaces P, D, and G are those with the smallest variation in G. For a given value of $\langle G \rangle$ these three surfaces differ further in their geometric features, namely their normalized surface to volume ratios, S^*, and the minimum distance d_{min} between neighbouring parallel sections of the minimal surface (this being relevant in scaling the interaction between surfactant sheets), based on which a system may choose between them.

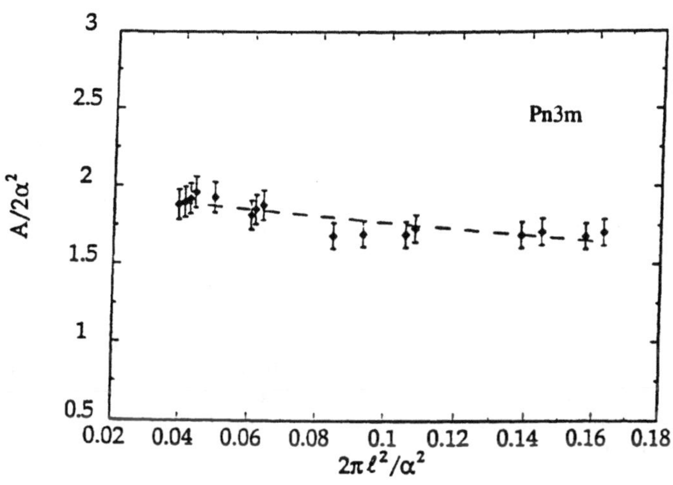

Figure 11. Master plot for cubic phases with double diamond symmetry in the system DDAB/D_2O/dodecane (see eqn. 13 and Table 2).

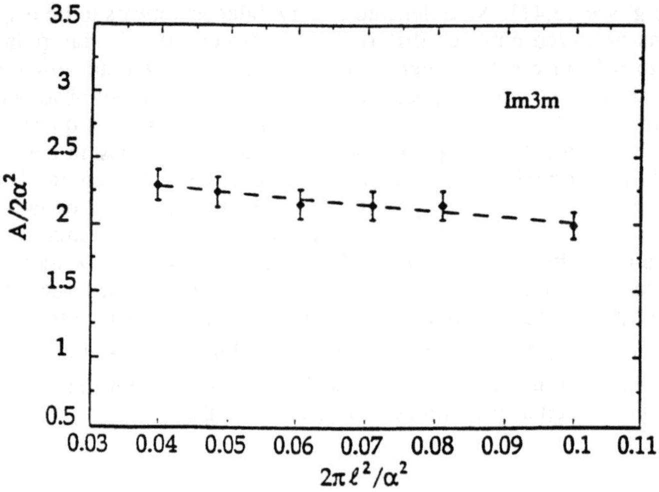

Figure 12. **Master plot for cubic phases with bcc symmetry in the system DDAB/D2O/toluene (see eqn. 13 and Table 2).**

2.2. RHEOLOGY OF TERNARY CUBIC PHASES

Inspection of the cubic phases immediately reveals large variations in rigidity between samples with different space groups (Their rigidity decreases in the sequence $P < D < G$). Within the same space group the samples also become less rigid on increasing the aqueous volume fraction. This seems reasonable, for the rigidity of these systems is largely imposed by the triply periodic surfactant membrane which tends to lose control as the aqueous labyrinths become increasingly swollen. The cubic samples further exhibit the rather interesting property of "ringing" [12,14,17], i.e. they ring at audible frequencies when struck gently. The frequency of ringing depends, of course, on the sample shape and volume, but if the latter are fixed by placing equal amounts of sample into identical containers, it is found that the ringing frequency varies inversely with the aqueous content. In the next sections we describe the main features of the rheology of surfactant (DDAB) cubic phases and a brief discussion of their dynamical behaviour.

2.2.1. *Strain Sweep*

Figure 13 [17] shows a strain sweep for a typical sample with Pn3m symmetry. The shear moduli are independent of strain λ, up to some critical value λ_c. For most cubic phases in the DDAB system $\lambda_c \sim 1-5\%$. It can be seen that both the storage modulus G' and the loss modulus G'' scale approximately with λ^{-1} at higher strains (see chapters of Rubinstein and Cates for a discussion of $G^*(\omega)$). This behaviour can be understood if one assumes that the network ruptures along planes parallel to the shear planes at some critical stress σ_c beyond which the stress increases only marginally for further strain [47]

$$\sigma_c \sim G'(\lambda, \omega)\lambda \qquad \text{or} \qquad G' \sim \lambda^{-1} \tag{14}$$

The dissipation per cycle for such a fluid is $\approx \lambda\sigma_c$ and this is related to the dissipative part of $G^*(\lambda, \omega)$ so that

$$\lambda\sigma_c \sim G''(\lambda, \omega)\lambda^2 \qquad \text{or} \qquad G'' \sim \lambda^{-1} \tag{15}$$

Thus both the real and imaginary parts of the complex non-linear modulus scale as λ^{-1} at large strains. This simple model allows an estimate of the rupture energy of the network.

$$E_r \leq NG\,\alpha^3 \tag{16}$$

where $G\alpha^3$ is the energy stored elastically per unit cell and N is the number of units separating dislocation planes. This gives (for $N \cong 1$) $E_r << 20kT$. Clearly, for the network to be stable we require

$$E_r >> kT \tag{17}$$

As T rises and G' falls the last inequality can no longer hold and thermal break-down of the network occurs.

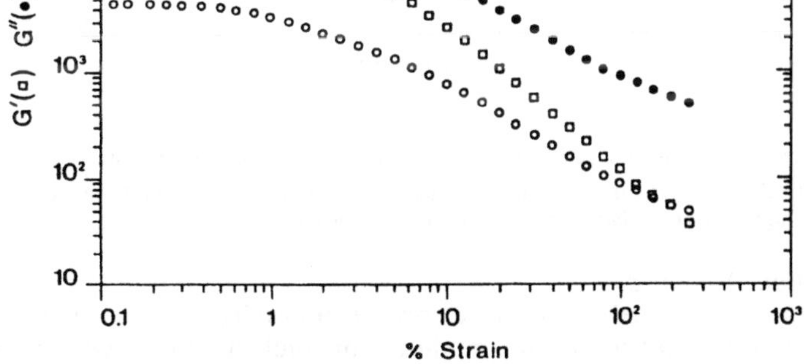

Figure 13. Dynamic strain sweep of a cubic sample with composition DDAB/D_2O/octane (32.99/60.26/6.75 weight %) and bcc symmetry (space group Im3m) at 27° showing the variation of the storage modulus G', loss modulus G'' and viscosity η as a function of strain at a frequency of 10rad/s.

2.2.2. *Frequency Dependence*

The frequency-dependent behaviour of surfactant cubic phases is seen in Figure 14 [17] which shows a frequency sweep for a sample with Pn3m symmetry. The behaviour of both $G'(\omega)$ and $G''(\omega)$ is somewhat reminiscent of polymer solutions. The storage modulus G' shows a high frequency plateau. Furthermore, the loss modulus G'' exhibits a maximum at $\omega_{max} \approx 10s^{-1}$, while at low ω, $G''(\omega) > G'(\omega)$. There appears to be a regime of scaling of the moduli down to a frequency of $\sim 3x10^{-2}s^{-1}$. Below this frequency there is Maxwell behaviour. The overall picture suggests a fluid with several relaxing structures with characteristic relaxation times in the range $0.1s - 30s$. Of course, since cubic phases are thermodynamically solids, they must have a finite modulus at $\omega \to 0$. So there must be a plateau in G' below the Maxwell regime, but it cannot be accessed in these experiments.

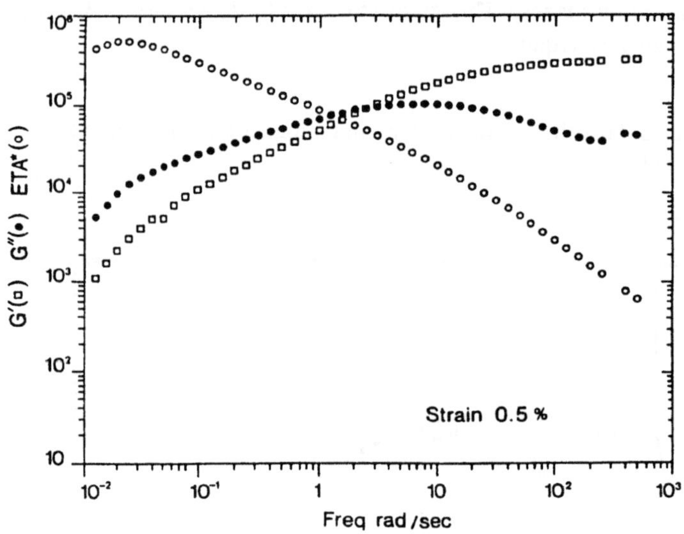

Figure 14. Frequency sweep of a cubic sample with composition DDAB/D $_2$O/octane (48.32/36.35/15.33 weight %) and diamond symmetry (space group Pn3m) at 25 °C. Note the maximum in G" and the high frequency plateau in G', while at low frequencies G'<G".

2.2.3. *Relaxation Dynamics*

It is clear that a simple model of an elastic network embedded in a viscous fluid *cannot* describe the ω-dependence of the moduli (for such a model we would have $G^*(\omega) = G_o + i\omega\eta_o$ where G_o is the elastic modulus of the network and η_o the viscosity of the solvent). Instead, to account for the observed non-trivial ω-dependence one must recognize the fluid nature of the surfactant membrane as well as of the water and oil regions. A shear-deformed minimal surface is no longer at a free-energy minimum but its original geometry may be partially regained through local transport of surfactant thus allowing the structure to relax to a new constrained minimum.

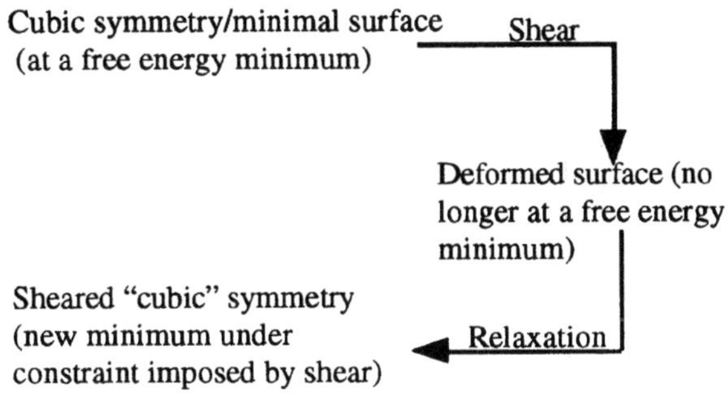

A rigorous characterization of the relaxation of the network would require identification of the eigenmodes of the surface in terms of a 2-d parametrization. The shear-stress relaxation depends on the coupling of a shear-induced deformation to the mode amplitudes. The shear - stress relaxation function $G(t)$ must be expressed in terms of these quantities [see ref. 17 for a more detailed discussion]. Although a detailed analysis on the basis of the scheme suggested above has yet to be carried out, consideration of the time constants involved leads to the conclusion that surfactant cubic phases behave as viscoelastic fluids in parallel with a weak elastic solid. The peak in $G''(\omega)$ may arise from the relaxation of shear-induced surface deformations through molecular flow in the surfactant membrane and surrounding bicontinuous channels.

Alternatively, one might invoke the interface between crystallites as a source of relaxation. The idea of such slip plane formation (introduced earlier in our discussion of the strain-dependent rheology) has been developed further by Jones and McLeish [18]. They propose a model that provides a means for relief of the residual stress arising from the sheared cubic symmetry we have considered above. This is a mechanism whereby the strain is accomodated by layers in the cubic lattice sliding over each other, the sliding taking place in the surfactant film; i.e. large deformations occur along certain planes in the cubic lattice while the rest of the structure sustains very little deformation. In this manner, the structure is disrupted locally while maintaining its global topological integrity. In such a picture, the surfactant skeleton does not break and re-form cell edges as would be the case in a dislocation mechanism. Instead, flow in the surfactant film forms the basis of the slippage process. Thus movement across the periodic potential may be possible at relatively low energy cost and relaxation may take place.

The calculations of Jones and McLeish show that the system behaves as a Maxwell liquid at low frequencies, but at high frequencies it displays Voigt cell characteristics and acts like a solid. An important result that emerges from these calculations is the existence of a *single* relaxation time τ even when there is polydispersity in the slip plane density, τ being a function of the mean density of slip planes.

2.2.4. *Melting Transition*

As the temperature is increased beyond a certain value, the cubic phases melt into free-flowing liquids. The temperature at which this occurs depends on the composition and increases with increasing aqueous content. The melting is characterized by the disappearance of higher-order Bragg reflections. At the melting temperature (more accurately a melting range of a few degrees °C) $G'(\omega)$ and $G''(\omega)$ become identical and scale with the frequency ω as can be seen in *Figure 15* [17].

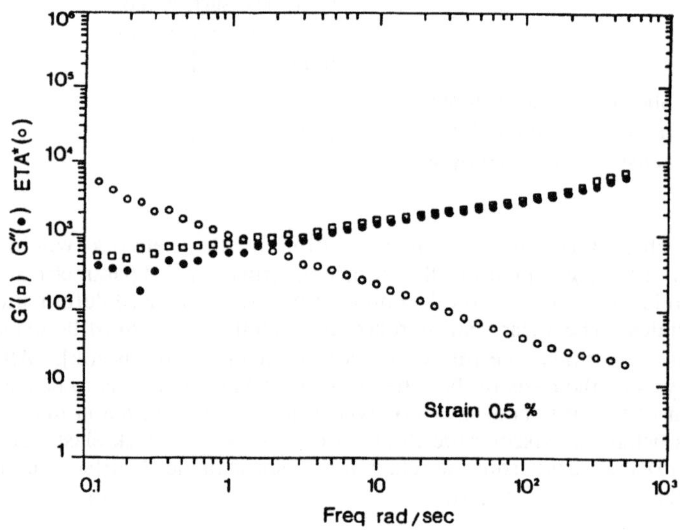

Figure 15. Frequency sweep of the same sample as in Fig. 14 at 40°C. Note the convergence of G' and G".

The situation is described by $G^* \sim \omega^u$ where $u = 0.55 \pm 0.05$, (see *Figure 16*). This is a little surprising since it suggests a 2nd-order transition, while one would expect 1st-order behaviour for a transition such as melting. Neutron-scattering measurements [48] suggest that the molten cubic phase has a sponge-like structure that maintains the bilayer geometry, while very recent, careful rheological measurements by Jones et al. indicate that the melting transition is fully reversible and highly reproducible [49]. Presumably, the melting transition is weakly first-order and this is further supported by differential scanning calorimetry which shows the absence of a large transition enthalpy, but the exact origin of the observed dynamical behaviour is difficult to understand and requires further investigation.

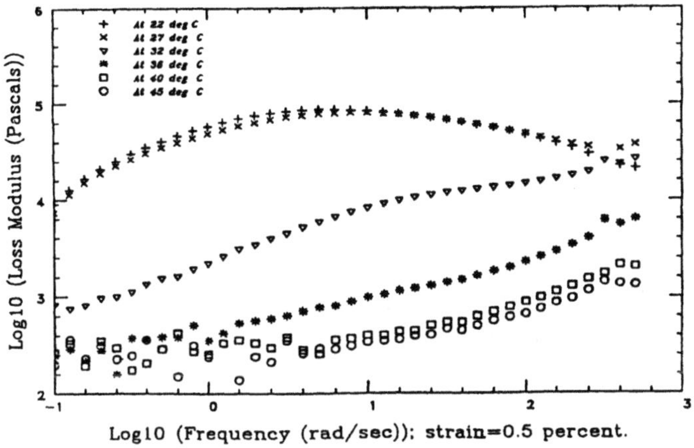

Figure 16. Frequency dependence of G" for the same sample as in Fig. 14 at various temperatures. The scaling of G" with ω at the melting transition yields an exponent of ca. 0.55.

2.2.5. *Ringing Behaviour*

As noted earlier many surfactant cubic phases exhibit the property of "ringing" (i.e. they ring like a bell when tapped gently). It might be reasonable to expect the ringing behaviour to be closely linked with the rheological properties of the system since both have an essentially mechanical origin. Ringing implies a resonator with high quality factor

$$Q = \frac{G'(\omega)}{G''(\omega)} \tag{18}$$

For ringing to occur we require $Q >> 1$ (see Figure 17). The ringing frequency is determined by the elastic modulus of the cubic phase and the characteristic sample size L and density ρ

$$\omega_n = k_n \left[\frac{G'(\omega)}{\rho L^2} \right]^{1/2} \tag{19}$$

where k_n are the ratios of modal frequencies to the fundamental frequency, determined by the geometry of the resonating cavity. We note straight away that compressional modes can be ruled out since they would give rise to frequencies that are far too high. The shear modes of the cubic phase can couple to compressional sound waves in air via the free surface of the samples.

As a trial function for $G'(\omega)$ we may take the Maxwell model with a single relaxation time τ.

$$G'(\omega)=\frac{G_o\omega^2\tau^2}{1+\omega^2\tau^2} \tag{20}$$

which leads to

$$\omega_n^2=\frac{k_n^2G_o}{\rho L^2}-\frac{1}{\tau^2} \tag{21}$$

where we impose the condition $G'(\omega_r) >> G''(\omega_r)$.
This has a solution only provided that

$$\frac{k_n^2G_o}{\rho L^2} >> \frac{1}{\tau^2} \tag{22}$$

which gives a lower cut-off to the observable ringing modes. Thus

$$\omega_r^2 \approx \frac{k_n^2G_o}{\rho L^2} \tag{23}$$

In fact, for shear modes with $k_n = 1$ we obtain $\omega_r \approx 1kHz$ in good accord with experiment.

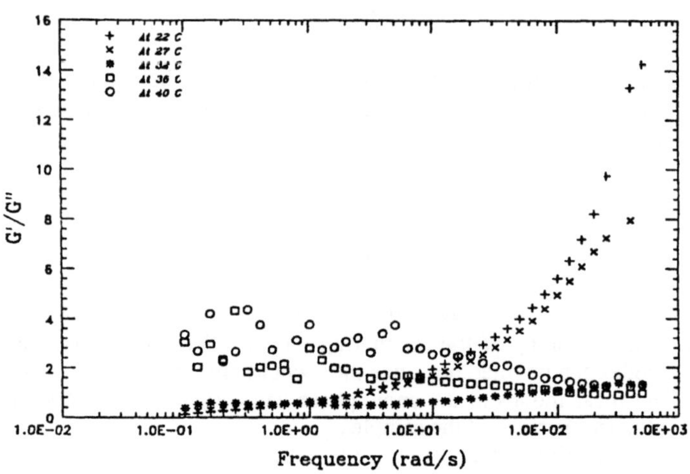

Figure 17. The quality factor, Q=G'/G" shown as a function of frequency at various temperatures. The sample is the same as that of Fig. 14.

3. Concluding Remarks

In recent years there has been a growing awareness of the ubiquity of bicontinuous cubic structures in condensed phases and the importance of minimal surfaces in describing them. Experimentally, the existence of cubic phases based on TPMS is firmly established; we now need to add the principal TPMS to the list of important crystallographic structures. Many hitherto unknown TPMS are awaiting mathematical discovery. The experimental challenge is to discover new structures possibly based on more exotic minimal surfaces. Physical theories too need to be refined so we may understand better the subtleties of phase transitions in these systems. Considerable progress has been made in our understanding of the rheological behavior of cubic phases, particularly through the work of Radiman, and of Jones and McLeish which has highlighted the important role played by the dynamics of the periodic surfactant layer. The way is open for more detailed calculations based on identification of the eigenmodes of specific TPMS structures which should allow full characterization of the relaxation of the network. The rich dynamical behavior promised by cubic phases is likely to attract new experimental studies, and it already appears that in addition to rheology, dynamic light scattering may develop into a useful tool in elucidating the dynamics of surfactant cubic phases.

4. References and Notes

1. Luzzati, V., Mariani, P., and Gulik-Krzywicki, T. (1987) *Physics of Amphiphilic Layers*; Springer-Verlag: Les Houches, France, 131.
2. Luzzati, V., and Spegt, P. A. (1967), *Nature* **215**, 701.
3. Luzzati, V., Gulik-Krzywicki, T., and Tardieu, A. (1968), *Nature* **218**, 1031.
4. Scriven, L. E. (1976), *Nature* **263**, 123.
5. Anderson, D. M. (1986) Ph.D. Thesis, Minnesota.
6. Charvolin, J., and Sadoc, J. F. (1987) *Physics of Amphiphilic Layers*; Springer- Verlag: Les Houches, France, 126.
7. Anderson, D. M., Gruner, S. M., and Leibler, S. (1988), *Proc. Natl. Acad. Sci.* **85**, 5364.
8. Hyde, S. T. (1989), *J. Phys. Chem.* **93**, 1458.
9. Anderson, D. M., Davis, H. T., Scriven, L. E., and Nitsche, J. C. C. (1990), *Adv. Chem. Phys.* **77**, 337.
10. Wang, Z.-G., and Safran, S. A. (1990), *Europhys. Lett.* **11**, 425.
11. Barois, P., Hyde, S. T., Ninham, B., and Dowling, T. (1990), *Langmuir* **6**, 1136-1140.
12. Radiman, S., Toprakcioglu, C., and Faruqi, A. R. (1990), *J. Phys. (Paris)* **51**, 1501.
13. Radiman, S., Toprakcioglu, C., Dai, L., Faruqi, A. R., Hjelm, R. P., Jr., and de Vallera, A. (1990), *J. Phys. (Paris)* **51**, *Colloque de Physique* **C7**, 375.
14. Maddaford, P.J., and Toprakcioglu, C. (1993), *Langmuir* **9**, 2868.
15. Strom, P., and Anderson, D. M. (1992), *Langmuir* **8**, 691.
16. Raedler, J.O., Radiman, S., de Vallera, A., and Toprakcioglu, C. (1989), *Physica B* **156**, 398.
17. Radiman, S., Toprakcioglu, C., and McLeish, T. (1994), *Langmuir* **10**, 61.
18. Jones, J.L. and McLeish, T.C.B. (1995), *Langmuir* **11**, 785.
19. Strictly speaking, the Gauss-Bonnet theorem applies to closed surfaces, but it is possible to form a closed surface from a unit cell of the interface by identifying surface elements linked by lattice translation vectors (see Ref. 8).
20. The genus of a compact surface S is the number of handles required to add to a sphere to produce a surface that is homeomorphic to S.
21. Hilbert, D., Cohn-Vossen, S. (1952) *Geometry and the Imagination*, Chelsea Publishing Co., New York.
22. Andersson, S. and Hyde, S.T. (1984), *Z. für Kristallogr.* **168**, 1.
23. Coxeter, H.S.M. (1969) *Introduction to Geometry*, John Wiley & Sons, New York.
24. Mackay, A. (1990), *J. Phys. (Paris)* **51**, *Colloque de Physique* **C7**, 399.

25. Weierstrass, K. (1864) *Untersuchungen über die Flachen, deren mittlere Krümmung überall gleich Null ist,* Monatsber. d. Berliner Akad., p. 612.
26. Terrones, H. (1990), *J. Phys. (Paris)* **51,** *Colloque de Physique* **C7,** 345.
27. Nitsche, J.C.C. (1989) *Lectures on minimal Surfaces,* Cambridge University Press, Vol. 1.
28. Kreyszig, E. (1959) *Differential Geometry*, University of Toronto Press.
29. Bonnet, O. (1853) *Note sur la theorie génerale des surfaces* , C.R. Acad. Sci. Paris, 529.
30. Lidin, S. (1990), *J. Phys. (Paris)* **51,** *Colloque de Physique* **C7,** 237.
31. Plateau, J.A.F. (1873) *Traité de Statique Expérimentale et Théorique des Liquides soumis aux seules Forces Moléculaires,* Gauthier-Villars, Trubner et cie, F. Clemm.
32. Schwarz, H.A. (1890) *Gesammelte Mathematische Abhandlungen*, Springer, Berlin.
33. Schoen, A. H. (1970), *NASA Technical Note* **TN D-5541** ; for recent progress on crystallographic aspects of minimal surfaces see also Fisher W. and Coch E. (1990), *J. Phys. (Paris)* **51,** *Colloque de Physique* **C7,** 131.
34. Kenmotsu, K. (1979), *Math. Ann.* **245**, 89.
35. Karcher, H. (1989), *Manuser. Math.* **64,** 291.
36. Gackstatter, F. (1990), *J. Phys. (Paris)* **51,** *Colloque de Physique* **C7,** 163.
37. Helfrich, W. (1973), *Z. Naturforsch.* **28C**, 693.
38. Safran, S.A. and Turkevich, L.A. (1983), *Phys. Rev. Lett.* **50,** 1930.
39. Safran, S.A., Roux, D., Cates, M., Andelman D. (1986), *Phys. Rev. Lett.* **57,** 491.
40. Wang, Z.-G., and Safran, S. A. (1990), *Europhys. Lett.* **11,** 425.
41. Turner, D. C., Wang, Z., Gruner, S. M., Mannock, D. A., and McElaney, R. N. (1992), *J. Phys. II* **2,** 2039.
42. Fontell, K., Ceglie, A., Lindman, B., and Ninham, B. (1986), *Acta Chem. Scand.* **A40**, 247.
43. Fontell, K., and Jansson, M. (1988), *Prog. Colloid Polym. Sci.* **76,** 169.
44. Fontell, K. (1974) Liquid Crystals and Plastic Crystals, in G.W. Gray and P. Winsor (eds.), Ellis Harwood, Chichester, Vol. 2, p.80.
45. Duesing, P. and Toprakcioglu, C., unpublished results.
46. Helfrich, W. and Rennschuh, H. (1990), *J. Phys. (Paris)* **51,** *Colloque de Physique* **C7,** 189.
47. It should be clear that in the discussion that follows ω refers to the frequency, and has nothing to do with the complex variable ω which appeared in the Weierstrass representation of minimal surfaces (see eqns. 7-9).
48. Maddaford, P. and Toprakcioglu, C. , manuscript in preparation.
49. Jones, J.L. and McLeish, T.C.B., manuscript in preparation.
50. The reader can benefit greatly from J. Phys. (Paris) **51,** Colloque de Physique **C7** (1990), which contains a wealth of information on minimal surfaces and the geometry of interfaces in general.

OF MICELLES AND MANY-LAYERED VESICLES

M. E. CATES
Department of Physics and Astronomy
University of Edinburgh
JCMB, Kings Buildings
Mayfield Road, Edinburgh EH9 3JZ, UK

Abstract. The major part of these lectures describe work done during the past decade or so on the flow behaviour of entangled micellar solutions. These can be described as equilibrium polymers, whose chains are subject to reversible scission and recombination "reactions". This leads to strong modifications of the reptation picture (which is the established description for conventional polymers in the entangled regime). The modified picture is sometimes called the "reptation-reaction" model. For rapid scission kinetics, the linear viscoelastic spectrum approaches a single-exponention (Maxwell) behaviour: small departures from this can be measured, and the model used to deduce information both on the micellar kinetics (lifetime of a typical micelle before breaking) and on the structure (the mean micelle length). These ideas work for several systems, but for others, unreasonable trends for such quantities are found. The most likely reason for this is micellar branching effects, which (as far as the reptation-reaction model is concerned) introduce an effective micellar length equal to the mean distance between branch points. Another possible discrepancy comes from the breakdown of mean-field averaging for the micellar reactions. The reptation-reaction model yields a nonlinear constitutive equation which shows a non-monotonic dependence of stress on strain rate, in simple steady shear. This leads one to expect flow instabilities, and (with further assumptions) suggests that a steady shear-banded flows should arise, in which macroscopic layers of fluid of different shear rates coexist. Several experimental observations support this general picture, although the instability could instead lead to wall-slip (in which the fast-flowing layers are microscopic), or unsteady flows. Some recent ideas on banding are reviewed. In the last part of the lectures, I describe some recent results on the rheology of a quite different system: that of close packed, monodisperse, many-layered vesicles

T McLeish (ed), Theoretical Challenges in the Dynamics of Complex Fluids, 257-283

made of surfactant bilayer. Unlike the case of micelles, such systems are metastable and this leads to serious difficulties with the description. Indeed, even the simplest quantity (the shear modulus in linear elasticity) appears to be influenced directly by quenched disorder effects.

1. INTRODUCTION

Surfactant molecules, in aqueous and other solvents, have a strong tendency to aggregate reversibly into extended structures. Depending on the local molecular geometry, the basic packing unit can range from a small sphere (conventional micelles), through rods, to very long, sometimes flexible cylinders (giant micelles, hexagonal phases), and on to bilayers (smectic phases, and the so called "sponge phase"). For a recent review, see [1]. The parameters controlling this sequence mainly involve the relative sizes of the polar head group and the hydrophobic tail [2]. In what follows, we take the preferred packing geometry as given, and discuss the flow properties of some relatively well-defined states of the self-assembled system. The main part of the lectures addresses the case of entangled giant micelles which can be viewed as reversibly breakable polymer chains. Like most isotropic phases of surfactants, these are in full thermodynamic equilibrium and their rheology (both linear and nonlinear) can be mainly understood in terms of the statistical mechanics of systems at or near equilibrium. In the last sections, however, we turn to another system of equal or greater technological importance: the "smectic dispersion" or "onion texture" which, as described in the lectures of Roux (this volume), arises when a smectic surfactant solution is gently sheared for a prolonged period. The resulting onions, although long-lived, are metastable and conventional equilibrium approaches are not necessarily applicable. Similar remarks apply to some closely related system, such as dense emulsions and foams (see the lectures of Lequeux, this volume) and some of these analogies are discussed.

2. ENTANGLED MICELLES

These can be viewed as one of a larger group of systems: "equilibrium polymers" (sometimes called "living polymers") which are reversibly polymerizing chain-like objects that arise in equilibrium through self-assembly of smaller units. In the case of giant micelles, the diameter of our "polymer" is of order 3 nm and a typical persistence length 15 nm. The contour length L is variable, governed by thermodynamic equilibrium, and in some systems extremely large (up to about 1 mm). For reviews, see [3, 4, 5, 6] which contain many important references omitted below.

2.1. SIZE DISTRIBUTION

Let the number per unit volume of chains of arc-length L be $c(L)$. We choose units so that the volume fraction is

$$\phi = \int_0^\infty L\, c(L)\, dL \tag{1}$$

In such units the gaussian size of a chain (which is a random walk) is given by $L^{1/2}$, with a prefactor (which we ignore) dependent on the local chain geometry (persistence length and diameter). The packing energy of a chain can be written as $AL + E$, where the term linear in L comes from the body of the chain and the constant term E is that required to create two chain ends (i.e., hemispherical end-caps in the case of micelles). This is typically $5 - 25\, kT$. We have in the Flory-Huggins approach the following (Helmholtz) free energy for the system: [7]

$$F = \int_0^\infty c(L)\, [kT \ln c(L) + E + AL] + F_{int}(\phi) \tag{2}$$

Where the final term arises from interactions between micelles. Insofar as this depends only on the total concentration ϕ it has no impact on the size distribution and may be dropped. The condition for equilibrium is then

$$\left(\frac{\delta F}{\delta c(L)}\right)_\phi = 0 \tag{3}$$

or equivalently

$$\frac{\delta}{\delta c(L)}(F - \mu\phi) = 0 \tag{4}$$

where the Lagrange multiplier (chemical potential) μ is fixed by applying Eq.(1).

Performing the minimization gives $c(L) = e^{-(E-\mu' L)/kT}$ with $\mu' = \mu - A$; this may be written using (1) as

$$c(L) \simeq e^{-E/kT}\, e^{-L/\bar{L}} \tag{5}$$

where

$$\bar{L} = \phi^{1/2} e^{E/2kT} \tag{6}$$

The dimensions of this equation may appear strange, but that is because we have made a certain choice for the unit of length, which is comparable to the diameter of a micelle (i.e. a molecular size). Suitable powers of this scale factor can be included to convert this equation (and similar ones below) into a form more directly comparable to experiment. From the above, we

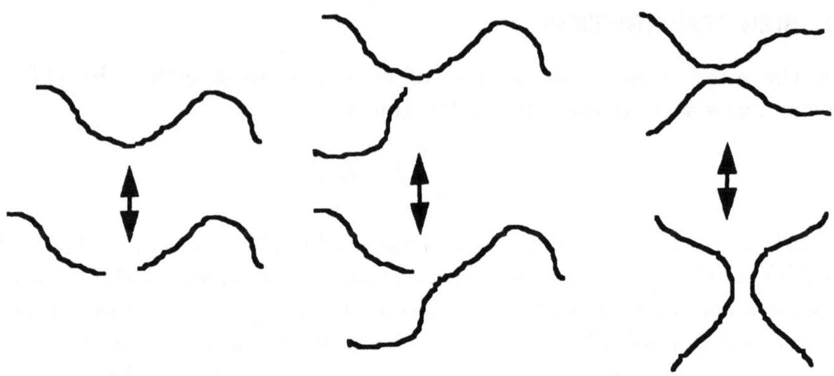

Figure 1. **The three kinetic schemes important for entangled micelles**

have a broad distribution of chain lengths with a mean that increases slowly with ϕ, and rapidly with E. These results are independent of the form of the mean-field interaction term $F_{int}(\phi)$. Different results are obtained if excluded-volume correlations are included, but in fact the modifications are quite weak and we do not discuss them further [8].

3. LIVING POLYMER KINETICS

In a system of equilibrium polymers, the chains can break and recombine reversibly. Various mechanisms are possible which we now consider, focussing attention on micellar systems.

3.1. KINETICS OF MICELLAR REACTIONS

Mechanisms for exchange of material include the following (Fig.1):

(i) Reversible scission: a chain breaks randomly anywhere along its length. The reverse reaction is end-to-end fusion.

(ii) End interchange: the end of one chain attacks the central part of another. The reverse reaction is the same process.

(iii) Bond interchange: two chains swap a central bond via a four-armed intermediate. The reverse reaction is the same process.

The kinetics can be probed experimentally by the Temperature Jump method [9, 10]. In equilibrium we have

$$c(L) = \frac{\phi}{\bar{L}^2} e^{-L/\bar{L}} \tag{7}$$

where $\bar{L} \sim e^{E/2kT}$ is the mean micelle length. If the temperature is suddenly changed (e.g. by a capacitor discharge), $\bar{L}$ is altered and $c(L)$ must relax to the new distribution. The relaxation is probed by light scattering which measures some functional $f[c(L)]$ of the size distibution.

Expanding perturbatively for a small jump ϵ gives the initial perturbation

$$\Delta c(L, t=0) = \epsilon\,(L - \bar{L}) e^{-L/\bar{L}} \tag{8}$$

To calculate the time evolution one must substitute this into suitable kinetic equations describing the mechanisms (i–iii) above. For example in reversible scission one has the equation of motion for the size distribution

$$\begin{aligned} \dot{c}(L) &= -k_1 L c(L) + 2k_1 \int_L^\infty c(L')\,dL' \\ &\quad + \frac{1}{2} k_2 \int_0^\infty c(L')c(L-L')\,dL' - k_2 \int_0^\infty c(L')\,dL' \end{aligned}$$

The first term represents scission of chains of length L, the second scission of chains of length $L' > L$ (to give a chain of length L), the third, nonlinear, term is the combination of two smaller chains to make one of length L, and the fourth term is loss of chains of length L by combination with another chain. The rate constants must obey $k_1/k_2 = e^{-E/kT}$; this is found by demanding that the steady state solution corresponds to thermodynamic equilibrium. Linearizing the kinetic equation and solving with the appropriate initial condition gives

$$\Delta c(L, t) = \Delta c(L, 0) \exp\left[-2t/\tau_b\right] \tag{9}$$

where $\tau_b = 1/k_1\bar{L}$ is the waiting time for a break on a chain of the mean length. The T-jump perturbation relaxes monoexponentially because it is an eigenfunction of the kinetic equation [10]. This behaviour has been observed unambiguously in experiment for chains near the overlap threshold (the signal becomes very small in the true semidilute regime). In fact, the T-jump perturbation is a *nonlinear* eigenfunction so the response can be found even for a large amplitude jump [11].

End-interchange and bond-interchange have similar kinetic equations. However, in both cases, the T-jump corresponds to a "zero mode": the perturbation $\Delta c(L, 0)$ does not decay in time [12, 13]. The presence of a zero mode is linked in each case to the fact that the reaction scheme preserves the total number of chains in the system. However, it is unclear why the T-jump experiment should coincide exactly with the relevant mode. All other perturbations to the size distribution relax with a longest decay time of order τ_b which we again define as the lifetime of a chain of length $\bar{L}$ before undergoing a reaction. This means that in these systems, the T-jump experiment does not offer useful information about the kinetics of micelles as they might affect flow behaviour (see below). For a predominantly end-interchange system, for example, T-jump might suggest a very long value of τ_b (associated with a low residual rate for reversible scission) whereas the actual breaking time for a typical chain, due to interchange, could be much shorter.

The kinetic scheme written above for reversible scission, and similar ones we have studied for the interchange processes [12, 13], are based on simple mean-field assumptions. In essence, we assume that a given chain-end, formed in (say) a scission reaction, is more likely to recombine in due course with an end of an independent chain, than with the very same end that it split away from in the original scission event. This can break down under certain conditions of concentration and τ_b [14], when ends are "hard to find" on the time scale of τ_b. Then, a strong "renormalization" of the rate constants occurs. We return to this below. For the other reaction schemes, however, since reactions do not involve two ends (but instead an end and a bond, or two bonds) a similar breakdown of the mean-field equations is not expected. For example, with end-interchange, the chances of a newly formed end attacking the bond just created by its formation are very remote since there are so many alternative bonds in the neighbourhood. Put differently, bonds are never "hard to find".

4. POLYMER DYNAMICS

The dynamics of unbreakable chains is described in the lectures of Rubinstein and McLeish (this volume). For entangled systems, the best description is based on the tube model. This describes motion at scales larger than the tube diameter a; at smaller scales, local modes (Rouse/Zimm modes) play a role. The latter are mainly ignored below. Here we give a brief summary of the tube model; in the next section we extend it to cover reversibly breakable chains.

4.1. THE TUBE MODEL

To model the entangled state, each chain is imagined confined to a tube of radius $a \sim L_e^{1/2}$ [15]. The tube has $N_T = L/L_e$ "tube segments", arranged in a random walk, each housing $\Delta s \simeq L_e$ monomers. The chain can diffuse only along the axis of the tube; as it emerges from the end of the tube, new tube is created by "reptation". There is a curvilinear diffusion constant $D_c(L) \sim 1/L$, this proportionality arising because the frictional drag on a chain is proportional to its length. The chain conformation is fully relaxed on the timescale for complete escape from the tube. This requires curvilinear diffusion over a length L, hence the relaxation time scales as $\tau_{rep} \sim L^3$. The prefactor can be estimated by demanding continuity with the "Rouse time", which is the orientational relaxation time for chains of length L_e and below (See the lectures of Rubinstein). This gives $\tau_{rep}(L_e) = \tau_{rouse}(L_e) \simeq \tau_0 L_e^2$ which requires finally the following scaling of the reptation time

$$\tau_{rep} \simeq \tau_{rouse} N_T \simeq \tau_0 L^3 / L_e \tag{10}$$

Typical reptation times for long, unbreakable chains are 0.1 s – 1 hour.

4.2. THE STRESS FORMULA

We consider a set of chains in a volume V and coarse-grain into subunits i of length Δs, each with an end-to-end vector $\Delta \mathbf{r}_i$. Each contributes $(p_i)_\beta (f_i)_\alpha$ to the elastic stress tensor $\sigma_{\alpha\beta}$ in our material, where α, β are cartesian indices. The quantity $(p)_\beta = \Delta r_\beta / V$ is the probability per unit area that a subunit crosses a given plane perpendicular to β, whereas $f_\alpha = kT \Delta r_\alpha / \Delta s$ is the thermodynamic force (spring tension) in the subunit. Hence the total contribution is

$$\sigma_{\alpha\beta}^{pol} = \frac{kT}{V} \sum_i \frac{\Delta r_\beta}{\Delta S} \frac{\Delta r_\alpha}{\Delta s} \Delta s \tag{11}$$

or, in the continuum limit,

$$\sigma_{\alpha\beta}^{pol} = C\,kT \int_0^L \langle \frac{\partial r_\beta}{\partial s} \frac{\partial r_\alpha}{\partial s} \rangle \; ds \tag{12}$$

where C is the number of chains per unit volume. For our purposes, the discrete formula is somewhat better, since there is a natural choice of Δs, namely L_e. That is, we can calcuate the stress for a discrete ensemble of elements consisiting of the tube segments in the system.

In practice, the stress tensor also has contributions from the isotropic pressure P and from velocity gradients in the solvent (often negligible); we write

$$\sigma_{\alpha\beta} = \sigma_{\alpha\beta}^{pol} + \sigma_{\alpha\beta}^{sol} + P\delta_{\alpha\beta} \tag{13}$$

The polymer contribution has 6 independent components of which one combination (the trace) is irrelevant since it can be absorbed into the pressure (itself a Lagrange multiplier to ensure fluid incompressibility). The remaining 5 components are the shear stresses, $\sigma_{xy}, \sigma_{yz}, \sigma_{xz}$ and the two normal stress differences $N_1 = \sigma_{xx} - \sigma_{yy}$ and $N_2 = \sigma_{xx} - \sigma_{zz}$.

4.3. LINEAR VISCOELASTICITY

Linear viscoelastic theory describes the response of a material to small deformations. Consider, for example, an imposed strain, $x \to x + \gamma y$, with γ the strain angle (assumed small) suddenly imposed at time zero and then held constant. The response is

$$\sigma_{xy}(t) = \gamma G(t) \tag{14}$$

where we can write $G(t) = G_0\mu(t)$ as the product of an instantaneous shear modulus (the plateau modulus) and a memory function $\mu(t)$. In a polymeric system, the stress will not fully decay until chains have had time to relax their contribution to the stress tensor by adapting their configurations to the strained shape of the sample. A related experiment is oscillatory shear, $\gamma = \gamma_0 e^{i\omega t}$, for which the stress is

$$\sigma_{xy}(t) = G^*(\omega)\gamma \tag{15}$$

with $G^*(\omega) = G'(\omega) + iG''(\omega)$ where G', G'' are called the storage modulus and the loss modulus. Linear response theory shows that

$$G^*(\omega) = i\omega \int_0^\infty G(t) e^{-i\omega t}\, dt \tag{16}$$

An example is the Maxwell fluid, $\mu(t) = e^{-t/\tau}$ which has

$$G'(\omega) = \frac{G_0\omega^2\tau^2}{1+\omega^2\tau^2} \qquad G''(\omega) = \frac{G_0\omega\tau}{1+\omega^2\tau^2} \tag{17}$$

When plotted in the "Cole-Cole" representation (a plot of G'' vs. G' with frequency parametrically eliminated) this gives a perfect semicircle.

4.4. STRESS DECAY IN THE REPTATION MODEL

The stress formula for gaussian chains can be written

$$\sigma_{\alpha\beta} = \frac{kT}{V} \sum \langle \frac{\Delta r_\beta}{\Delta s} \frac{\Delta r_\alpha}{\Delta s} \rangle \Delta s \tag{18}$$

where the sum is over tube segments, or

$$\sigma_{\alpha\beta} = C\, kT\, N_T \, \langle u_\alpha u_\beta \rangle \tag{19}$$

where $\mathbf{u}$ is a unit tangent to the tube. In an undeformed state, $\langle u_\alpha u_\beta \rangle_0 = \delta_{\alpha\beta}/3$ which contributes an irrelevant pressure term.

If a step strain γ is now applied, we have the instantaneous deformation $\mathbf{u} \to \mathbf{u} + \delta\mathbf{u}$ where $\delta\mathbf{u} = (\gamma u_y, 0, 0)$. Calculating

$$\langle u_\alpha u_\beta \rangle = \langle u_\alpha u_\beta \rangle_0 + \langle \delta u_\alpha u_\beta + u_\alpha \delta u_\beta \rangle \tag{20}$$

gives $\delta_{\alpha\beta}/3 + 4\gamma/15$. Hence

$$\sigma_{xy}(t = 0^+) = G_0 \gamma \tag{21}$$

where $G_0 = (4/15)\, C\, N_T$ is the plateau modulus. Since this can be measured, this serves as an operational definition of N_T and hence of the entanglement length.

As time proceeds, each chain escapes by curvilinear diffusion from its tube. New tube is created as the chain moves, but this is in isotropic equilibrium in the strained environment. Hence $\mu(t)$ is the fraction of the $t = 0$ tube through which neither chain end has passed, up to to time t (see the lectures of McLeish, this volume). This is found by imagining the tube diffusing relative to a fixed chain, and studying the survival probability of a labelled tube segment. This "particle" diffuses at a rate D_c and is killed when it meets either end of the chain on which it resides. A Fourier analysis gives

$$\mu(t) = \frac{8}{\pi^2} \sum_{p=odd} \frac{1}{p^2}\, e^{-tp^2/\tau_1} \tag{22}$$

where $\tau_1 \simeq \tau_{rep}$. This is a fairly monoexponential relaxation (most of the spectral weight is in the lowest mode). The theory accounts quite well for $G^*(\omega)$ although departures are seen at high frequencies. These have two origins:

(i) Breathing modes: one-dimensional Rouse motion within the tube causes the chain end positions to fluctuate through $\sqrt{N_T}$ tube segments on a time scale $\tau_{rouse} \ll \tau_{rep}$. This contributes to the high frequency relaxation spectrum;

(ii) Rouse modes within the tube: at frequencies larger than $\tau_{rouse}(L_e)^{-1}$ (the inverse Rouse time of a tube segment), the Rouse modes of unentangled subsections of the chain become important. These lead to a power law regime $G^*(\omega) \sim (i\omega)^{1/2}$.

Finally, if one has a distribution of chain lengths, the relaxation spectrum can be estimated (at the crudest level) by taking the weighted average

$$\bar{\mu}(t) \simeq \frac{\int Lc(L)\, \mu_L(t)\, dL}{\int Lc(L)\, dL} \tag{23}$$

For living polymer systems, with an exponential size distribution, this is an extremely nonexponential decay. A variety of more sophisticated estimates (along the lines discussed for polymer blends in Rubinstein's lectures, this volume) lead to the same basic conclusion.

5. LINEAR VISCOELASTICITY OF LIVING POLYMERS

On instantaneous deformation, living polymers behave just as "dead" ones do, and the plateau modulus is $G_0 = (4/15)\phi/L_e$ as before. (Note that in our units, ϕ is the total length of chain per unit volume so ϕ/L_e is the equivalent of CN_T for monodisperse chains.)

To describe the subsequent relaxation, we now have two time scales: τ_b and τ_{rep} characterising micellar reactions and diffusion respectively. When $\tau_b \gg \tau_{rep}$ there is no direct effect of breaking reactions and a formula like (23) can be used. This gives a strongly nonexponential relaxation, $\mu(t) \sim \exp[-(t/\tau_{rep}(\bar{L}))^{1/4}]$. This sort of decay has been reported in certain systems [16]. More commonly, however, entangled micelles approach the ideal limit of the Maxwell fluid: $\mu(t) \sim e^{-t/\tau}$. This corresponds to the fast breaking limit, $\tau_b \ll \tau_{rep}$.

In this limit, the relaxation function $\mu(t)$ can be calculated numerically from a one-dimensional stochastic model in which a randomly chosen tube segment diffuses relative to a chain. As in the unbreakable chain calculation, the particle is absorbed upon reaching a chain end and $\mu(t)$ is its survival probability. The chain length is now a function of time, however, and the chain ends make discrete jumps corresponding to reactions in which parts of the chain are removed and other parts added. The rate constants for these processes can be determined from the kinetic equations according to any of the three schemes discussed above.

For $\tau_b \ll \tau_{rep}$, a monoexponential relaxation is indeed approached with a relaxation time τ that varies as $(\tau_b\tau_{rep})^{1/2}$ for reversible scission and end interchange, but as $\tau_b^{1/3}\tau_{rep}^{2/3}$ for bond-interchange [17, 12, 13]. The laws for scission and end-interchange can be explained by estimating the characteristic distance along the tube that a chain can break for the newly created end to have an even chance of passing a given tube segment before being lost in a later reaction event. A new end is created by a "reaction" with an existing end; this happens somewhere along a chain with rate τ_b^{-1}. The new end diffuses along the tube but only covers a distance λ obeying $D_c\tau_b = \lambda^2$ before recombining with another end; the process relaxes a portion $\lambda/\bar{L}$ (on average) of the original tube. The rate of stress relaxation for any given tube segment is thus $\tau^{-1} \sim \tau_b^{-1}\lambda/\bar{L}$. But since $D_c\tau_{rep} \simeq \bar{L}^2$, this can be rewritten as $\tau \simeq (\tau_b\tau_{rep})^{1/2}$. (The mechanism for bond interchange is somewhat different since no ends are involved.) The physical origin of monoexponential behaviour is a separation of timescales: the stress relaxation time τ remains large compared to τ_b when $\tau_b \ll \tau_{rep}$. Many reactions occur on a given chain before a typical tube segment is relaxed; this gives a fast averaging effect on the timescale τ_b within which the various tube segments become mixed up with one another. Correspondingly, each sees an average decay and all tube segments relax with the same rate.

Corrections to the monoexponential behaviour can be computed numerically and yield direct information on the parameter $\bar{\zeta} \equiv \tau_b/\tau$. For values less than 0.1 the Cole-Cole plot is semicircular, with strong departures for values of order unity. Experimentally, measuring τ and the shape of the $G^*(\omega)$ spectrum allows τ_b to be estimated. For example in aqueous CTAB / KBr (0.3 M /0.25 M), τ = 95ms and the spectrum suggests $\bar{\zeta} = 1.2$. Thus τ_b is estimated as 115 ms, in good agreement with an independent estimate $\tau_b = 110 \pm 40$ ms from T-jump [9]. This shows that CTAB undergoes reversible scission reactions. For end- or bond-interchange, the same procedure gives the only possible estimate for τ_b since then the T-jump measurement does not couple to the kinetics, as discussed above.

Further corrections to the Maxwell spectrum arise at high frequencies. These can be analysed in terms of breathing and Rouse modes. The analysis of Granek [18] predicts a minimum value of the loss modulus, obeying (in the crudest approach) $G''_{min}/G_0 \simeq L_e/\bar{L}$. This can be used to get an absolute estimate of the chain length $\bar{L}$ from viscoelastic data. (No other method is known; for unbreakable chains the procedure is to dilute the system and perform light scattering, but for living polymers this alters the value of $\bar{L}$ that one is trying to measure.) The value of $\bar{L}$ found this way should show an increasing trend with concentration.

This trend is seen, for example in the system CTAC / NaSal / NaCl [19] with a fixed CTAC/NaSal ratio. At 0.1 M NaCl, $\bar{L}$ is in the range $0.5 - 1\mu$m and increasing with ϕ. However, at higher salinity (NaCl = 0.25 M) a decreasing trend is seen instead. It is expected that salt, which supresses end-caps, should also favour crosslinks and this data has been taken of evidence of network formation (see below). Various data on other systems suggest that this scenario is fairly common in systems at high salt (see e.g. [5]).

5.1. NETWORK DYNAMICS

The impact of reversible crosslinks on the reptation-reaction model has been examined by Lequeux [20]. He considered unsaturated networks (with more entanglements than crosslinks) for which a section of chain between crosslinks can still be considered as confined to a tube. Relaxation again requires the section of chain to break, at a point near enough to a given tube segment that the new end can pass through it before reacting with something else. Lequeux showed that the main effect of cross-linking was through the curvilinear diffusion constant D_c for a chain end. Remarkably, this is enhanced, rather than reduced, by the presence of reversible links:

$$D_c^{eff} \sim 1/\bar{L}_s \tag{24}$$

where $\bar{L}_s \leq L$ is the mean strand length between crosslinks. The enhancement can be understood in terms of "kinks" or stored length defects. (These kinks are nothing other than the "reptons" described in detail in Rubinstein's lectures, this volume.) In a linear chain, a kink must travel from one end of the chain to another before a chain displacement is established. In a network, it has to travel only a distance of order that to the nearest junction. The remainder of the network acts as a stored-length "reservoir" and absorbs the kink. To incorporate Lequeux's result, we need only replace $\bar{L} \rightarrow \bar{L}_s$ in our previous discussion. The latter is likely to be a decreasing function of ϕ at fixed salt, which would explain, in broad terms, the anomalous data on CTAC at high salinity mentioned above.

Obviously, this mechanism requires that crosslinks be freely mobile in the sense that a section of stored length (repton or kink) can pass right through a crosslink without dissipation. Since the surface of a micelle is fluid, this is a perfectly reasonable assumption though conceivably it might not always be the case. The fact that crosslinks *decrease* the viscoelastic relaxation time τ allows for an interesting speculation on the flow of "saturated" networks [21], which are networks in which the density of reversible cross-links is so large that no entanglements, in the proper sense, remain. (In other words, anything that might have been an entanglement is now a reversible link.) Here there is no tube, so the quantitative discussion does not apply; nonetheless one may argue that the saturated network is a freely flowing structure which will be nearly inviscid [22]. This may explain the anomalously newtonian rheology of the $CPyClO_3$ / $NaClO_3$ system.

Interesting questions arise if one looks into the dynamics of crosslinked networks in more detail: for example there is some time scale for equilibration of the *number* of crosslinks which may be separate from that for stress relaxation. To see this, note that to abolish a cross-link a chain end has to migrate all the way to its nearest branch-point, which may be unlikely on the time scale $\tau_b(\bar{L}_s)$ (i.e. this process may define a longer time scale, for relaxation of network topology). The kinetics of the reverse process whereby a new crosslink is formed by budding a free end out of the side of a chain is also interesting to consider: within the model just described, the principal of detailed balance dictates that this is how crosslinks are formed! Alternatively, there could be a separate dynamics in which free ends attack bonds to form three-fold junctions (which are the intermediates of an end-interchange reaction scheme for unbranched micelles) that could remain stable; conversely then network strands would preferentially break adjacent to a crosslink. (A bond-bond version involving four-fold junctions is also possible.) The details of these processes do not appear to affect the linear viscoelastic spectrum in any significant way; there could perhaps be effects on nonlinear flow behaviour but so far no large effect has yet been predicted [23].

5.2. BREAKDOWN OF MEAN-FIELD THEORY

The recent work of O'Shaughnessy and Yu [14] on correlation corrections to the reaction kinetics shows that the breakdown of mean field theory can lead to significant alterations to the basic picture.

These authors showed that correlated reactions occur, in reversible-scission systems, when τ_b (as calculated within mean-field theory) is less than or comparable to τ_h, the mean time taken to diffuse a distance h which is the mean separation between chain ends in the system. Since τ_b is a decreasing function of ϕ and τ_h an increasing one (as the chains become more entangled it takes them longer to sample the neighbourhood), one expects correlations to matter at high concentrations and when the intrinsic kinetics is fast. For $\tau_b \leq \tau_h$, ends repeatedly recombine with their previous partner (which to the first approximation prevents them from relaxing any stress) until a fully effective, uncorrelated break occurs on the time scale τ_h. Accordingly, τ_h takes over as the effective breaking time of the system. As mentioned alreadly above, similar corrections are not expected for the other reaction schemes.

This scenario leads to an effective breaking time which increases with concentration as ϕ^y with y an exponent between 1 and 2. Moreover, depending on the details (whether chains are breathing or reptating fully on the scale of the renormalized breaking time τ_h) one can expect anomalous stress relaxation in which the initial decay of $G(t)$ is extremely flat: $G(t) = G_0(1 - t^\alpha)$ with $\alpha = 5/4$ (compared to $\alpha = 1/2, 1/4$ for conventional reptation and breathing). Values of α bigger than unity, as predicted in Ref.[14], lead to an extraordinary "re-entrant" behaviour of the Cole-Cole plot which is not single valued at high frequencies (this corresponds to having a maximum in $G'(\omega)$). This fascinating prediction deserves careful experimental scrutiny.

6. NONLINEAR VISCOELASTICITY

The linear response treatment outlined above is restricted to the limit of small deformations. For general flows, we seek a *constitutive equation* or functional relationship

$$\sigma_{\alpha\beta}(t) = f\left[K_{\alpha\beta}(t' < t)\right] \tag{25}$$

where $K_{\alpha\beta} = \nabla_\alpha u_\beta$ is the velocity gradient or rate of deformation tensor. The deformation of a vector due to flow between times t' and t is

$$r_\alpha(t) = E_{\alpha\beta}(t', t) r_\beta(t') \tag{26}$$

where the tensor $E_{\alpha\beta}(t', t)$ will from now on be written $\mathbf{E}_{t't}$. This obeys

$$\mathbf{E}_{t't} = \exp\left[\int_{t'}^{t} K_{\alpha\beta}(t'')dt''\right] \tag{27}$$

so that $\dot{\mathbf{r}} = \mathbf{K} \cdot \mathbf{r}$.

6.1. CONSTITUTIVE EQUATION FOR LIVING POLYMERS

An accurate constitutive equation for unbreakable polymers, based on the reptation model, is given in Ref.[15]. Here we focus on the living case, which turns out to be somewhat simpler [24]. We assume that the deformation, which may be large on the scale of a chain, is not so large as to cause mechanical rupture of the micelles. This is reasonable since for strains of order unity (large), the chain tension is still of order kT/L_e. This is unlikely to rupture chains whose scission energies are of order 10-20 kT. Hence we assume the reaction rates for the micellar kinetics are unperturbed by flow.

The formula for the polymeric stress, coarse grained onto the scale of a tube segment, is given above as

$$\sigma^{pol}_{\alpha\beta} = \frac{kT}{V} \sum \langle \frac{\Delta r_\beta}{\Delta s} \frac{\Delta r_\alpha}{\Delta s} \rangle \Delta s \tag{28}$$

This applies for time scales large compared to $\tau_{rouse}(\bar{L})$ (so that breathing and local Rouse motions are not involved). It may also be written (cf. Eq.19)

$$\sigma^{pol}_{\alpha\beta} = (15/4)G_0\,(W_{\alpha\beta} - \delta_{\alpha\beta}/3) + P\delta_{\alpha\beta} \tag{29}$$

with P the pressure and

$$\mathbf{W} = \langle \mathbf{uu} \rangle \tag{30}$$

the second moment of the orientational distribution function, $P(\mathbf{u})$ for tube segments.

Consider now the birth and death of tube segments. In the linear region, we have the survival function $\mu(t) = e^{-t/\tau}$ corresponding to a death rate for tube segments $\mathcal{D} = 1/\tau$. By conservation, this is balanced by a birth rate $\mathcal{B} = 1/\tau$. These rates are modified, however, in nonlinear flows. The main effect is called *retraction* [15]. If a random walk is subject to a finite deformation, it increases in length by a factor

$$L'/L = \langle |\mathbf{E} \cdot \mathbf{u}| \rangle_0 \tag{31}$$

where the subscript 0 denotes average over an isotropic distribution, $P_0(\mathbf{u}) = 1/4\pi$. This stretching effect is compensated by a fast breathing-type motion in which the chain shrinks back down its tube to restore the original tube length. This causes segments to die at an additional rate

$$v = \frac{1}{L}\frac{\partial L}{\partial t} = \langle|\dot{\mathbf{u}}|\rangle = \langle \mathbf{u}\cdot\partial\mathbf{u}/\partial t\rangle = \mathbf{W} : \mathbf{K} \tag{32}$$

where $\dot{\mathbf{u}} = \mathbf{K}\cdot\mathbf{u}$ implies the last equality. The death rate by retraction depends on the current second moment $\mathbf{W}$ which in turn depends on the flow history. The rate v is positive in almost all flows [15], and we assume that applies here. For a nonlinear flow, we may approximate $\mathcal{B}$ and $\mathcal{D}$ by

$$\mathcal{B} = 1/\tau \tag{33}$$

$$\mathcal{D} = 1/\tau + v(t) \tag{34}$$

(Note that, were v to be negative, one would instead add a term $|v(t)|$ to $\mathcal{B}$ and omit the one in $\mathcal{D}$. This correction must be made if one wishes to consider, e.g., multiple step strain experiments.)

To get a constitutive equation we must find the stress contribution at the present time (t) due to a segment created earlier, at time t'. Such a segment was created isotropically (with probability distribution $P_0(\mathbf{u})$) but has been deformed by the flow. Its new length is $|\mathbf{E}_{t't}\cdot\mathbf{u}|$ and its unit tangent is

$$\mathbf{u}' = \frac{\mathbf{E}_{t't}\cdot\mathbf{u}}{|\mathbf{E}_{t't}\cdot\mathbf{u}|} \tag{35}$$

The mean contribution to $\mathbf{W}$ from this portion of tube, if it survives, is therefore

$$\hat{\mathbf{Q}}_{t't} = \langle\frac{\mathbf{E}_{t't}\cdot\mathbf{u}\,\mathbf{E}_{t't}\cdot\mathbf{u}}{|\mathbf{E}_{t't}\cdot\mathbf{u}|}\rangle_0 \tag{36}$$

To calculate the stress at time t, we need only compute $\mathbf{W}(t)$ as the sum over contributions from segments born earlier:

$$\mathbf{W}(t) = \int_{-\infty}^{t} \mathcal{B}(t')\,\exp\left[-\int_{t'}^{t}\mathcal{D}(t'')dt''\right]\,\hat{\mathbf{Q}}_{t't}\,dt' \tag{37}$$

where the exponential factor is the survival probability between times t' and t.

Bearing in mind the relation between the (polymeric) stress $\sigma_{\alpha\beta}$ and $W_{\alpha\beta}$ given in Eq.(29), this completes the constitutive equation we seek. As written above there is no approximation on top of the reptation-reaction model itself, except that the expressions for $\mathcal{B}, \mathcal{D}$ are slightly simplified from the exact forms (whose expression is very cumbersome [24]). However, since

$\mathbf{W}$ (via $v(t)$) occurs on both sides of the equation it can be unpleasant to solve in general. In some cases, it may be easier to use the following differential form, whose validity can be checked by explicit differentiation of Eq.(37) [25]:

$$\frac{d\mathbf{W}}{dt} = \mathbf{K} \cdot \mathbf{W} + \mathbf{W} \cdot \mathbf{K} - \mathbf{K} :\langle \mathbf{uuuu} \rangle - (\mathbf{K} : \mathbf{W})\mathbf{W} - \frac{\mathbf{W}}{\tau} + \frac{\mathbf{I}}{3\tau} \quad (38)$$

Though local in time, this form sacrifices autonomy in $\mathbf{W}$: the quartic average involves knowing the whole distribution function $P(\mathbf{u})$ and therefore requires some further decoupling approximation to express this in terms of $\mathbf{W}$ [15].

6.2. SHEAR BANDING SCENARIO

In any case, the full constitutive equation can be solved in simple flows, such as steady shear at strain rate $\dot{\gamma}$ [26]. For the polymeric shear stress $\sigma_{xy}^{pol}(\dot{\gamma})$ one finds a curve which shows a maximum shear stress, $\sigma_{max}^{pol} = 0.67G_0$ at shear rate $\dot{\gamma}_1 = 2.6/\tau$. At higher strain rates, the shear stress is a decreasing function of flow rate which corresponds to an unstable flow. The situation is saved (presumably) by a return to increasing stresses at very high flow rates; such an upturn is inevitable, if only from the contribution $\sigma_{\alpha\beta}^{sol}$ of the (newtonian) solvent which we have been ignoring. We denote by $\dot{\gamma}_2$ the shear rate at which σ_{xy} again equals σ_{max}; for faster flows than this, stable behaviour is recovered. The full stress vs. strain rate curve, therefore, should be of the form shown schematically in Fig.2. The actual behaviour of systems whose flow curve looks like this remains a classical and largely unsolved area of non-newtonian fluid mechanics [27].

Indeed, for $\dot{\gamma}_1 < \dot{\gamma} < \dot{\gamma}_2$, what happens in detail may depends on the geometry of the instrument. Typically a cone-and-plate rheometer is used, for which we might postulate [26, 28] the observed shear stress to obey $\sigma \equiv \sigma_{xy} = \sigma_{max}$ throughout this region. This is in excellent quantitative agreement with measurements of Rehage and Hoffmann on CPyCl / NaSal (100 mM / 60 mM) [16]. The prediction of a constant shear stress arises if one assumes a "shear-banding instability" where the system coexists between the upper and lower shear rates in bands whose relative volume fractions are chosen to match the imposed macroscopic shear rate $\dot{\gamma}$ (see Fig.3). There is, however, an additional assumption involved: that of "top-jumping" whereby one of the participating bands has a polymeric stress σ_{max}.

Although consistent with the data of Ref.[16], the top-jumping scenario is far from obvious and indeed cannot be justified in any rigorous way. On the contrary [29] a recent numerical solution of the constitutive equations in a parallel plate geometry suggests instead that on increasing shear

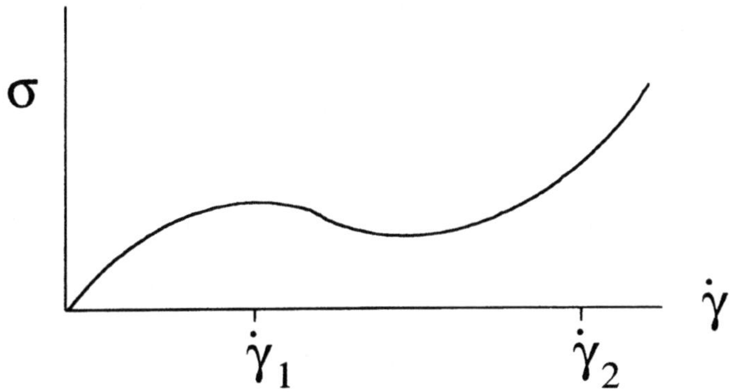

Figure 2. Schematic of the stress vs. flow rate curve for entangled micelles

Figure 3. Schematic of flow profiles (left) with shear bands (right) without. (Imagine the top plate to be moving and the bottom one stationary. Then the local flow velocity at a given height is given by the length of the horizontal line there.)

rate, the stress should rise to the maximum value σ_{max} but then, as the rate is increased further, drop down onto a lower plateau and remain constant. (Similar predictions are obtained for some simplified constitutive models which mimic the flow curve of Fig.2 [29].) Such behaviour has recently been seen in some experiments when controlled strain-rate rather than controlled-stress machinery is used [30]. It is quite possible that the original observation of an extended stress plateau at $\sigma = \sigma_{max}$ is a result of very slow transients, in other words, the data measured along this plateau does not truly correspond to steady state conditions. It would be understandable for the authors of Ref.[16] to miss this because they had no reason to suspect inhomogeneous flow and would therefore assume that a steady state was reached after a few Maxwell times. If the approach to steady state involves the migration of interfaces between bands [29], much longer timescales could be involved. Further careful studies with different rheometric protocols and flow histories will be necessary to validate or disprove the shear-banding scenario as manifested in flow behaviour [31].

It is therefore important to use other probes to establish the actual flow behaviour of materials in the apparent shear-banding regime. One of these is the birefringence microscope [32] which has been used to study an entangled micellar system for which shear-banding is expected (although in this case the high shear branch, $\dot{\gamma}_2$, may correspond to a metastable nematic mesophase). Using this technique fairly clear evidence for shear banding was found, albeit in a Couette geometry (a bright band is observed close to the inner cylinder of the Couette). Actually in the materials used for this study [33], there is a clear evolution of the observed stress plateau, with temperature and/or concentration, from values near the theoretical (top-jumping) prediction, $\sigma_{max} = 0.67G_0$, to substantially smaller values. (The latter could correspond to "bottom-jumping", or to some intermediate behaviour.) It has been argued that this is associated with the nucleation of nematic order in the high-shear branch, though since we do not understand yet the way in which any banded structure is selected, this remains unclear.

Another potentially vital probe is the direct measurement of flow profiles using NMR imaging [30]. So far this has been used mainly in pipe flow geometries, where, theoretically, there are additional hysteretic features associated with the spurt effect [34], because the shear stress is not constant across the width of the pipe. (Hence a flow instability, setting in in the high stress region near the wall, might propagate towards the centre giving a large hysteresis in throughput.) These features additionally complicate the interpretation of the data which at present cannot be said to support the shear-banding scenario in any definite way [30]. However, there is certainly strong evidence of complex and interesting behaviour at high enough stress.

An interesting alternative to shear-banding, partly suggested by the NMR studies, is that of wall-slip, in which very thin layers of highly deformed material lubricate the walls of the rheometer. The difference with the banding scenario is that these layers must, by definition, be directly altered in their properties by the presence of the wall. Accordingly, such slip depends on the type of wall (material, roughness...), and is generally thought of as a nonconstitutive property that is not predictable from the shape of the steady flow curve $\sigma(\dot{\gamma})$ alone. However, when the curve is nonmonotonic, as predicted for entangled micelles, it is possible that the *constitutive* instability ($d\sigma/d\dot{\gamma} < 0$) is the trigger that makes wall slip occur. In other words, once the steady homogeneous flow is unstable ($\dot{\gamma} \geq \dot{\gamma}_1$) a wall slip effect might "come to the rescue" before the intrinsic turnup in the $\sigma(\dot{\gamma})$ curve which arises from the newtonian solvent (or some other) contribution. These issues deserve detailed experimental and theoretical investigation.

In the discussion so far we have tacitly assumed that the turnup in the steady shear curve does indeed arise from the newtonian solvent contribution. In fact, at least for the system studied in [16], this is certainly not true, as evidenced by the fact that an extremely large, and steadily rising, first normal stress difference N_1 is measured throughout the postulated shear-banding regime (where $\sigma(\dot{\gamma})$ is constant). To explain this may require a more detailed model of the turnup at high shear, which may involve the local chain dynamics in a way not captured by the tube description. Some models along these lines have been proposed, in which the large values of N_1 are ascribed to a stretching of chains along the streamlines, due to terms left out of the simple reptation description. (At the level of description so far used, there is no discrimination between breakable and unbreakable chains.) However, although the simplest of these models seems to give a reasonable account of the normal stress data [26, 28], efforts to systematically improve it have tended to undermine these earlier results [35].

On reflection, perhaps, the extreme values of N_1 observed (several orders of magnitude above G_0 [16]) are more likely to reflect a local elasticity of fully-aligned micelles (perhaps in a hexagonal or nematic phase) than the entropic elasticity to which $\sigma^{pol}_{\alpha\beta}$ has above been entirely attributed. If so, this may be part of the explanation for why entangled micelles behave differently from high molecular weight entangled polymers, for which a very similar flow curve is, in principal, predicted [15, 34]. The existence of a single Maxwell mode in micelles makes them a stronger candidate for the shear-banding scenario, since the inflexion of the flow curve is easily removed by having a dispersion of relaxation times (via polydispersity in polymers), but this is not really enough to explain the fact that micelles generally show (apparent or real) plateaux in shear stress, strongly indicative of shear banding, whereas highly monodisperse, conventional polymers do not. An explanation may lie in improved modelling of the entanglement dynamics

[36], but may also involve a much clearer understanding of what exactly is happening to the micellar structures in the highly sheared parts of the sample.

7. LINEAR RHEOLOGY OF ONION TEXTURES

As shown by the above discussion, much remains to be understood in the rheology of entangled micellar systems, but a basic framework for comprehending the flow behaviour has at least been established and future progress will probably not lead to its demolition. A major aid, throughout these developments, has been that these systems are, at rest, in full thermodynamic equilibrium. We now turn away from this topic to a harder one: the rheology of onion textures. As described in detail in Roux's lectures (this volume), by shearing a smectic liquid crystal phase of surfactant for a prolonged period, one can make a structure consisting of close-packed, multilayer vesicles (onions) within which the local structural unit is a bilayer [37]. The resulting system is very similar to a dense emulsion or foam (Fig.4), containing polyhedral onion entities in close contact and strong deformation. (These become spherical if the system is diluted, a regime which we do not address here though initial theoretical studies are described by Lu [38].) All these materials are "soft solids" showing apparent yield behaviour and extreme rheological nonlinearity at very small strains (less than 1%). This is quite different from polymeric systems where rheological nonlinearity usually sets in at strains of order unity. The added difficulties in describing such materials under flow are manifold – in particular the physics of the *formation* of onions under shear is quite complex (see Roux's lectures). Accordingly we restrict ourselves to the simplest possible problem which is to predict the linear viscoelastic modulus $G^*(\omega)$, for a system which has been made by shear, but the shear then stopped so that small-amplitude, linear oscillatory measurements can be done.

Unfortunately, it is not even clear that this is a well-posed task! For a metastable system, there is no proof that a regime of linearity exists at small strains, and if it does exist, the usual laws of linear response theory (including for example the fact that $G(t)$ and $G^*(\omega)$ are related by a Fourier transform, Eq.16) have to be used with extreme caution. Indeed, careful measurements of $G^*(\omega)$, in what appears to be a linear regime, show formally "impossible" behaviour: the loss modulus G'' at low frequencies approaches a constant (or even a rising trend as frequency is lowered) [39]. This is puzzling: since $G(t) = 0$ for $t < 0$, $G''(\omega)$ is obliged by linear response theory to be an odd function. Very similar results are seen for dense emulsions and foams [40, 41]. We return to this below, but first address a less problematic issue, that of calculating the real part, $G'(\omega)$, which has

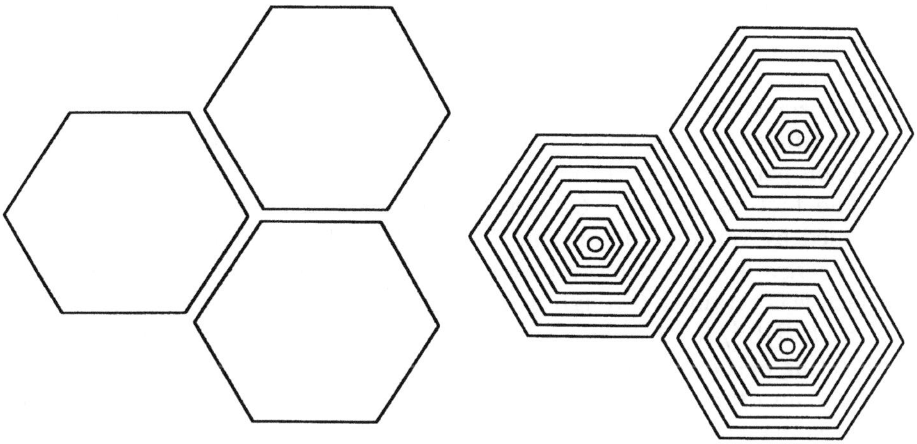

Figure 4. Schematic of a foam (left) and an onion texture (right).

a constant plateau extending to very low frequencies (less than 0.1 Hz). Onions can therefore be viewed as viscoelastic "solids", rather than fluids (although at lower frequencies still, creep may occur – a fact that also applies to conventional "solids", such as lead).

Viscoelastic measurements have in fact been made on a variety of onion-like textures in various systems [6, 44]. However, interpretation is difficult when the texture is not well controlled (for example, strongly polydisperse as arises when the onions form spontaneously without shear [45]). Here we restrict attention to data taken on samples with a uniform onion size.

7.1. ELASTIC MODULUS OF ONIONS

Our description is based on the elastic free energy density of a deformed smectic phase

$$2\Delta F = \bar{B}(\partial_z u)^2 + K(\partial_\perp^2 u)^2 \tag{39}$$

where the first term represents compression of layers and the second is layer bending. Here $\bar{B}$ is the smectic compression modulus at constant chemical potential, and K is the bending modulus (do not confuse this with $K_{\alpha\beta}$ used earlier for the strain rate). The variable u is a layer displacement from a local reference state consisting of flat layers perpendicular to the z direction.

From dimensional analysis of Eq.39, we can construct a length $\lambda = (K/\bar{B})^{1/2}$, called the "penetration" length. At scales larger than this, compression of the layers is costly and bending will usually occur in preference. This is why the onion texture consists of a dense polyhedral texture rather than spheres: the onions deform into close contact to maintain a uniform layer spacing throughout the medium. In fact, λ is rather small: it is one or two times the layer spacing, d. This means that the polyhedral shape of an onion persists right through to its centre.

The first estimate of G_0, the elastic modulus at low frequency, was provided by van der Linden and Droge [42], who calculated the deformation energy of an assembly of *spherical* onions of radius R whose outer surface was strained through a small angle γ. Their basic result was that $G_0 \simeq (K\bar{B})^{1/2}/R$ (they also gave a value for the prefactor). This result is interesting and can be interpreted as follows. Supposing each onion were to undergo an *affine* deformation, in which the inner layers followed the same shape as that imposed on the outermost one. This would impose an increase in the local layer spacing in some regions and a decrease in others, both of which are disfavoured by the compressional modulus; the overall elastic modulus of the material would then be of order $\bar{B}$. However, the inner layers need not deform affinely and instead will adjust their curvature to minimise the compressional distortion. This relaxation reduces the stored energy by a factor λ/R, giving the aforementioned scaling of G_0. The $1/R$ scaling is reminiscent of foams, which have an elastic modulus $G_0 \simeq \sigma/R$ where σ is the surface tension of the films. The onion result is equivalent to this if one recognizes that there is an effective surface tension $\sigma_{eff} = (K\bar{B})^{1/2}$, and indeed this plays the same role in other contexts [43]. However, unlike a true surface problem, the actual deformation energy, even after relaxation, is spread through the onion.

A better estimate for G_0 can be made by accounting for the polyhedral shape of the onions [39]. However, the basic scaling of G_0 turns out to be the same. The estimate can be made by noting that about half the free energy, in a polyhedral packing (prior to the application of a strain) is concentrated on the internal "facets" of the onion where there is high curvature of order $1/\lambda$. The area of such facets per unit volume scales as $1/R$ and so the corresponding energy is of order $K/(\lambda^2 R) \simeq (K\bar{B})^{1/2}/R$. On applying a small strain γ, the areas and curvatures of these regions is quadratically perturbed (since strains of opposite signs must be equivalent); the free energy shift therefore scales as $\gamma^2(K\bar{B})^{1/2}/R$, which gives $G_0 \simeq (K\bar{B})^{1/2}/R$ as before.

This prediction does not fit the data [39] which instead can be represented as

$$G_0 \simeq \alpha + \beta/R \tag{40}$$

with α and β constants (for typical R the two contributions are similar in magnitude). Note that the same functional form can be used for both an aqueous AOT system and an SDS-based system swollen with oil [39]. In broad terms, the scaling arguments given above explain the second term (and indeed the coefficient β is found to be comparable to $\sigma_{eff} = (K\bar{B})^{1/2}$ in magnitude [39]). However, they do not explain the first term. Note that this term is absent for foams, so the analogy via σ_{eff} is now breaking down.

7.2. DISORDER

A possible explanation for the α term comes from the intrinsic disorder in the foamlike structure (which is observable in freeze-fracture images, see the lectures of Roux, this volume). Recall that the relaxation of strains, described above, leads to a large reduction in elastic modulus from the naive estimate $\bar{B}$ that one would make on the basis of affine layer distortions. We now argue that this reduction cannot be effected completely when disorder is present [39].

Starting from an ensemble of polyhedral onions, we assume that each of these has some local "pre-strain" γ_0 which is equally likely to be positive or negative. The strain in a given onion, after an external shear through angle γ is imposed, is therefore not γ but $\tilde{\gamma} = \gamma + \gamma_0$. For an onion without pre-strain, an affine distortion of the onion would change the layer spacing as $d \rightarrow d' = d + O(\gamma)$, where the order γ term is positive in some parts of the onion and negative in others (with zero mean overall). The relaxation effect causes a redistribution of material, after which (to a sufficient approximation) one has instead $d' = d + O(\gamma^2)$ everywhere. This term leads to a compression energy density $\bar{B}(d' - d)^2/2$ which is quartic in the strain and therefore does not give any contribution of order $\bar{B}$ to the linear elastic modulus. However, if we now repeat this argument with γ replaced by $\tilde{\gamma}$, we find that after the relaxation stage, there is within any given onion a residual mean increase or decrease in the layer spacing of order $\gamma\gamma_0 d$ (with a random sign). Since, in contrast to the intra-onion relaxation, material cannot be reassigned from one onion to another, further relaxation of the layer spacing toward a globally uniform value is prohibited. Accordingly, there is in each neighbourhood a compression or expansion of the layers of order $d' - d = \gamma_0\gamma$, which gives a contribution to the elastic modulus $\Delta G_0 \simeq \bar{B}\gamma_0^2$ [39]. A more sophisticated estimate [38] suggests that the prefactor is close to unity.

This argument offers a reasonable explanation of the α term, though it is not very predictive! Specifically, to explain the constant offset in G_0, we have to assume that γ_0 is independent of the onion size R, which is far from obvious. If we also assume that it is independent of d, we recover a scaling $\alpha \sim d^{-3}$ which is observed experimentally [39]. For the systems studied so far, the magnitude of the α term can be explained by an r.m.s. prestrain of order 3% (in the SDS system) and 9% (in the AOT system). These values seem a little high, but are not unreasonable in the light of freeze-fracture images. A consequence of this prediction is that much more ordered onions, which can also be made (see Roux's lectures) should have much reduced α values. This remains to be tested experimentally.

7.3. CONSTANT LOSS MODULUS

We now return to the issue of why the loss modulus, G'', should appear to reach a plateau (or be rising gently) at low frequencies. One possibility is that there are slow relaxation modes arising from intrinsic dynamics of thin bilayer films, at frequencies far lower than those probed conveniently by oscillatory rheometry. However, as mentioned above, the same behaviour is seen for foams and dense emulsions; and in these systems, a rather exhaustive theoretical search has failed to uncover any such modes [46]. This suggests that the true mechanism is more likely to be a generic feature of disorder. Indeed, work of Bouchaud on spin-glasses [47] has revealed a very similar scenario in such systems (where there is quenched disorder affecting the interaction among spins), leading to a constant imaginary part of the susceptibility (analagous to the inverse of G^*) at low frequencies. The same work on spin glasses also reveals potentially deep deficiencies in linear response theory associated with ageing and related effects [47]. These may also be present in dense emulsions and onion phases; see the lectures of Lequeux (this volume). They remain the subject of current work [48, 49]. A barrier to our present understanding is to work out what is the analogue, in foams, emulsions and onions, of the individual spinflips in a spin glass. The most promising candidate might appear to be a local topological rearrangement; however, the energy scales for these are, at first sight, much too high to be accessible thermally (which would be required for the scenario of Ref.[47] to be truly analagous). The picture cannot yet be ruled out since it remains possible that the dynamics is dominated by relatively rare configurations where the energy barriers between topologies are untypically low. Alternatively there may be other local features of the morphology, as yet unrecognised, which play the required role.

8. CONCLUSION

The theoretical rheology of entangled micellar systems has progressed substantially in the past decade. Quantitative models exist for the linear viscoelastic spectrum and they work well in certain systems. As well as simple, unbranched micelles, these can now be taken to include lightly branched structures which apparently arise at high salt levels. Other systems, such as the lecithin-based reverse micelles studied extensively by Schurtenberger

[50], continue to show some baffling trends (stress relaxation times that decrease as concentration is raised, for example). We have not emphasized these systems above: their pedagogical value is limited because much less is understood! However, the author does not wish to leave the impression that there is no need for further studies, even in the realm of linear viscoelasticity. On the contrary, much remains to be clarified. The models describe here at least provide a baseline for comparison of new data, and when the data does not fit, that itself is interesting and may be a strong clue to new physics (branching is a good case in point).

The nonlinear rheology of entangled micellar systems is in a state of flux. The existence of a model constitutive equation is a major asset but this has, over the passed few years, raised more questions than it solves. The issue of shear banding is a fascinating one which has occupied rheologists for some decades. Entangled micellar systems offer a particularly clean and reproducible system to study such effects, partly because a constitutive equation is in place, but also because these materials are self-assembling and therefore do not degrade in use at high shear; they are also rather inexpensive, so that anyone with a rheometer can study them! It seems reasonable to hope for rapid further progress in understanding the relevant aspects of nonlinear flow behaviour, mainly because of the advent of new probes [32, 30] which allow flow patterns to be imaged directly rather than pieced together by detective work done on rheological data. Further theoretical enquiry is likely to be stimulated by these new experimental techniques.

Our understanding of the rheology of metastable surfactant materials, such as the onion texture of sheared smectic phases, remains in its infancy. It seems likely that quenched (or partially quenched) disorder plays a dominant role in even the most basic properties, such as the elastic modulus. This has, of course, been realized for a long time in the context of textured materials – see for example the lectures of Larson and Doi in this volume – of which defect-rich liquid crystals are a prime example. The importance of the onion texture, as a model system (both experimentally and theoretically) lies in the fact that the defect texture is under very much better control than usual.

8.1. ACKNOWLEDGEMENTS

The author wishes to thank the many colleagues and collaborators whose work has been described in these lectures, and apologise to those others whose work, of equal significance, has been arbitarily omitted. He is grateful for the hospitality of the Newton Institute where much of the preparation for these Lectures was done.

References

1. M. E. Cates, Phil. Trans. R. Soc. Lond. A 344 (1993) 339.
2. J. N. Israelachvili, D. Mitchell and B. Ninham, J. Chem. Soc. Faraday Trans. II, 72 (1976) 1525.
3. M. E. Cates and S. J. Candau, J. Phys. Condens. Matt. 2 (1990) 6869.
4. M. E. Cates, in Structure and Flow in Surfactant Solutions, C. A. Herb and R. K. Prud'homme, Eds., ACS Books, Washington (Symposium Series 578) (1994).
5. F. Lequeux and S. J. Candau, in Structure and Flow in Surfactant Solutions, C. A. Herb and R. K. Prud'homme, Eds., ACS Books, Washington (Symposium Series 578) (1994).
6. H. Hoffmann, in Structure and Flow in Surfactant Solutions, C. A. Herb and R. K. Prud'homme, Eds., ACS Books, Washington (Symposium Series 578) (1994).
7. P.-G. de Gennes, Scaling Concepts in Polymer Physics, Cornell, Ithaca, 1979.
8. M. E. Cates, J. Physique 49 (1988) 1593.
9. S. J. Candau, F. Merrikhi, G. Waton and P. Lemarechal, J. Physique 51 (1990) 977.
10. M. S. Turner and M. E. Cates, J. Physique 51 (1990) 307.
11. C. M. Marques and M. E. Cates, J. Physique II, 1 (1991) 489.
12. M. S. Turner and M. E. Cates, Langmuir 7 (1991) 1590.
13. M. S. Turner and M. E. Cates, J. Physique II, 2 (1992) 453; M. S. Turner, C. M. Marques and M. E. Cates, Langmuir, 9 (1993) 695.
14. B. O'Shaughnessy and J. Yu, Phys. Rev. Lett. 744 (1995) 4329.
15. M. Doi and S. F. Edwards, The Theory of Polymer Dynamics, Clarendon, Oxford, 1986.
16. H. Rehage and H. Hoffmann, Molecular Phys. 74 (1991) 933; H. Rehage and H. Hoffmann, J. Phys. Chem. 92 (1988) 4712.
17. M. E. Cates, Macromolecules 20 (1987) 2289.
18. R. Granek and M. E. Cates, J. Chem. Phys. 96 (1992) 4758; for a recent improvement see, R. Granek, Macromolecules 28 (1995) 5370.
19. F. Kern, R. Zana and S. J. Candau, Langmuir 7 (1991) 1344.
20. F. Lequeux, Europhys. Lett. 19 (1992) 675.
21. T. J. Drye and M. E. Cates, J. Chem. Phys. 96 (1992) 1367.
22. A. Khatory, F. Kern, F. Lequeux, J. Appell, G. Porte, N. Morie, A. Ott and W. Urbach, Langmuir 9 (1993) 933; J. Appell, G. Porte, A. Khatory, F. Kern and S. J. Candau, J. Physique II, 2 (1992) 1045.
23. T. C. B. McLeish and M. E. Cates, unpublished.
24. M. E. Cates, J. Phys. Chem. 94 (1990) 371.
25. This equation was derived in unpublished work by the author and independently by C.-Y. D. Lu.
26. N. A. Spenley, M. E. Cates and T. C. B. McLeish, Phys. Rev. Lett. 71 (1993) 939.
27. G. V. Vinogradov, Rheol. Acta 12 (1973) 273; T. C. B. McLeish and R. C. Ball, J. Polym. Sci. Polym. Phys. Edn, 24 (1986) 1735 and references therein.
28. M. E. Cates, T. C. B. McLeish and G. Marrucci, Europhys. Lett. 21 (1993) 451.
29. N. A. Spenley, X.-F. Yuan and M. E. Cates, J. Physique II, 6 (1996) 375.
30. P. T. Callaghan, M. E. Cates, C. J. Rofe and J. B. A. F. Smeulders, J. Physique II, 6 (1996) 375.
31. C. Grand, J. Arrault and M. E. Cates, work in progress.
32. R. Makhloufi, J. P. Decruppe, A. Ait-Ali, and R. Cressely, Europhys. Lett. 32 (1995) 253.
33. J. F. Berret, D. C. Roux, G. Porte, and P. Lindner, Europhys. Lett. 25 (1994) 521; J. F. Berret, D. C. Roux and G. Porte, J. Phys. II, 4 (1994) 1261; see also V. Schmitt, F. Lequeux, A.Pousse and D. Roux, Langmuir 10 (1994) 955.
34. T. C. B. McLeish and R. C. Ball, J. Polym. Sci. Polym. Phys. Edn., 24 (1986) 1735; T. C. B. McLeish J. Polym. Sci. Polym. Phys. Edn., 25 (1987) 2253, and references therein.

35. N. A. Spenley and M. E. Cates, Macromolecules, 27 (1994) 3850.
36. See, for example, G. Marrucci, J. Non-Newtonian Fluid Mech., 62 (1996) 279; G. Ianniruberto and G. Marrucci, to be published.
37. O. Diat and D. Roux, J. Physique II, 3 (1993) 9.
38. C.-Y. D. Lu, to be published; Ph.D. Thesis, Cambridge University (1995).
39. P. Panizza, D. Roux, V. Vuillaume, C.-Y. D. Lu and M. E. Cates, Langmuir, 12 (1996) 248.
40. T. Mason, Ph.D. Thesis, Princeton University (1995).
41. H. M. Princen, J. Colloid Sci. 91 (1983) 160; S. A. Khan and R. C. Armstrong, J. Non Newtonian Fluid Mech. 22 (1986) 1.
42. E. van der Linden and J. H. Droge, Physica A, 193 (1993), 439.
43. P. G. de Gennes, "the Physics of Liquid Crystals", Clarendon, Oxford (1976).
44. H. Hoffmann, C. Thunig, P. Schmiedel and U. Munkert, Nuov. Cimento del Soc. Ital. di Fis. D, 16 (1994) 1373.
45. M. Dubois and T. Zemb, Langmuir, 7 (1991) 1352; M. Dubois, T. Gulik-Krzywicki and B. Cabane, Langmuir, 9 (1993) 673.
46. D. M. A. Buzza, C.-Y. D. Lu and M. E. Cates, J. Physique II, 5 (1995) 37.
47. J.-P. Bouchaud, J. Physique I, 2 (1992) 1705.
48. P. Sollich and M. E. Cates, work in progress.
49. F. Lequeux and colleagues, work in progress.
50. See, e.g, P. Schurtenberger, R. Scartazzini and P. Luisi, Rheol. Acta 28 (1989), 372; P. Schurtenberger and C. Cavaco, J. Phys. II, 3 (1993) 1279; J. Phys. II, 4 (1994) 305; Langmuir 10 (1994) 100.

EXCESS DISSIPATION IN DENSE EMULSIONS

SRIRAM RAMASWAMY
Department of Physics
Indian Institute of Science, Bangalore 560 012, India
sriram@physics.iisc.ernet.in

Abstract. A new model for the structure, elastic properties and dynamics of foams and concentrated emulsions is presented, based on the idea of local regions lacking shear-rigidity in one or more directions which vary randomly through the medium. It is shown to lead naturally to slow ($t^{-1/2}$) stress-relaxation, implying a piece of the dynamic modulus scaling with frequency ω as $\omega^{1/2}$. Striking experimental confirmation of this prediction using a novel experimental technique is reported, and challenges for the theoretician are offered. This work was done in collaboration with Andrea Liu, Tom Mason, Hu Gang, and David Weitz [1].

1. Introduction and results

Emulsions and foams are metastable mixtures of droplets of a fluid (A) suspended in a liquid background (B), with the tendency to phase separate suppressed by the addition of surfactant. The term 'foam' is applied to the case where A is a gas, while for emulsions, sometimes called biliquid foams, it is a liquid. The case of interest for this lecture is that of a concentrated emulsion, in which (without losing the droplet character of the dispersion) the volume fraction of A far exceeds that of B. This state is achievable because of the highly deformable nature of the droplets and the stability of the surfactant films separating them. As anyone who has shaken a beer-foam knows, foams and concentrated emulsions, although composed of fluids alone, have a shear modulus [2, 3] as well as, of course, a viscosity. In general, the magnitudes of the elastic and viscous responses depend on the frequency at which the measurement is made. The relevant quantity measured in small-amplitude shear experiments is the frequency-dependent viscosity $\eta(\omega) = \eta'(\omega) + i\eta''(\omega) \equiv \sigma(\omega)/\dot{\gamma}(\omega)$ which is the ratio

T McLeish (ed), Theoretical Challenges in the Dynamics of Complex Fluids, 285-292

of stress σ to strain-rate $\dot{\gamma}$ at frequency ω, or equivalently the complex dynamic shear modulus $i\omega\eta^*(\omega) \equiv G^*(\omega) = G'(\omega) + iG''(\omega)$, where $G'(\omega)$ and $G''(\omega)$ are called the storage and loss modulus respectively. For a solid with a well-defined shear modulus and viscous damping, $G'(\omega) \sim$ constant and $G''(\omega) \sim \omega$ for $\omega \to 0$. François Lequeux and Mike Cates (this volume) discuss foams and emulsions as well, particularly the role of aging in their dynamics. The work I am presenting here [1], done in collaboration with Andrea Liu, Tom Mason, Hu Gang, and David Weitz, is a speculative attempt to explain the observation that G'' is much larger at low frequencies, and G' much smaller, than reasonable models for viscous flow [4] and droplet deformation [2, 3, 5, 6, 7, 8] would lead one to expect. Our model predicts an excess $\omega^{1/2}$ piece in $G^*(\omega)$ on top of a background of the conventional viscoelastic solid behaviour described above. There is substantial evidence for this [1] in experiments by Mason, Gang, and Weitz, using a remarkable reinterpretation [9] of light-scattering measurements which I describe below. Details of the comparison with experiment can be found in [1].

2. The model

Our model [1] is based on a prejudice: that a structure which is everywhere well-packed, with a solid-like elastic response in all directions in all regions, cannot display anomalous dissipation. Ultimately, shear waves must travel in such a medium as they do in any solid, undergoing at most some multiple scattering from the disorder. To produce a large timescale of the sort that Cates (this volume) and [4] talk about requires some unusually slow or soft mode. The simplest way to produce such a mode is to eliminate one of the the elastic moduli locally here and there in the material, in a random manner. It seems reasonable to suppose that a random packing of deformable droplets would have "weak regions" which are elastically ultra-weak in some direction; it is also known [10] that an extreme example of such a random structure, *viz.*, a polydomain *lamellar* phase, has power-law stress relaxation. If a structure with the aforementioned weak regions were subjected to a macroscopic shear strain it would not deform affinely; some regions would slip rather than deform, resulting in an increased dissipation and a weakened macroscopic elasticity. In this section I present a model which implements these speculations, and work out its implications for the frequency- and packing-fraction-dependence of the dynamic modulus.

The model consists of an elastic medium containing domains each of which lacks an elastic modulus for shear deformations with gradient direction normal to a certain plane (the "weak plane"). The normal, which we will call the *local c-axis* varies randomly from one domain to the next, depending on the nature of the local packing of droplets. Clearly, the response of a given region to an externally applied shear strain will be quite different for gradients along the local c-axis from that to gradients perpendicular to it. The elastic free-energy F [11] of each weak region centred around a point $\mathbf{r}$ is written in terms of the symmetric strain tensor $u_{ij} = \frac{1}{2}[\nabla_i u_j + \nabla_j u_i]$, where u_i is the displacement field of the medium. In a local Cartesian frame $\{abc\}$ assuming, for simplicity alone, that the region is isotropic in the ab-plane and symmetric under $c \to -c$:

$$\begin{aligned} F &= \frac{1}{2}\lambda_1(\mathbf{r})u_{cc}^2 + \lambda_2(\mathbf{r})\left[(u_{aa} - u_{bb})^2 + 4u_{ab}^2\right] \\ &\quad -\lambda_3(\mathbf{r})u_{aa}u_{bb} + \lambda_4(\mathbf{r})(u_{ac}^2 + u_{bc}^2), \end{aligned} \tag{1}$$

plus a bulk modulus term whose effects we can ignore by taking the incompressible limit and disregarding vacancies [12]. Since the ab plane is "weak", $\lambda_4(\mathbf{r}) = 0$. This free-energy has also been used to describe anisotropic glasses [13] and "decoupled" lamellar phases of tethered membranes [14]. Strictly speaking, the limit of vanishing λ_4 is a pathological one unless terms at higher order in the gradient expansion are included. There are three candidates for such terms: (i) a bending energy of the sort that stabilises smectic or discotic liquid crystals [15], which is unlikely to be appropriate here; (ii) an exceedingly small but nonzero λ_4; (iii) the intervention of coarsening and rearrangements at very long times. All three mechanisms will introduce a low-frequency cutoff ω_{min} into the problem, whose value depends on the details of the mechanisms; the predictions we make for $G^*(\omega)$ from a $\lambda_4 = 0$ model are asymptotic results, *independent* of ω_{min} for $\omega \gg \omega_{min}$. Experiments [1] suggest $\omega_{min} \simeq 10$ Hz for emulsions.

3. Calculating the macroscopic modulus

Dynamic moduli can be calculated by correlation or response function methods: see, e.g., [16]. Response-function methods for the present problem have not completely been sorted out [17], while a correlation function approach [10] proves more tractable. Let us first define the spatiotemporal correlation function

$$S(\mathbf{r},t) = \langle \sigma_{xy}(\mathbf{r},t)\sigma_{xy}(\mathbf{0},0)\rangle \tag{2}$$

of the local shear-stress field $\sigma_{xy}(\mathbf{r},t)$, referred to a fixed xyz coordinate frame (the "lab frame"). Then the fluctuation-dissipation theorem tells us that

$$G^*(\omega) = i\omega\eta(\omega) = \frac{i\omega}{k_B T}\int_0^\infty dt e^{-i\omega t}\tilde{S}(k=0,t), \tag{3}$$

where $\tilde{S}(k = 0,t) \equiv \langle \tilde{\sigma}_{xy}(k = 0,t)\tilde{\sigma}_{xy}(k = 0,0)\rangle$ is the autocorrelation of the shear-stress at wavenumber $k = 0$. In (3),the angle-brackets denote averages over time (or an equilibrium canonical ensemble) and the static random distribution of the orientations of the weak planes.

The lab-frame xy-component of the local stress, obtained by differentiating (2) with respect to the strain, gets contributions from all nonzero pieces of the shear-stress in the abc frame:

$$\begin{aligned}\sigma_{xy}(\mathbf{r},t) &= a_x(\mathbf{r})a_y(\mathbf{r})\frac{\delta F}{\delta u_{aa}(\mathbf{r},t)} + b_x(\mathbf{r})b_y(\mathbf{r})\frac{\delta F}{\delta u_{bb}(\mathbf{r},t)} \\ &\quad + c_x(\mathbf{r})c_y(\mathbf{r})\frac{\delta F}{\delta u_{cc}(\mathbf{r},t)} \\ &\quad + \frac{1}{2}(a_x(\mathbf{r})b_y(\mathbf{r}) + a_y(\mathbf{r})b_x(\mathbf{r}))\frac{\delta F}{\delta u_{ab}(\mathbf{r},\mathbf{t})}, \end{aligned} \tag{4}$$

where $a_i(\mathbf{r})$, $b_i(\mathbf{r})$ and $c_i(\mathbf{r})$ are the lab-frame components of the local basis $\{abc\}$. It is clear from (4) that the macroscopic shear-stress at zero wavenumber involves, through convolutions, the local abc stress and frame-fields at all wavenumbers. Inserting (4) in (2), we can express the stress autocorrelation as a simultaneous thermal and disorder average of products of four frame fields and two abc-frame components of strain fields (here and hereafter I eschew detailed algebra in favour of sketchy arguments). To move any further we must resort to some kind of decoupling, and the simplest is that used by [10] for polydomain smectics, *viz.*, to break each such term up into a product of the average of four frame-fields and the average of a pair of strain fields. The strain autocorrelations remain difficult to compute because they too are referred to the local frames, which vary from region to region. So let us assume, as do [10], that the major contribution to the autocorrelation is from regions over which the $\{abc\}$ frame is highly correlated, i.e., from wavenumbers $q \gg \xi^{-1}$, where ξ is the size of the typical weak region of a given orientation. Then we can treat the response of each such region as if that region were unbounded and independent of the others, and add incoherently the contributions from distinct regions. The result is an expression for $\tilde{S}(k=0,t)$ expressed as a sum of integrals over wavevector $\mathbf{q}$ of products of disorder averages of the form $\langle \tilde{a}_x(\mathbf{q})\tilde{a}_y(\mathbf{q})\tilde{a}_x(-\mathbf{q})\tilde{a}_y(-\mathbf{q})\rangle$ and strain-strain correlation functions of the form $\langle \tilde{u}_{aa}(\mathbf{q},t)\tilde{u}_{aa}(-\mathbf{q},0)\rangle$[10]. The coordinate axes $\{abc\}$ are correlated only over the coherence length ξ, the size of the local region of slip:

$$\langle \tilde{a}_x(\mathbf{q})\tilde{a}_y(\mathbf{q})\tilde{a}_x(-\mathbf{q})\tilde{a}_y(-\mathbf{q})\rangle = f(q^2\xi^2), \tag{5}$$

and the local axes are uncorrelated with each other, so that correlation functions of the form $\langle \tilde{a}_x(\mathbf{q})\tilde{a}_y(\mathbf{q})\tilde{b}_x(-\mathbf{q})\tilde{b}_y(-\mathbf{q})\rangle$ vanish.

The strain autocorrelations themselves can be obtained from fluctuating hydrodynamic equations for the displacement fields $u_i(q)$ at wavenumber q, in close analogy to those used for crystals [12] or liquid crystals

[15, 10]. We proceed as follows. On timescales long compared to $\tau_o = \rho/\eta(\xi/2\pi)^2 \approx 10^{-6}$ s (the viscous damping time on the scale of a slip region), where $\rho \approx 1\text{gm/cm}^3$ is the mass density, $\eta \approx 1cp$ is the film viscosity, and ξ has been taken to be several droplet diameters $\approx 3\mu$m, the *overdamped* limit is appropriate. The equations of motion, after longitudinal modes have been projected out by incompressibility then take the schematic form $\eta q^2 \partial_t u_i(q) = -B_{ij}(\theta, \phi) q^2 u_j(q)+$ noise, where B_{ij} is a combination of elastic constants of typical magnitude B and angular factors (from the anisotropic elasticity as well as the projection). Time correlations therefore decay as e^{-t/τ_B} where $\tau_B \equiv \eta/B$, is the characteristic time in the problem: we see that in this overdamped limit the decay is *independent* of the *magnitude* of the wavevector, and depends only on its direction. The strain autocorrelations follow in a straightforward from those of the u_i. There are only two modes due to incompressibility. Using the resulting expressions to get the stress autocorrelations yields

$$\tilde{S}(k=0,t) = \int d\varphi d\cos(\theta) \left[f_1(\theta,\varphi)\, e^{-\gamma_1(\theta,\varphi)t} + f_2(\theta,\varphi)\, e^{-\gamma_2(\theta,\varphi)t} \right], \quad (6)$$

where $\gamma_i(\theta, \varphi)$ are the relaxation rates (related to the eigenvalues of B_{ij}) for the two modes with wavevector $\mathbf{q}$ oriented at an angle θ to the local c axis and and an azimuthal angle φ in the ab plane, and the f_i are functions whose detailed properties do not concern us here. We are interested in the long time behavior of $\tilde{S}(k = 0, t)$, which we can obtain by an asymptotic estimate of (6) for $t \to \infty$. The integral (6) will then be dominated by the minima of the γ_i. One of them, say γ_2, is nonzero, $\sim 1/\tau_B$ for all directions, reflecting the isotropic elasticity of the ab planes, while γ_1 vanishes analytically as $\tau_B^{-1} cos^2\theta$ for $\theta \to \pi/2$, as a result of the existence of a weak plane. The integral $\tilde{S}(k = 0, t) \sim \int dx\, e^{-x^2 t/\tau_B} \sim (t/\tau_B)^{-\frac{1}{2}}$, for $t \gg \tau_B$, so that $G^*(\omega) \sim (i\omega\tau_B)^{\frac{1}{2}}$ as asserted in the Introduction and, curiously, exactly as for polydomain smectics [10]. This result should hold for $\omega_{min} \ll \omega \ll \tau_B^{-1}$, where ω_{min} is the lower cutoff mentioned in Section 2, and the high frequency cutoff is because of our asymptotic estimate. For concentrated emulsions, the scale of the elastic constants B should be set by the ratio Γ/a of the surface tension to the droplet radius. The emulsions on which the experiments discussed below were done had [18] $\Gamma \sim 10$ dyn cm^{-1}, $a \sim 1\mu$m, and film viscosity $\eta \sim 1$ cp, so that $\tau_B^{-1} \sim 10$ MHz.

4. Discussion of experiment

The experimental method of [9] extracts information about the dynamic shear modulus over a frequency range much higher than that accessible to conventional rheometers, from the thermal mean-square displacement

(MSD) of particles. The principle behind the method is a remarkable, empirically based generalisation of the Stokes-Einstein relation to finite frequncy measurements. Here is how it works. Imagine a particle (an emulsion droplet or colloidal sphere) of radius *a* at *thermal equilibrium* in a medium with a nonlocal mobility (inverse friction) kernel $\hat{\mu}(t-t')$. Such a particle would execute a non-Brownian random walk with MSD, in three dimensions

$$\langle \Delta r^2(t) \rangle = 6k_BT \int \hat{\mu}(t-t')t'dt', \tag{7}$$

where I've assumed a fluctuation-dissipation theorem relating (timescale-dependent) mobility and diffusivity. Such nonlocal mobilities arise naturally even at equilibrium (see, e.g., my first lecture) in concentrated dispersions where particle motion is highly constrained and hence correlated. Laplace-transforming (7) relates the frequency-dependent mobility $\bar{\mu}(s)$ to the transform of the MSD:

$$\bar{\mu}^{-1}(s) = \frac{6k_BT}{s^2\langle \Delta \bar{r}^2(s) \rangle}. \tag{8}$$

where overbars denote a real Laplace transform, as a function of the frequency s. For a simple fluid governed by the Stokes equations, the mobility μ (a local quantity) is related to the viscosity η (a macroscopic quantity): $\mu = 1/6\pi\eta a$. If we *assume* that the relation between local mobility and macroscopic viscosity continues to hold for a correlated medium like our emulsion, then we can turn (8) into a relation between the dynamic viscosity $\bar{\eta}(s)$ (or the dynamic modulus $\bar{G}(s)$) and the measured MSD:

$$\bar{G}(s) = s\bar{\eta}(s) = \frac{k_BT}{\pi a s \langle \Delta \bar{r}^2(s) \rangle}. \tag{9}$$

Empirically, this relation works remarkably well [9], suggesting that on the short timescale of the experiments on the emulsions, self and collective motions (see my first lecture) are very similar. Note that we can get back to the complex dynamic modulus $G^*(\omega)$ from $\bar{G}(s)$ through analytic continuation: $s = i\omega$ [9]. Thus, nearly s-independent behavior in $\bar{G}(s)$ reflects a large elastic contribution since $G'(\omega)$ dominates, while linear s-dependence implies viscous behaviour.

The experiments used monodisperse silicone-oil-in-water emulsions [18] at packing fractions $\phi > 0.64$, the close-packing value, and measured the thermal MSD of the droplets themselves [9] by dynamic light scattering. Since the samples multiple-scatter a great deal, the method of choice is diffusing-wave spectroscopy (DWS) [19, 20]. The dynamic modulus of the emulsion was extracted by numerically Laplace transforming the MSD time-trace and using Eq. 9. For figures and experimental details I refer the reader to [1], giving only a summary here. Measurements were made over seven decades in frequency, and found nearly s-independent behaviour at low frequencies and linear s dependence at high frequencies, as expected for a viscoelastic solid. However, if these two conventional behaviours, with coefficients adjusted to match the high and low frequency ends of the data, are subtracted out, what remains is $s^{1/2}$ (and hence, for $G^*(\omega)$, $(i\omega)^{1/2}$) over the entire range! It should be pointed out that a small and concentration-independent $s^{1/2}$ piece is seen [21] for $\phi < .64$ as well, and is generally attributed [22] to the high-frequency tail of *free* Brownian motion. In the concentrated regime, however, this latter mechanism cannot possibly apply, since the droplets do *not* diffuse freely at all. Moreover, the coefficient of the $s^{1/2}$ piece increases by a factor or 4 or 5 as the packing fraction ϕ goes from .64 (the close-packing value) to .85, suggesting strongly that this is behaviour intrinsic to well-packed emulsions. A more detailed quantitative comparison between experiment and theory is found in [1].

5. Challenges

While the observation of the predicted behavior of the dynamic modulus is strong support for the idea of weak regions, direct observation of the process of slip in sheared foams or emulsions, either in experiment or perhaps simulation, is the single major challenge for the simulator or experimenter. For the theoretician, it is essential to put the weak-regions idea to a more stringent test by extracting as many predictions from it as possible. An inviting problem is a detailed calculation of the *storage* modulus at zero frequency, which experiments [3] find to be much smaller than a periodic packing of cells would predict. It seems likely that the idea of elastically weak regions will prove important in understanding this anomaly. Including the effects of coarsening and rearangements in the present picture remains an outstanding problem.

References

1. A.J. Liu, S. Ramaswamy, T. G. Mason, H. Gang, and D. A. Weitz, Phys. Rev Lett. **76**, 3017 (1996).
2. H.M. Princen and A.D. Kiss, J. Colloid Interface Sci. **112**, 427 (1986).
3. T. G. Mason, J. Bibette, D. A. Weitz Phys. Rev Lett. **75**, 2051 (1995).
4. D. M. A. Buzza, C.-Y. D. Lu and M. E. Cates, J. Phys. II France **5**, 37 (1995).
5. A. M. Kraynik, Ann. Rev. Fluid. Mech. **20**, 325 (1988).
6. S. A. Khan, C. A. Schnepper and R. C. Armstrong, J. Rheol. **32**, 69 (1989).
7. D. Weaire, F. Bolton, T. Herdtle and H. Aref, Phil. Mag. Lett. **66**, 293 (1992).
8. D.J. Durian, Phys. Rev. Lett. **75**, 4780 (1995).
9. T. G. Mason and D. A. Weitz, Phys. Rev. Lett. **74**, 1250 (1995).
10. K. Kawasaki and A. Onuki, Phys. Rev. A **42**, 3664 (1990).
11. L. D. Landau and E. M. Lifschitz, "Theory of Elasticity", (Pergamon Press, NY, 1986), Ch. 1.

12. P.C. Martin, O. Parodi, and P.S. Pershan, Phys. Rev. A **6**, 2401 (1972).
13. L. Golubović and T. C. Lubensky, Phys. Rev. Lett. **63**, 1082 (1989).
14. L. Golubović and T. C. Lubensky, Phys. Rev. A **43**, 6793 (1991).
15. P.-G. de Gennes and J. Prost, *The Physics of Liquid Crystals*, Clarendon, Oxford (1994).
16. R. Kubo, *Lectures in Theoretical Physics*, (Interscience, New York, 1959), vol. 1.
17. S. T. Milner and G. H. Fredrickson, preprint (1995).
18. J. Bibette, J. Colloid and Interface Sci. **147**, 474, (1991).
19. D.J. Pine, D.A. Weitz, E. Herbolzheimer and P.M. Chaikin, Phys. Rev. Lett. **60**, 1134 (1988).
20. D.J. Pine, D.A. Weitz, J.X. Zhu and E. Herbolzheimer, J. Phys. (Paris) **51**, 2101 (1990).
21. T. G. Mason and D. A. Weitz, Phys. Rev. Lett. **75**, 2770 (1995).
22. R. A. Lionberger and W. B. Russel, J. Rheol. **38**, 1885 (1994).

DYNAMICS OF DOMAINS AND TEXTURES

M. DOI
Department of Applied Physics,
Nagoya University,
Chikusa-ku, Nagoya 464 JAPAN

1. Introduction

Many complex fluids involve structures of mesoscopic length scale ranging from 0.1μ to 100μ. For example, liquid crystals have various textures resulting from spatially varying molecular order [1, 2](see chapter of R.G.Larson, present volume). Self assembling molecules such as surfactants or block copolymers form various structures of spherical, rodlike or lamelar shapes[3, 4]. Colloidal particles form large aggregates or gels when the particle interaction becomes attractive[5]. These mesoscopic structures are very fragile, and easily deformed by weak external forces. As a consequence, these systems show strongly non-linear rheological responses which are quite important in technological application of these materials.

Predicting the rheological properties of materials which have structures of mesoscopic length scale is a theoretical challenge. The random nature of the structure of these materials requires a statistical description. On the other hand, standard methods of statistical mechanics cannot be used since the thermal motion plays only a minor (or completely negligible) role in these systems. At present we do not have any general theoretical means to handle such systems. However, attempts have been made in related areas, many of which are discussed in this book. By accumulating this knowledge and experience, I hope that we can construct some sort of theoretical paradigm for such systems.

In this chapter, I describe the classical knowledge of domain dynamics. It is obtained in the study of the mixture of simple liquids. In the first part, I discuss the phase separation kinetics [6, 7]. The aim here is to give an introduction to the transport equations which are used in the other parts of this book (see e.g. the chapters by S.Mil ner and A. Semenov). In

T McLeish (ed), Theoretical Challenges in the Dynamics of Complex Fluids, 293-314

the second part, I discuss the rheology of two immiscible fluids [9]. The aim here is to show a characteristic feature of the system governed by mesoscale dynamics.

2. Phase Separation Kinetics

2.1. PHASE SEPARATION

Let us consider the phenomenon that upon changing the temperature, a homogeneous solution consisting of two fluid components A and B separates into two phases each having different compositions. From the thermodynamic point of view, the phenomenon can be completely described by the mixing-free-energy function $f(\phi)$ which is defined as follows.

For the sake of simplicity, we consider the case that the volume of the fluid does not change upon mixing. Thus if the pure fluid A of volume V_A is mixed with the pure fluid B of volume V_B, the solution has a volume $V = V_A + V_B$. If the solution is homogeneous, the composition of fluid A is expressed by the dimensionless parameter

$$\phi = \frac{V_A}{V_A + V_B} \tag{1}$$

The function $f(\phi)$ is defined as the free energy per unit volume of the solution which has uniform composition ϕ. If the two fluids do not mix uniformly, we consider a hypothetical state where the two fluids are forcefully mixed by some external means, and determine the free energy. Alternatively, we can obtain the function $f(\phi)$ theoretically. For example, a simple mean field theory gives,

$$f(\phi) = \frac{k_B T}{a^3}\left[\phi \ln \phi + (1-\phi)\ln(1-\phi)\right] + \frac{\epsilon}{a^3}\phi(1-\phi) \tag{2}$$

where a^3 is the molecular volume and ϵ is the excess interaction energy for the $A - B$ molecules in contact. The actual form of $f(\phi)$ is not important in the following discussion.

Given $f(\phi)$, one can judge whether the solution remains uniform or not. Suppose that the solution of volume V separates into two phases each having composition ϕ_1 and ϕ_2. The mass conservation law tells that the volume of each phase is $V_1 = V(\phi_2 - \phi)/(\phi_2 - \phi_1)$ and $V_2 = V(\phi - \phi_1)/(\phi_2 - \phi_1)$, so that the change of free energy of the whole system is

$$\begin{aligned}\Delta F &= V_1 f(\phi_1) + V_2 f(\phi_2) - V f(\phi) \\ &= V\left[\frac{\phi_2 - \phi}{\phi_2 - \phi_1} f(\phi_1) + \frac{\phi - \phi_1}{\phi_2 - \phi_1} f(\phi_2) - f(\phi)\right]\end{aligned} \tag{3}$$

For the system to remain uniform at composition ϕ, ΔF must be positive for all possible choices of ϕ_1 and ϕ_2. From this criteria, it can be easily shown that

(i) If $f''(\phi) > 0$ (the temperature T_1 in Fig.1), the system is stable and the composition remains uniform.

(ii) If $f''(\phi) < 0$, the system is locally unstable (the temperature T_3), and separates into two phases with composition ϕ_A and ϕ_B, where ϕ_A and ϕ_B are the equilibrium compositions obtained by the common tangent construction shown in Fig.1. (Note that ϕ_A and ϕ_B represent the compositions of A molecules in A rich phase and B rich phase respectively.)

(iii) If $f''(\phi) > 0$ (the temperature T_2), the system is locally stable, but it will phase separate if ϕ is between ϕ_A and ϕ_B.

The phase separation in the case (ii) is called spinodal decomposition. In the following we shall consider this case.

Let us now consider the case that the composition is not uniform. Let $\phi(\boldsymbol{r})$ be the composition at point $\boldsymbol{r}$. In this case, the free energy of the whole system is given by

$$F[\phi] = \int d\boldsymbol{r} \left[f(\phi(\boldsymbol{r})) + \frac{A}{2}(\nabla\phi)^2 \right] \tag{4}$$

The first term represents the contribution from each volume element. The second term represents the cost of the free energy when the composition is not uniform. It is needed to account for the interfacial tension. Minimizing eq.(4), it can be shown that when the two phases of compositions ϕ_A and ϕ_B coexist, the thickness of the interface ξ and the interfacial tension Γ is given by

$$\xi \simeq \sqrt{\frac{A}{|f''(\phi_C)|}} \quad , \quad \Gamma \simeq \sqrt{A|f''(\phi_C)|} \tag{5}$$

where ϕ_C is the maximum value of $|f'(\phi)|$ for $\phi_B < \phi < \phi_A$.

2.2. EARLY STAGE KINETICS

Let us now consider that a solution of uniform composition ϕ_0 is suddenly brought to a temperature where $f''(\phi_0) < 0$. Then the phase separation will take place. We shall discuss how the process proceeds, i.e., how $\phi(\boldsymbol{r}, t)$ changes in time.

The initial stage of the phase separation is governed by the molecular diffusion: the A molecules tend to cluster together by exchanging their position with the neighbouring B molecules. This process is described by a transport equation for $\phi(\boldsymbol{r}, t)$.

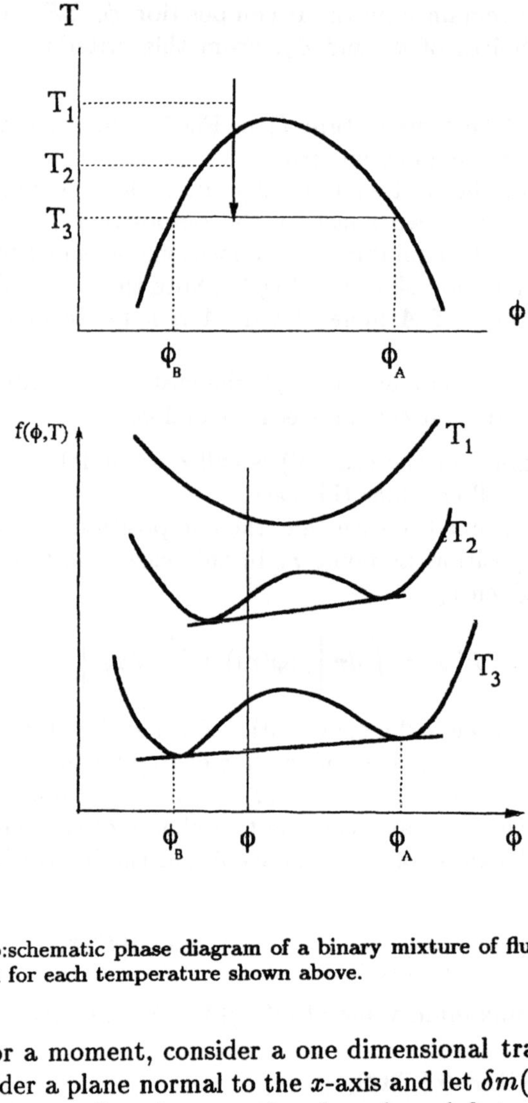

Figure 1. Top:schematic phase diagram of a binary mixture of fluids. Bottom:the free energy function for each temperature shown above.

Let us, for a moment, consider a one dimensional transport along the x-axis. Consider a plane normal to the x-axis and let $\delta m(x)$ be the number of A molecules transferred across the plane from left to right in time δt . Since the total volume is constant, $\delta m(x)$ is also equal to the number of B molecules transferred across the plane from right to left. (Here and in the subsequent discussion, I assume that both A and B molecules have the same volume a^3.) To obtain the rate of such transportation , we calculate how the operation (i.e., exchanging A molecules with B molecules across the plane) changes the free energy. Since the operation changes the composition at x as

$$\delta\phi = -\frac{\partial \delta m}{\partial x} \tag{6}$$

it changes the free energy as

$$\delta F = \int dx \frac{\delta F}{\delta \phi} \delta \phi = \int dx \underline{\frac{\partial}{\partial x} \left(\frac{\delta F}{\delta \phi} \right)} \delta m \tag{7}$$

The underlined term corresponds to the force conjugate to the variable $m(x)$. Therefore we may assume that the rate of exchange, $\dot{m}(x)$, which is the flux $j(x)$ across the plane, is proportional to this force. Thus

$$j(x) = \dot{m}(x) = -L(\phi) \frac{\partial}{\partial x} \left(\frac{\delta F}{\delta \phi} \right) \tag{8}$$

where $L(\phi)$ is a kinetic coefficient called the Onsager coefficient. $L(\phi)$ can depend on ϕ. Eqn.(8) has a simple physical meaning. The functional derivative

$$\mu(x) = \frac{\delta F}{\delta \phi(x)} \tag{9}$$

represents the change of the free energy when a A molecule at position x is replaced by a B molecule. This is equal to the difference between the chemical potential of A molecule and that of B molecule. In the jargon of phase separation kinetics μ is simply called the chemical potential. Eq.(8) therefore states the usual diffusion law that the flux is proportional to the gradient of the chemical potential.

Eq.(8) and the mass conservation of A molecules give the following time evolution equation for ϕ (in the three dimensional case);

$$\frac{\partial \phi}{\partial t} = -\nabla \cdot \boldsymbol{j} = \nabla \cdot L \nabla \left(\frac{\delta F}{\delta \phi} \right) \tag{10}$$

or using eq.(4)

$$\frac{\partial \phi}{\partial t} = \nabla \cdot L \nabla \left(f'(\phi) - A \nabla^2 \phi \right) \tag{11}$$

Eq. (10) is a non-linear equation for ϕ, and solving it generally requires numerical computation. However, in the very early stage , where $\phi(\boldsymbol{r}, t)$ is not very much different from the initial value ϕ_0, it can be solved easily.

By expanding the right hand side with respect to $\delta\phi = \phi(\boldsymbol{r}, t) - \phi_0$, we have

$$\frac{\partial}{\partial t} \delta \phi = L(\phi_0) \nabla^2 \left(f''(\phi_0) - A \nabla^2 \right) \delta \phi \tag{12}$$

In terms of the Fourier component, eq.(12) gives

$$\frac{\partial \delta \phi_{\boldsymbol{k}}}{\partial t} = \Lambda_{\boldsymbol{k}} \delta \phi_{\boldsymbol{k}} \tag{13}$$

where

$$\Lambda_k = L(\phi_0)k^2 \left(-f''(\phi_0) - Ak^2\right) \tag{14}$$

In the unstable region where $f''(\phi_0) < 0$, Λ_k becomes positive in a center k region and has a maximum at

$$k^* = \sqrt{\frac{|f''(\phi_0)|}{2A}} \tag{15}$$

Therefore, in the initial stage, the compositional inhomogeneity with the wave vector k^* grows predominantly. Experimentally, this is seen that the structure factor $S(k) =< \phi_k \phi_{-k} >$ grows in time with a peak at $k = k^*$.

2.3. LATE STAGE KINETICS

The compositional inhomogeneity is small in the early stage, but as it grows, there appear distinct domains each having the composition nearly equal to the equilibrium composition ϕ_A and ϕ_B. In this late stage, the time evolution is described by the change of the size and shapes of domains. The time evolution of domains is governed not only by diffusion but also by hydrodynamic flow. Complete kinetic equations which describe these effects will be given later. Here the kinetics are discussed in simple physical terms.

It has been proposed that three processes are important in the late stage kinetics.

(a) Lifshitz-Slyozov Process

The diffusion, which is dominant in the early stage kinetics, slows down as the composition of each domains approaches the equilibrium value ϕ_A or ϕ_B. However it does not cease until the true equilibrium is reached. Lifshitz and Slyozov pointed out that even if the compositions in each domain become equal to the equilibrium value, there is a spatial gradient of the chemical potential $\mu = \delta F/\delta\phi$ due to the interfacial tension.

Consider a simple case that two droplets of radius r_1 and r_2 ($r_1 > r_2$) are placed with the distance R (See Fig.2). Suppose we take out a A molecule from the second droplet and put it into the first droplet. Then the radius of each droplet changes as $\delta r_1 \cong a^3/4\pi {r_1}^2$ and $\delta r_2 \cong -a^3/4\pi {r_2}^2$ respectively, and the interfacial area changes as

$$\delta A = 8\pi(r_1\delta r_1 + r_2\delta r_2) \cong \frac{1}{2}\left[\frac{a^3}{r_1} - \frac{a^3}{r_2}\right] \tag{16}$$

This changes the free energy of the system by $\Gamma\delta A$. This means that the chemical potential in the second droplet is larger than that in the first

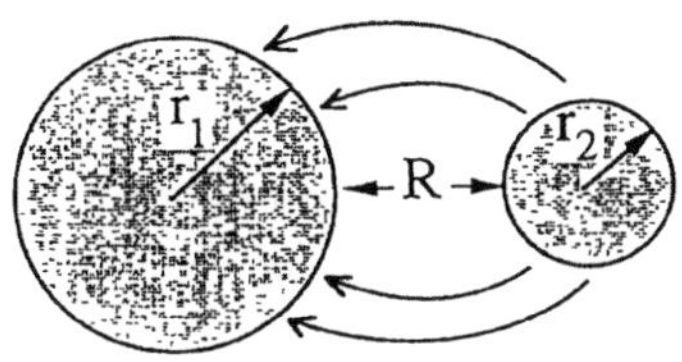

Figure 2. Lifshitz-Slyozov process in the case of two droplets

droplet. The difference in the chemical potential is

$$\Delta\mu \simeq \Gamma a^3 \left(\frac{1}{r_2} - \frac{1}{r_1}\right) \tag{17}$$

Therefore there is a diffusion flux of A molecules from the second droplet to the first. The diffusion is driven by the difference in the chemical potential. This process is called the Lifshitz-Slyozov process.

In the Lifshitz-Slyozov process, the molecules in small droplets diffuse through the B rich phase and move into larger droplets. The process is often called evaporation-condensation process. This process is very slow if ϕ_B (the equilibrium composition of A molecule in the B rich phase) is small. The Lifshitz-Slyozov process takes place not only among droplets, but also within a droplet: if the surface of a droplet is not spherical, the molecule sitting close to the surface with large curvature diffuses to the region near the surface with smaller curvature.

The growth rate of the droplet size is estimated by the following argument. Let R be the characteristic droplet size. (Since we are considering the case of $\phi \cong 0.5$, R represents the characteristic distance between the droplets.) Then the difference of the chemical potential between the droplets $\Delta\mu$ is of the order of $\Gamma a^3/R$, and the flux j is estimated by

$$j \simeq L|\nabla\mu| \simeq \frac{L\Delta\mu}{R} \simeq \frac{L\Gamma a^3}{R^2} \tag{18}$$

The number of A molecules coming into the droplet of size R per unit time is equal to $\dot{n} = jR^2$. Therefore the volume change of the droplet is

$$\frac{dR^3}{dt} \simeq \dot{n}a^3 \simeq L\Gamma a^6 \tag{19}$$

Therefore

$$R(t) \simeq \left(L\Gamma a^6\right)^{1/3} t^{1/3} \tag{20}$$

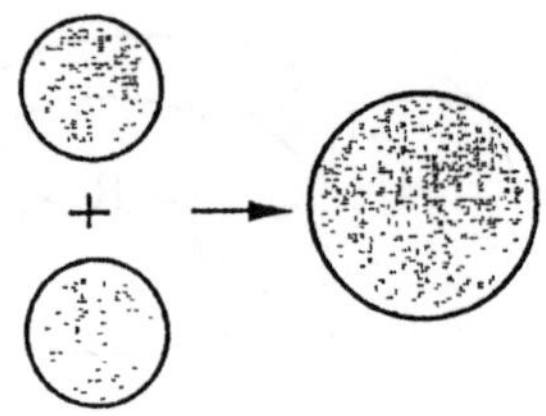

Figure 3. **Coagulation of droplets**

The power law $R(t) \sim t^{1/3}$ is often observed experimentally. However, the power law does not mean that the underlying mechanism is the Lifshitz-Slyozov process since the droplet coagulation described in the following also gives the same power law.

(b)Brownian Coagulation

The small droplets are moving by Brownian motion. When they collide, they coalesce (see Fig. 3). The time variation of the average droplet size due to this mechanism can be estimated as follows. We consider the case that the froplets of A rich phase are moving in the B rich phase.

According to the theory of diffusion controlled aggregation, the number of droplets colloiding with a given droplet of size R per unit time is

$$\dot{n} \simeq DRc \tag{21}$$

where D is the diffusion constant of the droplet,

$$D \simeq \frac{k_B T}{\eta(\phi_B) R} \tag{22}$$

and c is the number density of the surrounding droplets. The expression for $\dot{n}$ can be derived by dimensional analysis, or see [10].

$$c \simeq \frac{\phi_0}{R^3} \tag{23}$$

Since an event of coalescence increases the volume of the droplet by R^3, the increase of the average droplet volume is described by

$$\frac{dR^3}{dt} \simeq \dot{n} R^3 \simeq \frac{k_B T}{\eta(\phi_B)} \phi_0 \tag{24}$$

This also gives the growth law $R(t) \sim t^{1/3}$.

Figure 4. Bicontinuous structures. In this case, the domain size grows due to the hydrodynamic flow induced by interfacial tension.

The droplet coagulation mechanism involves the flow of fluids. (Notice that the viscosity of the surrounding media $\eta(\phi_B)$ appears in eq.(24).) Therefore this process cannot be described by the diffusion equation (10).

(c)Relaxation of droplet shape by hydrodynamic flow

This process, first suggested by Siggia [11], is especially important for the domains with bicontinuous structure such as shown in Fig.4. In such structures, the hydrodynamic flow caused by the interfacial tension decreases the average curvature of the system and increases the characteristic size of the domains. The rate of the change of the characteristic length R due to this mechanism can be estimated by a dimensional analysis. The rate dR/dt must be determined by the viscosity η, the interfacial tension Γ and the characteristic length of domains R. Dimensional analysis gives

$$\frac{dR}{dt} \simeq \frac{\Gamma}{\eta} \tag{25}$$

i.e.,

$$R(t) \simeq \frac{\Gamma}{\eta} t \tag{26}$$

This growth law $R(t) \sim t$ dominates in the late stage kinetics.

2.4. DIFFUSION-HYDRODYNAMIC COUPLING

As it has become clear in the previous arguments, the hydrodynamic flow induced by the interfacial tension plays an important role in the kinetics of phase separation. Let us now modify the kinetic equation to take this effect into account.

Since the Reynolds number is very small, we can use the Stokes equation to determine the fluid velocity $\boldsymbol{v}$.

$$\nabla\eta(\phi)\cdot\left[\nabla\boldsymbol{v}+(\nabla\boldsymbol{v})^{t}\right]-\nabla P=\boldsymbol{g} \tag{27}$$

where P is the pressure which is determined by the incompressible condition;

$$\nabla\cdot\boldsymbol{v}=0 \tag{28}$$

and $\boldsymbol{g}$ stands for the force arising from the interfacial tension. For sharp interface, $\boldsymbol{g}$ is given by

$$\boldsymbol{g}=-2H\Gamma\nabla\Omega(\boldsymbol{r}) \tag{29}$$

where the function $\Omega(\boldsymbol{r})$ takes 1 in the A rich domain and 0 in the B rich domain, and H denotes the mean curvature at the interface. In the problem of phase separation, the interface is not always sharp; it is fuzzy, especially, in the early stage. The expression for $\boldsymbol{g}$ in such a case is again obtained from the free energy functional $F[\phi]$. Let us suppose that we displace the fluid element at $\boldsymbol{r}$ to $\boldsymbol{r}+\delta\boldsymbol{u}(\boldsymbol{r})$. This changes the composition at $\boldsymbol{r}$ by $\delta\phi=-\nabla\cdot(\phi\delta\boldsymbol{u})$, and changes the free energy of the system as

$$\delta F=\int d\boldsymbol{r}\frac{\delta F}{\delta\phi}\delta\phi=\int d\boldsymbol{r}\delta\boldsymbol{u}\cdot\underline{\phi\nabla\left(\frac{\delta F}{\delta\phi}\right)} \tag{30}$$

The underlined part is the force conjugate to the displacement, and can be regarded as the force acting on the fluid element at $\boldsymbol{r}$. Therefore the equation which determines $\boldsymbol{v}$ is

$$\nabla\eta(\phi)\cdot\left[\nabla\boldsymbol{v}+(\nabla\boldsymbol{v})^{t}\right]-\nabla P=-\phi\nabla\left(\frac{\delta F}{\delta\phi}\right) \tag{31}$$

Eqs. (28) and (31) determine $\boldsymbol{v}$ for given $\phi(\boldsymbol{r})$.

The time evolution equation for ϕ now must include the convection term:

$$\frac{\partial\phi}{\partial t}=-\nabla\cdot(\phi\boldsymbol{v})+\nabla\cdot L\nabla\left(\frac{\delta F}{\delta\phi}\right) \tag{32}$$

Eqs.(31) and (32) are the basic equations that describe the phase separation kinetics of simple liquids. These equations have been used to study the effect of shear flow [12, 13], and also have been used in numerical simulation [14, 15].

2.5. EFFECT OF ELASTICITY

Extension of the above theory to take into account various new effects has been done (see for example, [6, 7, 16]). Here I discuss the effect of elasticity on the phase separation kinetics. The reason for this is that elasticity introduces a dramatic change in the kinetics.

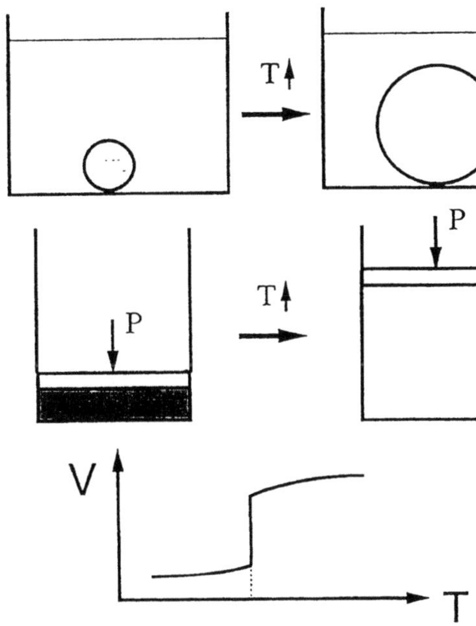

Figure 5. **Comparison between the volume transition of gels and the gas-liquid transition of simple fluids**

To demonstrate how the elasticity changes the kinetics, I take an example of volume transition of gels[17]. This is a phenomenon that a gel immersed in an abundant solvent changes its equilibrium volume discontinuously when the temperature is changed continuously (see Fig. 5) . The phenomenon appears to be very much like the usual gas-liquid phase transition: under a constant pressure, a fluid confined in a piston changes its equilibrium volume discontinuously at a certain temperature at which the fluid transforms from liquid to gas. Indeed the volume transition of gels has been discussed in the same framework as that of the gas-liquid transition. Recently, however, it has been pointed out that such analogy is very misleading[19, 20].

In the case of gas-liquid transition, the equilibrium volume of the fluid under given temperature T and pressure P is determined by the volume V which minimizes the following quantity:

$$F(V,T) = \tilde{F}(V,T) + PV \tag{33}$$

where $\tilde{F}(V,T)$ is the usual Helmholtz free energy for the homogeneous

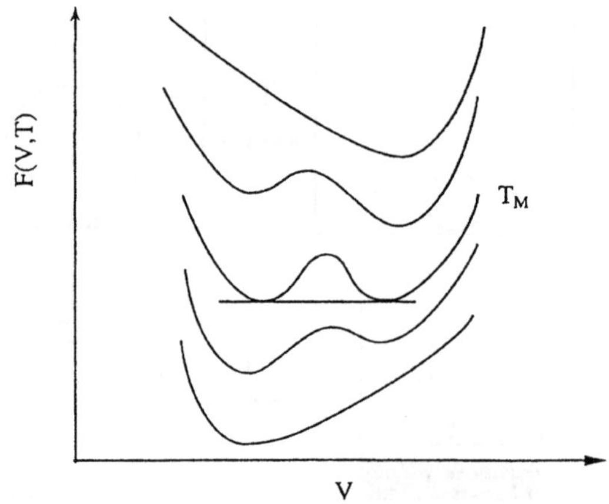

Figure 6. **The free energy $F(V,T)$ of a fluid undergoing gas-liquid transition is plotted against V for various temperature T.**

state of volume V and temperature T. (The function $F(V,T)$ depends on P, but since P is kept constant throughout the present discussion, it is not listed in the argument of F.) Fig. 6 is a schematic drawing of $F(V,T)$ as a function of V when T is varied. At high temperature, $F(V,T)$ has a single minimum, which corresponds to the equilibrium gas phase. As the temperature is lowered, another local minimum appears at a smaller volume, which corresponds to the metastable liquid phase. At a certain temperature T_M, which we call the Maxwell temperature, the height of the two minima become equal to each other. Below the temperature T_M, the free energy of the liquid phase becomes less than that of the gas phase. Therefore the gas-liquid transition takes place at the temperature T_M.

A similar argument has been used for gels. The free energy of the system (gel plus solvent) is calculated as a function of the gel volume V, and it is claimed that the volume transition takes place at the temperature T_M at which the two local minima of the function $F(V,T)$ become equal to each other. This argument is wrong.

At the temperature T_M, the free energy of the small-volume phase (which we call collapsed phase) is equal to the free energy of the large-volume phase (swollen phase). However, in gels, when the two phases coexist within a sample, both phases are deformed from their equilibrium state, and the free energy of the system becomes higher than that of the single phase.

To make the argument clear, let us introduce the minimal free energy

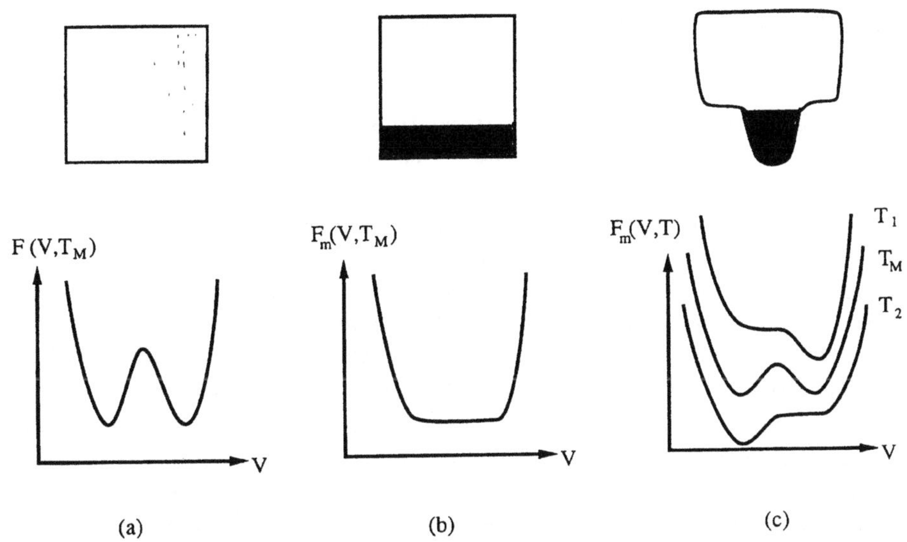

Figure 7. **The free energy $F(V,T_M)$ and theminimal free energy $F_m(V,T)$ is plotted against V for the Maxwell temperature T_M. (a) the uniform state, (b) the coexistent state of a fluid undergoing gas-liquid transition, and (c) the coexistent state of a gel undergoing volume transition.**

function $F_m(V,T)$ which represents the minimum value of the free energy of the system for given volume V and temperature T among all possible states (which of course include the state that the two phases coexist within the sample). The behaviour of this function at the Maxwell temperature is shown in Fig. 7. For the fluid system, $F_m(V,T_M)$ becomes a horizontal straight line connecting the two local minima of $F(V,T_M)$. Strictly speaking, $F_m(V,T_M)$ is slightly larger than this line since there is an interfacial energy. However the interfacial energy is proportional to $V^{2/3}$, and is totally negligible compared with the bulk energy. Therefore, in the fluid system, the liquid phase spontaneously transforms to the gas phase when the temperature is slightly increased above T_M. Conversely, the gas phase spontaneously transforms to the liquid phase when the temperature is slightly lowered below T_M.

The situation is entirely different for gels. In gels, the free energy of the coexisting state must involve the elastic energy of deformation, which is an extensive quantity comparable to the bulk energy. Therefore, in order to cause the spontaneous transition from the collapsed to the swollen state, one has to increase the temperature above T_M so that $F_m(V,T)$ monotonously decreases as the gel transforms from the collapsed phase to the swollen phase. This is realized at the temperature T_1 shown in Fig 7 . Conversely, to cause the transition from the swollen state to the collapsed state, one has to decrease the temperature to T_2 shown in the same figure. Therefore the temperature at which the collapsed to swollen transition takes place is different from the temperature at which the swollen to collapsed takes place, i.e., there is a hysteresis.

Notice that the argument given here has nothing to do with the nucleation rate or any kinetics. The temperatures T_1 and T_2 are theoretical temperatures determined by thermodynamics: the temperature T_1 is determined in such a way that the collapsed to swollen transition is not possible below the temperature. In practice, one has to go to temperatures higher than T_1 in order to cause the transition with reasonable speed.

The above argument demonstrates an anomalous feature in the phase transition kinetics caused by the effect of elasticity. In the volume transition of gels, many other anomalous effects are observed and predicted, such as the surface instability [24], the existence of the incubation time which depends on the gel-size [20] and the trapped phases [21, 22]. Similar effects are also reported for the transition in metals [23].

The elasticity is expected to have an important effect in the phase separation kinetics of polymer solutions and polymer melts since these materials are viscoelastic, meaning that they have elasticity in short time scale. Experiments of Tanaka [24] show the anomalous morphological development during the process of phase separation in polymer solutions. Generalization of the transport equations to account for the viscoelasticity has been done [25, 26, 27]. (See the chapter of Milner and that of Fredrickson).

3. Rheology of Mixtures of Two Immiscible Fluids

3.1. THE PROBLEM

Having discussed the structural evolution in a quiescent state, let us now turn to a new problem, i.e., the structural change in a shear flow. Consider a mixture of two immiscible Newtonian fluids with same density. When a shear flow is applied, the domains of each phase deform, rupture and coagulate. The problem is how the structural evolution takes place and how it affects the rheological properties.

The problem may be attacked by using the transport equations discussed in the previous sections. Here, however, we use a different approach. We set up a phenomenological constitutive equation for such system using simple argument. The present theory is crude, and not very successful in describing the actual complex rheological behaviours observed experimentally. Nevertheless, I discuss it here to demonstrate one way of attacking the problem of mesoscopic systems, and also to illustrate a characteristic aspect of such systems.

The time evolution of the domain patterns can, in principle, be calculated by solving the hydrodynamic equations (eqs.(27), (28) and (29)). For a given configuration of domains, these equations can be solved for $\boldsymbol{v}$, and its value at the interface determines the time evolution of the interface. Such rigorous procedure has been carried out for single droplet, but it is difficult to execute for complex interfaces [28, 29].

As an alternative, we take a phenomenological approach. This approach is motivated by the following formula for the stress tensor for the mixture of fluids which have equal viscosity ($\eta_A = \eta_B = \eta$).

$$\sigma_{\alpha\beta} = \eta(K_{\alpha\beta} + K_{\beta\alpha}) - \frac{\Gamma}{V}\int dS(n_\alpha n_\beta - \frac{1}{3}\delta_{\alpha\beta}) - P\delta_{\alpha\beta} \tag{34}$$

where $K_{\alpha\beta} = \partial\bar{v}_\alpha/\partial r_\beta$ is the macroscopic velocity gradient, $\boldsymbol{n}$is the unit vector normal to the interface, and the integral is done over the whole interface in the system of volume V.

Equation (34) can be derived as follows (compare the derivation of the stress in a polymer fluid in the chapter by T.McLeish, this volume). Suppose we apply an instantaneous deformation $\boldsymbol{\delta\epsilon}$ to the system, and let δF be the change of the free energy caused by this deformation, then the extra stress arising from the thermodynamic force (called the elastic stress in ref. [30]) is given by

$$\delta F = V\boldsymbol{\sigma}^{(E)} : \boldsymbol{\delta\epsilon} \tag{35}$$

Now in the present system, the free energy is only the interfacial energy:

$$F = \Gamma\int dS \tag{36}$$

If the viscosities of the two fluid are equal to each other, the instantaneous deformation causes an affine deformation: i.e, the point $\boldsymbol{r}$ is displaced to $\boldsymbol{r} \to \boldsymbol{r}' = (1 + \boldsymbol{\delta\epsilon}) \cdot \boldsymbol{r}$. It is easy to show that this deformation changes the area of a surface element ΔS as

$$\Delta S \to \Delta S' = (1 - \boldsymbol{\delta\epsilon} : \boldsymbol{nn})\Delta S \tag{37}$$

Consequently, the change of the free energy is

$$\delta F = -\Gamma\int dS\boldsymbol{\delta\epsilon} : \boldsymbol{nn} \tag{38}$$

From eqs.(35) and (38), eq.(34) is obtained.

Equation(34) indicates that the stress is determined by the tensor

$$\boldsymbol{q} = \frac{1}{V}\int dS(\boldsymbol{nn} - \frac{1}{3}\boldsymbol{I}) \tag{39}$$

which is called the interface tensor. The interface tensor represents the mesoscale structure in the system. It plays a role of the order parameter tensor in the theory of liquid crystals (see the lecture of Marrucci). To see the meaning of the this tensor, let us introduces one more "order parameter" which is a scalar:

$$Q = \frac{1}{V}\int dS \tag{40}$$

$\boldsymbol{q}$ and Q represents the orientation and size of the domains in the system. We aim at deriving the time evolution equation of this tensor for given macroscopic velocity field.

Crudely speaking, there are two factors which determine the time evolution of $\boldsymbol{q}$ and Q: one is the flow field $\boldsymbol{K}$ which enlarges and orients the interface, and the other is the interfacial tension Γ which opposes these effects. If one takes into account these effects separately, one can write

$$\frac{d\boldsymbol{q}}{dt} = \left.\frac{d\boldsymbol{q}}{dt}\right|_{\boldsymbol{K}} + \left.\frac{d\boldsymbol{q}}{dt}\right|_{\Gamma} \quad \text{and} \quad \frac{dQ}{dt} = \left.\frac{dQ}{dt}\right|_{\boldsymbol{K}} + \left.\frac{dQ}{dt}\right|_{\Gamma} \tag{41}$$

We shall now discuss these terms separately.

3.2. DEFORMATION BY MACROSCOPIC FLOW

To evaluate the first term in eq.(41), let us assume that the interfacial tension is negligibly small and consider how a surface element at the interface changes under flow. If the two fluids have the same viscosity and the interfacial tension is negligible, every point in the material moves with the macroscopic velocity given by $\boldsymbol{v}(\boldsymbol{r}) = \boldsymbol{K}\cdot\boldsymbol{r}$. Then, it is easy to calculate the change of $\boldsymbol{q}$ and Q. For example, the change of Q is given by(see eq.(37)),

$$\left.\frac{dQ}{dt}\right|_{\boldsymbol{K}} = \frac{1}{\Delta t}\int \frac{dS}{V}(-\boldsymbol{\delta\epsilon} : \boldsymbol{nn}) = -\boldsymbol{q} : \boldsymbol{K} \tag{42}$$

The change for $\boldsymbol{q}$ can be calculated in the similar way , and it has been shown that [8]

$$\begin{aligned}\left.\frac{d}{dt}q_{\alpha\beta}\right|_{\boldsymbol{K}} = & - q_{\alpha\gamma}K_{\gamma\beta} - q_{\beta\gamma}K_{\gamma\alpha} + \frac{2}{3}\delta_{\alpha\beta}K_{\mu\nu}q_{\mu\nu} \\ & - \frac{Q}{3}(K_{\alpha\beta} + K_{\beta\alpha}) + \frac{q_{\mu\nu}K_{\mu\nu}}{Q}q_{\alpha\beta}\end{aligned} \tag{43}$$

3.3. RELAXATION BY INTERFACIAL TENSION

The second term in eq.(41) is not easy to assess. Here a simple phenomenological argument is used. Generally speaking, the interfacial tension has two effects:(i) it decreases the area of the interface, and (ii) it makes the system isotropic. The area of the interface is Q, and the degree of the anisotropy can be expressed by $\boldsymbol{q}/Q$. Thus the simplest relaxation equation is:

$$\begin{aligned} \frac{d}{dt}Q\Big|_{\Gamma} &= -r_1 Q \\ \frac{d}{dt}\left(\frac{\boldsymbol{q}}{Q}\right)\Big|_{\Gamma} &= -r_2\left(\frac{\boldsymbol{q}}{Q}\right) \end{aligned} \qquad (44)$$

where r_1 and r_2 are the relaxation rates with r_1 representing the decrease of the interfacial area (size relaxation), and r_2 the relaxation of anisotropy (shape relaxation). In general, these relaxation rates are determined by the viscosity η , the interfacial tension Γ and the configuration of the interface. The last can be characterized by Q and $\boldsymbol{q}$. If we disregard the dependence on $\boldsymbol{q}$ and assume that r_1 and r_2 are determined by η , Γ and Q, we can show by dimensional analysis,

$$r_1 = c_1\frac{\Gamma Q}{\eta} \qquad r_2 = c_2\frac{\Gamma Q}{\eta} \qquad (45)$$

where c_1 and c_2 are positive numbers, which may depend on the volume fraction ϕ.

Equation(45), however, has an unphysical feature. According to eqs.(44) and (45), in a quiescent state, the system approaches the state of $Q = 0$, which is a macroscopically phase separated state. This can happen for the bicontinuous structure shown in Fig. 4. However, for the usual islands-in-sea structure, the final state will be the state that all droplets are spherical (here the Brownian coagulation and the Lifshitz-Slyozov process are neglected). Such situation is perhaps better described by

$$r_1 = c_1\frac{\Gamma}{\eta}(\boldsymbol{q}:\boldsymbol{q})^{1/2} \qquad (46)$$

which guarantees that the relaxation stops when all droplets become spherical, i.e., when $\boldsymbol{q} = 0$.

3.4. CONSTITUTIVE EQUATION

Combining eqs.(44) , (45) and (46) with eqs.(42) and (43), we have the time evolution equation for $\boldsymbol{q}$ and Q.

$$\frac{d}{dt}q_{\alpha\beta} = -q_{\alpha\gamma}K_{\gamma\beta} - q_{\beta\gamma}K_{\gamma\alpha} + \frac{2}{3}\delta_{\alpha\beta}K_{\mu\nu}q_{\mu\nu} - \frac{Q}{3}(K_{\alpha\beta} + K_{\beta\alpha})$$

$$+ \quad \frac{q_{\mu\nu}K_{\mu\nu}}{Q}q_{\alpha\beta} - \lambda[(1-\mu)Q + \mu q]q_{\alpha\beta} \tag{47}$$

and

$$\frac{dQ}{dt} = -K_{\alpha\beta}q_{\alpha\beta} - \lambda\mu q Q \tag{48}$$

where

$$q = (\Sigma q_{\alpha\beta}^2)^{1/2} \tag{49}$$

Equations (47) and (48) are the time evolution equations for the shape and size of the interface. Given $q_{\alpha\beta}$ and Q, the stress can be calculated by

$$\sigma_{\alpha\beta} = \eta(K_{\alpha\beta} + K_{\beta\alpha}) - \Gamma q_{\alpha\beta} - P\delta_{\alpha\beta} \tag{50}$$

Thus the set of equations, eqs.(47) (48) and (50) give the rheological constitutive equation.

3.5. STEADY SHEAR FLOW

The constitutive equation we have derived has properties quite different from those of the usual viscoelastic materials like polymeric liquids. The most important difference is that the present constitutive equation has no intrinsic time constant. This follows from the dimensional analysis: the present system is characterized by Γ, η and ϕ, from which one cannot construct a quantity that has a dimension of time. This leads to various curious rheological properties.

Consider, for example, the steady shear flow with shear rate $\dot{\gamma}$.

$$v_x = \dot{\gamma} y \qquad v_y = v_z = 0 \tag{51}$$

It can be shown from the constitutive equation that the shear stress and the first normal stress differences are proportional to the shear rate:

$$\sigma_{xy} = \eta(\phi)\dot{\gamma} \qquad \sigma_{xx} - \sigma_{yy} = \eta_N(\phi)|\dot{\gamma}| \tag{52}$$

Notice also that the first normal stress difference is proportional to $\dot{\gamma}$, and is singular at $\dot{\gamma} = 0$. This is like nematic liquid crystals. The singularity comes from the fact that even an infinitesimally small shear rate can change the structure drastically. This behaviour has been confirmed experimentally[31].

3.6. SCALING PROPERTY

The lack of the intrinsic time scale leads to an interesting scaling property. As an example, consider a stepwise change in the shear rate from $\dot{\gamma}_1$ to $\dot{\gamma}_2$

at time $t = 0$. Then it can be shown from the constitutive equation that the stress for $t > 0$ is written as

$$\sigma(t, \dot{\gamma}_1, \dot{\gamma}_2) = \sigma_1 f\left(\dot{\gamma}_1 t, \frac{\dot{\gamma}_2}{\dot{\gamma}_1}\right) \tag{53}$$

where σ_1 is the steady state value of the stress for the shear rate $\dot{\gamma}_1$. Thus for a fixed ratio of $\dot{\gamma}_1/\dot{\gamma}_2$, the plot of $\sigma(t, \dot{\gamma}_1, \dot{\gamma}_2)/\sigma_1$ against the shear strain $\dot{\gamma}_2 t$, becomes independent of the magnitude of $\dot{\gamma}_1$. Such scaling law has been confirmed experimentally [31].

The origin of the scaling law is that since there is no intrinsic time scale in the system, the characteristic time of the system is determined by the external conditions such as the shear rate applied for $t < 0$.

There are other classes of materials which satisfy the same condition and show the scaling relation. An example is the nematic liquid crystals which involve many defects and have the polydomain structures. The basic equation of nemato-hydrodynamics (i.e., the Ericksen-Leslie equation) includes only the viscosity (the Leslie coefficients) and the Frank elastic constants. Thus there is no intrinsic time constant, and therefore the same scaling law holds. Actually, the scaling law was first found for liquid crystals [32, 33]. A phenomenological constitutive equations has been proposed for polydomain nematics[34].

3.7. TRANSIENT FLOWS

Fig. 8 shows the shear stress calculated by the present constitutive equation. Here $\dot{\gamma}$ is changed as $1 \to 4 \to 1 \to -1$ as it is indicated in the figure.

The figure reproduces some characteristic features of the rheological responses observed experimentally[31] such as (i) when the shear rate is increased stepwisely, the first normal stress difference shows an overshoot. (ii) when the magnitude of the shear rate is increased, it takes a long time for the stress to reach the steady state value, while when the flow is reversed ($\dot{\gamma} \to -\dot{\gamma}$), the stress immediately reaches the steady state value, and (iii) when the shear flow is stopped, the stress immediately drops to zero.

On the other hand, the figure fails to reproduce the overshoot of the shear stress observed in the step-wise increase of the shear rate experiments. The failure may be removed by using a better model for the relaxation dynamics. Since experimental data is now available for the evolution of the droplet size and orientations [35], it will be possible to have such equations in near future.

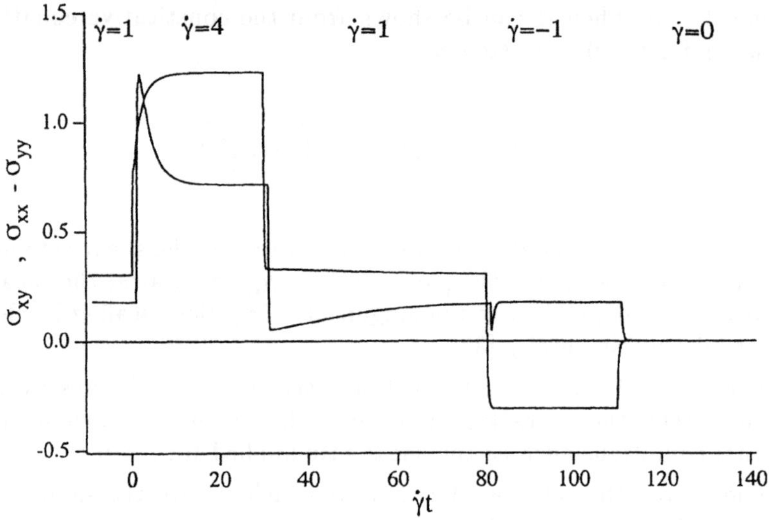

Figure 8. **The shear stress σ_{xy}, and the first normal stress difference $\sigma_{xx} - \sigma_{yy}$ calculated by the constitutive equation for the step change of the shear rate $\dot{\gamma}$ is plotted against the time. They show a complex rheological properties due to the deforming the domains.**

4. Theoretical Challenges

There are many theoretical challenges in this area. The theory I described here is for the simplest case: the mixture of two simple fluids. since the phase separation takes place very fast for such systems. Most experiments are done for metals or polymeric systems. This is because (i) the slow kinetics of these materials allows various experimental means to study the structural evolution and (ii) these materials are technologically much more important than simple fluids.

The problems of how the mesoscale structures are determined and how they can be controlled have an immense technological importance. However, the theoretical model described here cannot answer many simple questions.

(i) When the phase separation takes place, one phase forms continuous sea and the other phase forms islands (or droplets). Which component forms the continuous phase?
(ii) It is known that the mesoscale structure is strongly affected by the third components (surfactant, block copolymers etc.). How do they affect the mesoscale structure?
(iii) How do solidification or gelation affects the mesoscale structure?

To answer these questions, we need new theoretical models and also efficient computational methods to solve the equations.

The phenomenological approach described in section 3 should be pushed further. The drawback of the equation is that it does not have clear connection to the process of coagulation and ruptures of droplets. A better description is clearly needed. It is also interesting to study other systems with mesoscale structure from the scaling view point. The experimental data in colloids (see the chapter by P.Pusey, this volume) and surfactant systems (see the chapter of Roux) may have a hidden scaling relation. Challenges!

References

1. de Gennes, P.G. and Prost, J. (1993) *Physics of Liquid Crystals*. Oxford Univ. Press, Oxford.
2. Marrucci, G. and Greco, F. (1994) , *Adv. Chem. Phys.*, **86**, pp. 331.
3. Israelachvili, J. (1991) *Intermolecular and Surface Forces*. Academic Press, New York
4. Fredrickson, G.N. and Bates, F.S. Dynamics of Block copolymers; Theory and Experiments *Ann. Rev. Material Sci.*, , to be published.
5. Russel,R.B., Saville D.A.S. and Schowalter W.R.S. (1989) *Colloidal Dispersions*, Cambridge Univ. Press, Cambridge.
6. Gunton, J.D., Miguel, M.S. and Sahni, P.S. (1983) in *Phase Transition and Critical Phenomena*, vol **8**, ed C. Domb and J.L. Lebowitz, Academic Press, N.Y.
7. A. Onuki (1992) *Polymer Physics and Phase Transition Dynamics* Iwanami, Tokyo (in Japanese).
8. Doi, M. and Ohta, T. (1991) , *J. Chem. Phys.*, **95**, pp. 242.
9. Doi, M. (1993) Rheology of Textured Materials in *Complex Fluids, Lecture Notes in Physics*, ed by L. Garrido, Springer Verlag, pp. 221-242.
10. Chandrasekhar, S. (1943), *Rev. Mod. Phys.*,**15**, 1-89.
11. Siggia, E.D. (1979) , *Phys. Rev. A*, **20**, pp. 595.
12. Onuki, A. and Kawasaki, K. (1979) , *Ann. Phys.*, **121**, pp. 456.
13. Onuki, A. (1987) , *Phys. Rev. A*, **35**, pp. 5149.
14. Shinozaki, A. and Oono, Y. (1991) , *Phys. Rev. A*, **66**, pp. 173 and (1993) , *Phys. Rev. E*, **48** pp. 2622.
15. Koga, T. and Kawasaki, K. (1993) , *Physica A*, **196**, pp. 389.
16. Glotzer, S. C. (1995) , *Ann. Rev. Comp. Phys.* , **II**, World Sci. page 1
17. K. Dusek Ed, *Responsive Gels* (1993) Advances in Polymer Science 109 and 110 Springer
18. Tanaka, T. Sun,S.T.Hirokawa,Y.Katayama.S.,Kucera J., Hirose Y., and Amiya, T. (1987) *Nature* **325** pp. 796.
19. Sekimoto, K. (1993) , *Phys. Rev. Lett*, **70**, pp. 4154 see also (1995), *Int. J. Mod. Phys.*.
20. Tomari, T. and Doi, M. (1995) , *Macromolecules*, **28**, pp. 8334 .
21. Sekimoto, K, Suematsu, N. and Kawasaki, K. (1989) *Phys. Rev. B.*, **39**, pp.4912.
22. Kuroki,Y. and Sekimoto, K. (1994) *Europhys. Lett.* **23** pp.113
23. Nishimori, H. and Onuki, A. (1991), *J. Phys. Soc. Jpn.*, **60**, pp.1208.
24. Tanaka, H. (1993) , *Phys. Rev. Lett.*, **71**, pp. 3158.
25. Helfand, E. and Fredrickson, G.H. (1992) , *Phys. Rev. Lett*, **62**, pp. 2468.
26. Doi, M. (1990) *in Dynamics and Patterns in Colplex fluids*. ed by Onuki, A. and Kawasaki, K., Springer pp.100 .
27. Doi, M. and Onuki, A. (1992) , *J. Phys. II France*, **2**, pp. 1631-1656.
28. Rallison, J. M. (1984) , *Ann. Rev. Fluid Mech.*, **16**, pp. 45 .
29. Stone, H. A., Bently, B.J. and Leal, L.G. (1986) , *J. Fluid Mech.*, **173**, pp. 131.
30. Doi, M. and Edwards, S.F. (1986) *The Theory of Polymer Dynamics*. Oxford Univ.

Press, Oxford.
31. Takahashi, T, Noda, I., Doi, M. (1994) , *J. Rheol.*, **38**, pp. 699.
32. Larson, R. G. and Mead, D.W. (1989) , *J. Rheology*, **33**, pp. 1251.
33. Moldenaers, P., Yanase, H. and Mewis, J. (1990) , *ACS Symposium Miami*, chap **26**, pp. 370-380.
34. Larson, R.G. and Doi, M. (1991) , *J. Rheology*, **35**, pp. 539.
35. Takahashi, T., Noda, I., (1995) , *in Flow Induced Structure in Polymers*. ed Nakatani, A.I., Dadnum, M.S., ACS Symposium Series 597 page 140.

CURRENT TOPICS IN DIFFUSION-CONTROLLED REACTIONS OF POLYMERS

GLENN H. FREDRICKSON
Departments of Chemical Engineering and Materials
University of California, Santa Barbara, CA 93106, USA

Abstract.
Recent progress in the theory of diffusion-controlled reactions is described, with particular emphasis on developments relevant to reactive blending operations. Topics covered include the role of convection on intermolecular reactions in polymer melts and diffusion-controlled coupling at polymer-polymer interfaces.

1. Introduction

Reactive blending is a very versatile method of strengthening and emulsifying incompatible polymer blends[1]. The technique involves melt blending two polymeric components (A and B), each pure component containing a fraction of chains with functional reactive groups. If the reactive groups in the A phase are designed to couple only with complimentary groups in the B phase, then the reaction is confined to the interfacial regions between the two phases. The interfacial reaction product is an A-B graft or block copolymer and is conveniently located where it is most effective–in the interfacial zone. Thus, the need for dispersing a copolymer surfactant is eliminated (e.g. if it was added as a third component). Once at the interface, the copolymer serves a dual role: (1) to control droplet size (through interfacial tension reduction[2] and by suppressing droplet coalescence[3]) and, (2) to increase interfacial adhesion[4]. Applications of reactive blending technology are very widespread, including toughening/modification of nylons, polyolefins, epoxies, etc.

In view of the importance of this technology, it is surprising that *theoretical* studies of reactive blending processes are extremely limited in scope. However, on closer examination, realistic blending processes are extremely

T McLeish (ed), Theoretical Challenges in the Dynamics of Complex Fluids, 315-329

involved, with non-Newtonian fluid mechanics, mass transport, and interfacial reaction kinetics intertwined in a complex fashion. Here, I describe a few recent steps towards unraveling this complexity.

2. Review of Diffusion-Controlled Reactions of Small Molecules

The classical description of diffusion-controlled reactions in small molecule (or colloidal) systems is due to Smoluchowski[5]. This approach is appropriate for a *dilute* solution of reacting particles in which two particles couple *irreversibly* when they approach each other to a separation less than some "capture radius" a. Moreover, the intrisic rate of reaction is assumed to be extremely fast (instantaneous reaction at separations $r \leq a$) to ensure diffusion-control. The Smoluchowski theory is based on a simple diffusion equation for a pair distribution function, $\psi(\mathbf{r}_1, \mathbf{r}_2; t)$, representing the joint probability density of observing two reactive particles at positions $\mathbf{r}_1$ and $\mathbf{r}_2$ at time t. (This focus on two-particle correlations is justified by the assumption of dilute reactive species.) The equation can be written

$$\partial_t \psi(\mathbf{r}_1, \mathbf{r}_2; t) = D_0(\nabla_1^2 + \nabla_2^2)\psi \tag{1}$$

where D_0 is the self-diffusion coefficient of a reactive particle. Due to the translational invariance in a macroscopic system, ψ is actually a function only of the distance between the particles, $r_{12} = |\mathbf{r}_1 - \mathbf{r}_2|$, and at large distances the particle positions are assumed to be uncorrelated: $\lim_{r_{12}\to\infty} \psi(r_{12}; t) = \rho^2$, where ρ is the spatially-averaged number density of reactive particles. Furthermore, the instantaneous reaction of particles at distances $r_{12} \leq a$ can be easily accomodated by a simple absorbing boundary condition at $r_{12} = a$, $\psi(a; t) = 0$. The steady state solution of Eq. (1) satisfying both boundary conditions is given by

$$\psi_0(r) = \rho^2(1 - a/r) \quad r \geq a \tag{2}$$

The reaction rate is easily calculated by integrating the diffusive flux over the surface of the "capture sphere"; i.e. $4\pi a^2(2D_0)\partial_r\psi_0|_{r=a}$, where $2D_0$ is the diffusion coefficient describing the dynamics of *relative* displacement of two particles. Defining an effective second order rate coefficient, k_S, by the expression

$$-d\rho/dt \equiv k_S\rho^2 = 4\pi a^2(2D_0)\partial_r\psi_0|_{r=a} \tag{3}$$

leads immediately to the Smoluchowski formula:

$$k_S = 8\pi a D_0 \tag{4}$$

In subsequent sections, I will have frequent occasion to refer to this fundamental formula, expressing the rate coefficient for diffusion-controlled reaction as the product of a particle diffusivity and a capture radius.

The Smoluchowski theory has been generalized in a number of important ways over the years[6], including modifying the "absorbing" boundary condition to allow for a finite intrinsic reaction rate and incorporating hydrodynamic interactions between particles (important if they are colloidal). More important for the present work are extensions to include *convective*, as well as diffusive, transport of particles. For example, if a steady, linear shear flow is imposed on the dilute suspension of reactive particles, described by a velocity field $u_x(\mathbf{r}) = \dot{\gamma}y$ ($\dot{\gamma}$ is the shear rate), then it is natural to consider the following "modified Smoluchowski" problem[7]:

$$2D_0\nabla^2\psi - \dot{\gamma}y\partial_x\psi = 0 \tag{5}$$

$$\psi(a;t) = 0 \quad \psi(\infty;t) = \rho^2 \tag{6}$$

Notice that the above equations imply that the linear flow is unperturbed on the scale of the capture radius, a. While this assumption is perhaps not appropriate for colloidal systems, its validity will subsequently be argued for polymer melts.

I am unaware of analytical solutions to the above equations in the literature, but asymptotic results can be inferred from closely related studies of convective heat transfer caused by flow past a sphere[8, 9]. The relative importance of convective to diffusive transport can be described by a dimensionless Peclet number, Pe, which in the present application is defined by

$$Pe \equiv \dot{\gamma}a^2/D_0 \tag{7}$$

Weak flow and strong flow situations are thus characterized by $Pe \ll 1$ and $Pe \gg 1$, respectively, and the general solution to Eqs. (5) and (6) leads to an effective reaction rate coefficient k of the form $k = 8\pi D_0 a\, \Phi(Pe)$, where $\Phi(x)$ is a dimensionless scaling function.

For weak flows, the heat transfer studies indicate that the leading correction to the Smoluchowski formula is *non-analytic* in Peclet number:

$$k = 8\pi D_0 a[1 + c\, Pe^{1/2} + \cdots] \quad Pe \ll 1 \tag{8}$$

with c a numerical constant. This unusual result is a consequence of the slow *power law* decay of the unperturbed Smoluchowski solution $\psi_0(r)$ given in Eq. (2). At any finite shear rate $\dot{\gamma}$, convection will dominate diffusion at scales r exceeding $(D_0/\dot{\gamma})^{1/2}$. As a result, the power law tail of ψ_0 is fundamentally changed by the imposition of flow; i.e. flow is a singular perturbation.

The situation is more complicated for strong flows. Strict translation of the heat transfer results to the present problem would imply that $\Phi(Pe) \sim 1$ for $Pe \to \infty$, but this result was obtained for a neutrally buoyant sphere free to rotate in the shear flow. This leads to closed streamlines around the sphere, which is not implied by Eq. (5) above. In more typical situations of no closed streamlines, heat transfer studies[10] indicate that the strong flow asymptotic behavior is changed to $\Phi(Pe) \sim Pe^{1/3}$. I suspect (guess!) that this is the correct asymptotic behavior of the modified Smoluchowski problem stated above:

$$k = 8\pi D_0 a(c' \, Pe^{1/3}) \qquad Pe \gg 1 \tag{9}$$

with c' a second numerical constant. The physics behind this scaling is associated with the onset of steep concentration boundary layers around the capture sphere in which ψ varies from 0 to ρ^2 over a boundary layer thickness $\xi \sim a \, Pe^{-1/3}$.

3. Diffusion-Controlled Reactions in Polymers

Next, I turn to consider a generalization of the Smoluchowski problem relevant to reactive coupling in polymer melts. Suppose that a low concentration ρ of chains with a single *terminal* reactive group and degree of polymerization N are dispersed into a melt of unreactive, but otherwise identical, linear chains. This situation is depicted in Figure 1. The theoretical description of intermolecular diffusion-controlled reaction in such systems is quite mature[11, 12, 13, 14]; here I will focus on a particularly convenient formulation due to de Gennes[13].

Guided by earlier studies of Fixman-Wilemski and Doi, de Gennes' approach amounts to replacing Eq. (1) by the following equation:

$$\partial_t \psi = \mathcal{L}\psi - Q_0 a^3 \delta(\mathbf{r}_1 - \mathbf{r}_2)\psi \tag{10}$$

where $\mathcal{L}$ is a diffusive operator describing the independent dynamics of two *reactive chain ends* located near $\mathbf{r}_1$ and $\mathbf{r}_2$. For simple diffusing particles, $\mathcal{L}$ is simply the differential operator given on the right hand side (RHS) of Eq. (1); in the present case of polymer dynamics, $\mathcal{L}$ is generally a non-Markovian operator. The last term on the RHS of the above equation is a crude model of a reaction sink and replaces the absorbing boundary condition of the Smoluchowski calculation. In this regard, Q_0 describes the intrinsic reaction rate of two reactive chain ends that fall within a microscopic capture volume a^3 of each other. Effective absorbing boundary conditions (instantaneous reaction on contact) are generated by taking $Q_0 \to \infty$ at a later stage in the calculation.

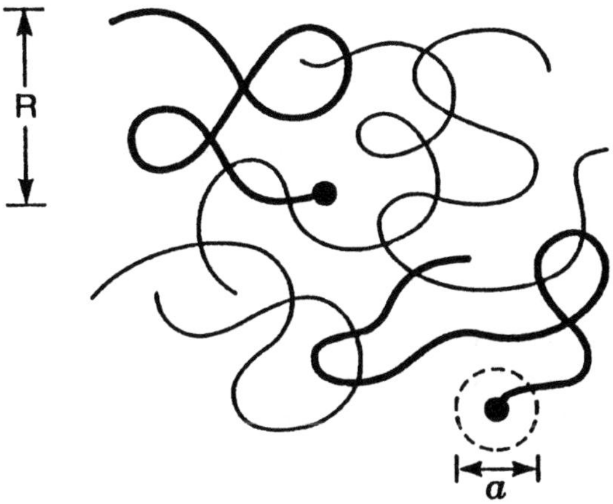

Figure 1. **The reacting polymer system under consideration consists of a melt of linear chains, with a small fraction of the chains possessing one terminally attached reactive site (solid dot). When two sites approach within a "capture radius" a, an instantaneous and irreversible reaction occurs. The melt is assumed monodisperse with degree of polymerization N and radius of gyration R.**

While the non-Markovian operator $\mathcal{L}$ is difficult to construct for any reasonable model of chain dynamics (e.g. Rouse or reptation), it turns out that the inverse operator $\mathcal{L}^{-1}$ is actually much simpler to deal with. The kernel of this inverse operator is a two-particle Green's function, which can be factored into a product of two one-particle Green's functions, each factor $G(\mathbf{r}, t)$ describing the probability density for the end of a reactive chain to undergo a displacement $\mathbf{r}$ over a time interval t. By focusing just on the diagonal elements of the pair distribution function, $\Psi(t) \equiv \psi(\mathbf{r}, \mathbf{r}; t)$, i.e. the "contact" probability of two reactive chain ends, and imposing the initial condition $\Psi(0) = \rho^2$, Eq. (10) can be rewritten as the following Volterra integral equation:

$$\Psi(t) = \rho^2 - Q_0 a^3 \int_0^t d\tau \; S(t - \tau)\Psi(\tau) \tag{11}$$

where the kernel $S(t)$ is defined by

$$S(t) \equiv \int d\mathbf{r} \; [G(\mathbf{r}, t)]^2 \tag{12}$$

While several physical interpretations can be ascribed to $S(t)$[13], a particularly useful one is to note that $1/S(t)$ scales as the *volume swept out by a diffusing chain end in a time t*. More precisely,

$$S(t) \sim 1/[\bar{x}(t)]^3 \tag{13}$$

where $\bar{x}(t)$ is the root-mean-squared (RMS) displacement of a diffusing chain end. The omitted numerical prefactor in this scaling expression can be restored by insertion of appropriate expressions for $G(\mathbf{r}, t)$ in Eq. (12)[13].

The above Volterra equation is easily solved by Laplace transformation (carets are used to denote transformed quantities) and a frequency-dependent reaction rate coefficient can be defined by $\hat{k}(s) \equiv Q_0 a^3 \rho^{-2} \hat{\Psi}(s)$, i.e. proportional to the probability of contact of two sites times their intrinsic reactivity. A particularly interesting result is that this expression reduces in the instantaneous reaction limit ($Q_0 \to \infty$) to $\hat{k}(s) = 1/[s\hat{S}(s)]$. Since the RHS of this expression is independent of both Q_0 and a, I arrive at the result, previously obtained by Doi and de Gennes, that the rate of intermolecular diffusion-controlled reactions in polymer melts is *insensitive to the microscopic details of the capture event!* Finally, if $\hat{S}(0)$ is finite (the so-called "non-compact" exploration regime[13] appropriate for all situations considered in this paper), a simple Kubo-like formula can be derived for the long-time effective rate coefficient:

$$k \equiv \lim_{t \to \infty} k(t) = \left[\int_0^\infty dt \; S(t) \right]^{-1} \tag{14}$$

3.1. UNENTANGLED MELTS

Next, I discuss the application of Eq. (14) to *unentangled* polymer melts ($N < N_e$), where the Rouse model is appropriate. The Green's function $G(\mathbf{r}, t)$ can be computed exactly for Rouse chains and is found to be a Gaussian. Carrying out the integral in Eq. (12) then leads to an expression for the omitted prefactor in Eq (13). Moreover, an explicit expression for the mean-squared displacement $\bar{x}^2(t)$ of a chain end in the Rouse model can be found in the text by Doi and Edwards[15]. For times larger than the longest Rouse time, $\tau_R \sim N^2$, the RMS displacement corresponds to classical diffusion: $\bar{x}^2(t) \sim D_0 t$, where $D_0 \sim 1/N$ is the center of mass diffusivity of a reactive chain. In contrast, at times shorter than τ_R, the displacement is sub-diffusive: $\bar{x}^2(t) \sim b(D_m t)^{1/2}$. Here, $D_m \sim N D_0$ is a monomeric diffusivity and b is the statistical segment length. Since $S(t)$ thus scales as $t^{-3/4}$ at short times and as $t^{-3/2}$ at long times, the Kubo integral in Eq. (14) is convergent. Numerical analysis leads to[7]

$$k \approx 50.3 \; D_0 R \sim N^{-1/2} \tag{15}$$

where $R^2 = Nb^2/6$ is the mean-squared radius of gyration.

This result, originally due to Doi[12] and de Gennes[13], is remarkable in that the *microscopic* (i.e. monomer-scale) capture radius of the Smoluchowski formula Eq. (4) is replaced by the *mesoscopic* length R. Evidently the physics underlying Eq. (15) is that the rate limiting step for combination of a pair of reactive ends separated by a large distance ($\gg R$) involves getting the centers of mass (COM) of the two coils to overlap. Once this has occured, the actual end combination event proceeds by faster sub-diffusive internal Rouse dynamics. Provided that the initial concentration of reactive chains, ρ, is sufficiently small that the typical separation of a reactive pair greatly exceeds R ($\rho R^3 \ll 1$), the conditions required for the above mechanism are met and the equation $d\rho/dt = -k\ \rho^2$ faithfully describes the decay of the reactive chain population.

3.2. ENTANGLED MELTS

Similar results can be obtained for *entangled* ($N > N_e$) polymer melts[13]. In fact, if the RMS displacement of a chain end in the *reptation model* is expressed in the form $\bar{x}^2(t) = 6D_0 t\ \Phi_d(t/\tau_d)$, where $D_0 \sim 1/N^2$ is the COM diffusivity of a reptating chain, $\Phi_d(x)$ is a scaling function, and $\tau_d \sim N^3$ is the disengagement time of the reptation model, then the reptation model results can be written in a form surprisingly similar to those of the Rouse model. This relies on the fact that for times $t \gg \tau_R$, the functional form of $\Phi_d(x)$ is *precisely* the same[15] as the corresponding scaling function for the Rouse model, $\Phi_R(x)$, describing the crossover between COM and internal mode-dominated dynamics. If one further approximation is made, namely that $G(\mathbf{r}, t)$ is approximated by a Gaussian function with the proper second moment, $\bar{x}^2(t)$, then even the numerical coefficient is preserved in the final result for reptating chains[7]:

$$k \approx 50.3\ D_0 R \sim N^{-3/2} \tag{16}$$

The difference in the molecular weight scaling from the Rouse result Eq. (15) thus derives solely from the N-dependence of D_0.

4. Shear Effects on Polymeric Diffusion-Controlled Reactions

Recently, in collaboration with L. Leibler[7], I examined the effect of linear flow fields on the rate of intermolecular diffusion-controlled reactions in polymer melts. The model is precisely the same as that considered in the previous section, except that a steady linear velocity field of the form $u_i(\mathbf{r}) = K_{ij} r_j$ is imposed on the homogeneous melt. (K_{ij} is the transpose of a uniform velocity gradient tensor.) Retracing the steps of the de Gennes procedure outlined in the last section leads to an identical Volterra equation as Eq. (11), but with a modified definition of $S(t)$:

$$S(t) = \int d\mathbf{\Delta}\ [G(\mathbf{\Delta}, t)]^2 \tag{17}$$

where $\mathbf{\Delta} \equiv \mathbf{r}(t) - \exp(\mathbf{K}t) \cdot \mathbf{r}(0)$ is the "diffusive displacement" of a chain end, i.e. the end position at time t minus the advected initial position of the end. In place of Eq. (13), the diffusive displacement in the presence of flow is *anisotropic*, leading to the following scaling relation:

$$S(t) \sim [\det \mathbf{M}(t)]^{-1/2} \tag{18}$$

where $\mathbf{M}(t)$ is the mean-squared diffusive displacement tensor, $M_{ij}(t) \equiv \langle \Delta_i(t)\Delta_j(t)\rangle$. As in the previous section, the numerical prefactor in the above scaling relation can be established by resort to specific models for $G(\mathbf{\Delta}, t)$ and the rate coefficient k in flow is still connected to $S(t)$ by the Kubo-like relation Eq. (14).

4.1. UNENTANGLED MELTS

Due to the linearity of the Rouse equations, the mean-squared diffusive displacement tensor $\mathbf{M}(t)$ can be *exactly* computed for *any* linear flow. Here I will only be concerned with unentangled melts and simple shear flow ($\dot{\gamma} = K_{12}$). From the previous discussion of flow effects in Section 2, it is expected that a Peclet number will emerge to gauge the flow strength. However, the capture radius a entering the definition of Pe in Eq. (7) should be replaced with the *mesoscopic* radius of gyration R, as discussed in Section 3. Thus,

$$Pe \equiv \dot{\gamma}R^2/D_0 \sim \dot{\gamma}\tau_R \equiv De \tag{19}$$

indicating that the Peclet number is actually a Deborah number, De, for the present case of polymer melts.

Asymptotic results for the rate coefficient in cases of weak ($De \ll 1$) and strong ($De \gg 1$) shear flows are straightforward to obtain from the integral

$$k^{-1} = 2^{-3}\pi^{-3/2} \int_0^\infty dt\ [\det \mathbf{M}(t)]^{-1/2} \tag{20}$$

where the exact numerical prefactor has been restored. For weak shear flows

$$k = 50.3\ D_0R[1 + 0.807\ De^{1/2} + \cdots] \quad De \ll 1 \tag{21}$$

while for strong shear we found[7]

$$k \approx 52.9\ D_0R\ De^{1/3} \sim N^{1/6}\dot{\gamma}^{1/3} \quad De \gg 1 \tag{22}$$

It is interesting to see that these equations have the same form as the weak and strong flow expressions for the "modified Smoluchowski" problem with $Pe \sim De$, Eqs. (8) and (9). A particularly nice feature of sheared *polymer melts* is that these nonanalytic results follow by direct asymptotic expansion of the integral in Eq. (20). In contrast, sophisticated matched asymptotic expansion techniques were required to deduce Eqs. (8) and (9) in the analogous heat transfer problem[8, 9, 10]. Moreover, the assumption of an unperturbed linear flow inside the "capture sphere" is evidently a good one in the present case of a mesoscopic capture radius of order $R \sim N^{1/2}$.

4.2. ENTANGLED MELTS

For chains above the entanglement threshold ($N > N_e$), the mean-squared diffusive displacement tensor $\mathbf{M}(t)$ should be computed with the reptation model. Weak and strong flows are delineated by a modified Deborah number, $De \equiv \dot{\gamma}\tau_d$, where $\tau_d \sim N^3$ is again the disengagement time. For weak shear flow, advection of the reactive chain COM's gives rise to the leading nonanalytic flow correction, so it is not difficult to argue for a form identical to the Rouse model[7]:

$$k = 50.3\ D_0 R[1 + 0.807\ De^{1/2} + \cdots] \quad De \ll 1 \tag{23}$$

In this expression, however, both De and the COM diffusivity, $D_0 \sim 1/N^2$, have new definitions (and N-dependence).

For strong flows, $De \gg 1$, calculation of $\mathbf{M}(t)$ proves considerably more difficult. The details of the calculation are beyond the scope of this article, so I only provide an outline here and refer the interested reader to the original paper[7]. Instead of directly computing $\mathbf{M}(t)$, it turns out to be easier to derive equations of motion for a related tensor $\mathbf{Q}(s, s'; t)$ defined by

$$\mathbf{Q}(s, s'; t) \equiv \langle [\mathbf{R}(s,t) - e^{\mathbf{K}t} \cdot \mathbf{R}(s', 0)][\mathbf{R}(s,t) - e^{\mathbf{K}t} \cdot \mathbf{R}(s', 0)] \rangle \tag{24}$$

where $\mathbf{R}(s,t)$ is the position of the sth segment of the primitive chain at time t. The desired tensor $\mathbf{M}(t)$ is obtained by simply setting s and s' to zero, i.e. $M_{ij}(t) = Q_{ij}(0,0;t)$. Employing the "earthworm" equations of Doi and Edwards for the primitive chain dynamics and invoking the "independent alignment approximation"[15], leads to a linear PDE for $\mathbf{Q}$:

$$\partial_t \mathbf{Q} = D_c \partial_s^2 \mathbf{Q} + \mathbf{K}\mathbf{Q} + \mathbf{Q}\mathbf{K}^T \tag{25}$$

subject to the boundary conditions

$$\partial_s Q_{ij}|_{s=0} = -\partial_s Q_{ij}|_{s=L} = -(1/3) a_t \delta_{ij} \tag{26}$$

In the above equations, the quantity $D_c \sim 1/N$ is the curvilinear tube (Rouse) diffusion coefficient and a_t is the tube diameter. The initial condition, $U_{ij}(s,s') \equiv Q_{ij}(s,s';0)$, satisfies a related PDE in the independent alignment approximation:

$$D_c(\partial_s + \partial_{s'})^2\mathbf{U} + \mathbf{K}\mathbf{U} + \mathbf{U}\mathbf{K}^T = 0 \tag{27}$$

and is subject to the same boundary conditions as $\mathbf{Q}$ on *both* the s and s' variables. It should be noted that the above equations describing the dynamics of the primitive chain are valid only for times exceeding the Rouse time τ_R, so the subsequent flow calculations will be restricted to $1 \ll De \ll \tau_d/\tau_R \sim N$.

The strategy for obtaining strong shear results for entangled chains thus consists of the following: (1) solve Eq. (27) to obtain the initial condition $\mathbf{U}$, (2) solve Eqs. (25)-(26) to obtain the tensor $\mathbf{Q}$, (3) deduce $\mathbf{M}$ from $M_{ij}(t) = Q_{ij}(0,0;t)$, and (4) asymptotically expand the Kubo integral Eq. (20) for $De \gg 1$. Carrying out this procedure leads to[7]

$$k \approx 32.2\ D_0 R\ De/\ln De \quad 1 \ll De \ll \tau_d/\tau_R \tag{28}$$

where the numerical coefficient is approximate due to the fact that the Gaussian prefactor in Eq. (20) was employed, even though $\mathbf{\Delta}(t)$ is not strictly a Gaussian random variable in the reptation model. This striking asymptotic scaling of the reaction rate coefficient, $k \sim \dot{\gamma}N^{3/2}/\ln(\dot{\gamma}N^3)$, is quite interesting because it suggests that the reaction rate can be more readily enhanced by flow in entangled melts than for unentangled ones [c.f. Eq. (22)]. Further studies to examine the accuracy of the independent alignment approximation in the present application would be quite useful.

5. Diffusion-Controlled Reactions at Polymer-Polymer Interfaces

Very recently, O'Shaughnessy and Sawhney[16] and Fredrickson[17] independently extended the above ideas to diffusion-controlled coupling at polymeric interfaces. Here I will summarize the latter approach, which is a natural extension of the formalism described above.

The system of interest is depicted in Figure 2; namely, a symmetric, flat interface[18] between bulk A and B homopolymer phases. These bulk phases have small concentrations, ρ_A and ρ_B, respectively, of end-functionalized chains. Type A reactive groups can react only with type B groups in the interfacial region of thickness ξ. For simplicity, all chains have degree of polymerization N. Furthermore, it is assumed that the narrow interface regime of Helfand and Tagami[18] is appropriate, so that $\chi \ll 1$, $\chi N \gg 1$, and $\xi = 2b/(6\chi)^{1/2}$, where χ is the A-B Flory interaction parameter. Thus,

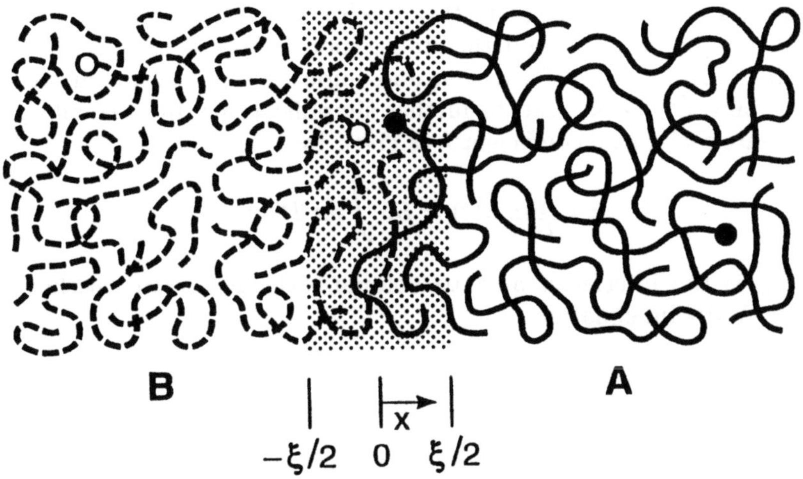

Figure 2. Symmetric A-B polymer interface and coordinate system employed. The interfacial region (shaded), corresponds to the region $-\xi/2 \le x \le \xi/2$ in which loops of A and B polymers overlap. The bulk A phase ($x \to \infty$) consists of type-A homopolymer chains (solid curves) mixed with a low concentration ρ_A of end-functionalized (solid circle) type-A chains. The bulk B phase ($x \to -\infty$) consists of type-B homopolymer chains (dashed curves) mixed with a low concentration ρ_B of end-functionalized (open circle) type-B chains. Solid circles react with open circles irreversibly on contact in the interfacial region.

ξ is an N-independent length larger than the monomer size b, but much smaller than the radius of gyration R.

The formalism of de Gennes, described in Section 3 above, can be straightforwardly extended to the present situation[17]. One important difference is that the "contact" probability of an A and a B reactive end, $\Psi(x,t) \equiv \psi(\mathbf{r},\mathbf{r};t)$, is no longer homogeneous, but depends on the coordinate normal to the interface, x. The integral equation Eq. (11) is replaced by

$$\begin{aligned}\Psi(x,\quad t\quad) &= \rho_A\rho_B \exp[-V_A(x) - V_B(x)] \\ &- \quad Q_0 a^3 \int_0^t d\tau \int d\mathbf{r}'\, G_A(\mathbf{r},\mathbf{r}';t-\tau) G_B(\mathbf{r},\mathbf{r}';t-\tau)\Psi(x',\tau) \quad (29)\end{aligned}$$

where $V_K(x)$ is the potential (units of k_BT) experienced by a type-K (A or B) chain with its reactive end placed at location x within the interface. $G_K(\mathbf{r},\mathbf{r}';t)$ is the Green's function for the displacement of a type-K reactive chain end. The above equation can be simplified considerably by factorizing the Green's functions and invoking translational invariance in the transverse coordinates $\mathbf{r}_\perp = (y, z)$: $G_K(\mathbf{r}, \mathbf{r}'; t) = H_K(\mathbf{r}_\perp - \mathbf{r}_\perp'; t) g_K(x, x'; t)$. (This is exact for simple diffusion and the Rouse model, approximate for reptating chains.) Laplace transformation and imposition of the instantaneous reaction limit, $Q_0 \to \infty$, lead to further simplifications. In particular, the effective interfacial rate coefficient k is given by

$$k = \lim_{t\to\infty} k(t) = S_V \int dx\ P(x) \tag{30}$$

where S_V is the interfacial area to volume ratio and $P(x)$ satisfies the following Fredholm equation

$$\exp[-V_A(x) - V_B(x)] = \int dx'\ \hat{K}(x, x'; 0) P(x') \tag{31}$$

The kernel $\hat{K}(x, x'; 0)$ appearing in this equation is the $s \to 0$ limit of the object

$$\hat{K}(x, x'; s) \equiv \int_0^\infty dt\ e^{-st} g_A(x, x'; t) g_B(x, x'; t) S_\perp(t) \tag{32}$$

where

$$S_\perp(t) \equiv \int d\mathbf{r}_\perp\ H_A(\mathbf{r}_\perp; t) H_B(\mathbf{r}_\perp; t) \tag{33}$$

With a specified potential model for the $V_K(x)$, the above equations permit a quantitative assessment of the rate of interfacial reaction described by

$$d\rho_A(t)/dt = d\rho_B(t)/dt = -k\ \rho_A(t)\rho_B(t) \tag{34}$$

It is perhaps surprising at first glance[19] that a *steady state* rate coefficient exists for the present problem, since viewed globally it is a one-dimensional diffusion-reaction system. The above kinetic description holds only for times that exceed the terminal relaxation time τ and that are smaller than the characteristic timescale τ_ρ for reactants to be depleted in the interfacial region. The latter time can be shown to scale as $\tau \ln^2 N/(\rho R^3)^2$, where $\rho \sim \rho_A \sim \rho_B$ is the initial reactive end concentration. Since τ_ρ greatly exceeds τ for low reactant concentrations, $\rho R^3 \ll 1$, it is evident that Eq. (34) will hold for a significant range of times.

5.1. UNENTANGLED MELTS

Here I restrict consideration to the Rouse model and continue with the assumption of symmetric interfaces and equal-length chains, $N = N_A = N_B$. Under such conditions, there is only one independent potential to be specified, $V(x) \equiv V_A(x) = V_B(-x)$, and only one Green's function, $g(x, x'; t) \equiv g_A(x, x'; t) = g_B(-x, -x'; t)$. An expression for $V(x)$ follows from the Helfand-Tagami theory[18]: $V(x) = (1/2)\ln[1 + \exp(-4x/\xi)]$. In addition, the integral in Eq. (33) can be carried out explicitly to yield the scaling relation: $S_{\perp}(t) \sim 1/[\bar{x}(t)]^2$.

I will subsequently argue that for $N \gg 1$, expressions derived for k are independent of the detailed interfacial structure. Thus, in place of the Helfand-Tagami expression, it is permissible and convenient to choose a simpler representation of $V(x)$. The choice of a linear potential field inside the interfacial region, $V(x) = 1/2 - x/\xi$ for $-\xi/2 < x < \xi/2$, is particularly convenient, because it renders the left hand side of the Fredholm Eq. (31) independent of x. Furthermore, in the bulk A phase, $x > \xi/2$, it is convenient to choose $V(x) = 0$, while a reflecting boundary will be imposed at $x = -\xi/2$, i.e. $V(x) = \infty$ for $x < -\xi/2$. The last condition has the added advantage that both x and x' are restricted to the *finite* interval $(-\xi/2, \xi/2)$ in Eqs. (30) and (31).

With the above simple potential model, it is easy to extract the large N asymptotic behavior of k. An important timescale is τ_0, the time required for a chain end that is initially in the interface to escape that region. Since $\xi \ll R$, τ_0 is estimated using the sub-diffusive Rouse regime, i.e. $\bar{x}(\tau_0) \sim b^{1/2}(D_m\tau_0)^{1/4} \sim \xi$. This leads to $\tau_0 \sim b^2/(D_m\chi^2)$, an N-independent timescale. Next, consider the $s \to 0$ limit of Eq. (32). Breaking the integral at a time of order τ_0 yields an N-independent contribution arising from the time interval $(0, \tau_0)$ and an N-dependent contribution from the interval (τ_0, ∞). In the second integral, for times $t \gg \tau_0$ and x and x' inside the interface, the Green's function $g(x, x'; t)$ can be approximated by the simple form $g(x, x'; t) \sim \exp[-V(x)]/\bar{x}(t)$, since the end position has equilibrated over a scale of order $\bar{x}(t) \gg \xi$ at such times. Thus, the N-dependent contribution to the kernel is

$$\hat{K}(x, x'; 0) \sim e^{-1} \int_{\tau_0}^{\infty} dt\, [\bar{x}(t)]^{-4} \sim (D_0R^2)^{-1} \ln(\tau_R/\tau_0) \sim \ln N \tag{35}$$

which clearly dominates the N-independent contribution for $N \gg 1$. Finally, solving Eqs. (30) and (31) and restoring neglected numerical prefactors leads to

$$k \approx 50.3\, D_0R(2RS_V)/\ln N \sim 1/\ln N \tag{36}$$

This interesting result, obtained independently by Fredrickson[17] and O'Shaughnessy and Sawhney[16], can evidently be expressed in the form $k \approx k_{homog}\phi_a$, where $k_{homog} = 50.3\, D_0R$ is the rate coefficient for a homogeneous system [c.f. Eq. (15)] and $\phi_a = 2RS_V/\ln N$ is a measure of the volume fraction of the inhomogeneous system that is *accessible for reactive coupling*. It is important to note that $\phi_a \neq S_V\xi$, which had previously been assumed in the literature[1]. Furthermore, a simple physical interpretation can be given: the rate limiting step for interfacial reaction involves transporting the COM's of reactive A and B coils to within R of the interface on either side. Thus, $2RS_V$ is a natural choice for the volume fraction accessible to the reaction. Apparently, however, there is a residual N-dependence associated with the internal Rouse dynamics that subsequently couple the two chain ends; this results in the extra factor of $1/\ln N$.

5.2. ENTANGLED MELTS

For chains exceeding the entanglement threshold, the above analysis can be repeated using the reptation model[17]. Regardless of whether the interfacial thickness ξ is smaller or larger than the tube diameter a_t, it is straightforward to argue that the timescale τ_0 for end equilibration within the interfacial region is N-independent. At times exceeding τ_0, parallel and perpendicular diffusion approximately decouple, even for reptating chains, so the formalism presented in the last section is preserved. Retracing those steps, it can be argued that the dominant contribution to the kernel for $N \gg 1$ is

$$\hat{K}(x, x'; 0) \sim e^{-1} \int_{\tau_R}^{\infty} dt\, [\bar{x}(t)]^{-4} \sim (D_0 R^2)^{-1} \ln(\tau_d/\tau_R) \sim N \ln N \quad (37)$$

This leads to the following expression for the interfacial rate coefficient:

$$k \approx 100.6\, D_0 R (2RS_V)/\ln N \sim 1/(N \ln N) \quad (38)$$

The factor of 2 difference in the prefactor from Eq. (36) is explained by the fact that $\ln(\tau_d/\tau_R) = (1/2)\ln(\tau_R/\tau_0)$ for $N \to \infty$.

6. Summary and Discussion

The above examples serve to illustrate that in spite of the seeming complexity of high molecular weight polymer melts, the slow COM diffusion of reactive chains (in comparison with their faster internal mode dynamics) imparts a surprising degree of universality to the rate of intermolecular diffusion-controlled reactions in such systems. This is true for reactions occuring in a homogeneous single phase (Section 3), as well as for end-coupling reactions at polymer-polymer interfaces (Section 5). Moreover, the universality is preserved even on imposition of flow (Section 4), allowing for predictions to be made that transcend particular classes of polymers or end-group chemical reactivity. Because relaxation times are long in polymer melts, it should also be feasible to experimentally access regimes where convective enhancement of the reaction rate is large (i.e. $De \sim Pe \gg 1$).

While we have made relatively modest progress towards a complete theory of reactive blending, several of the ingredients are now in place. Carefully designed experiments to test the predictions outlined in Sections 3-5 would go far towards confirming our understanding of bulk and interfacial diffusion-controlled coupling, as well as the role of convection in homogeneous reacting systems. In addition, parallel experimental and theoretical studies on well-characterized inhomogeneous systems in well-defined flows would be extremely valuable. Only by developing an understanding of the competing effects of diffusion, convection, and reaction in such idealized situations can we hope to be able to tackle realistic cases of reactive blending. Much remains to be accomplished in this exciting field!

This work was supported by the National Science Foundation under grant NSF-DMR 9505599.

References

1. Guegan, P., Macosko, C. W., Ishizone, T., Hirao, A. and Nakahama, S. (1994) *Macromolecules* **27**, 4993.
2. Leibler, L. (1988) *Makromol. Chem., Macromol. Symp.* **16**, 1.
3. Milner, S. T., Xi, H. (1996) *J. Rheology*, (in press).
4. Brown, H. R., Char, K., Deline, V. R. and Green, P. F. (1993) *Macromolecules* **26**, 4155.
5. Smoluchowski, M. V. (1916) *Phys. Z.* **17**, 557; Chandrasekhar, S. (1943) *Rev. Mod. Phys.* **15**, 1.
6. Noyes, R. M. (1961) Effects of Diffusion Rates on Chemical Kinetics, in G. Porter (ed.), *Progress in Reaction Kinetics*, Pergamon Publishers, New York, Vol. 1, pp. 130-160.
7. Fredrickson, G. H. and Leibler, L. (1996) *Macromolecules*, (in press).
8. Batchelor, G. K. (1979) *J. Fluid Mech.* **95**, 369.
9. Acrivos, A. (1980) *J. Fluid Mech.* **98**, 299.
10. Leal, L. G. (1992) *Laminar Flow and Convective Transport Processes: Scaling Principles and Asymptotic Analysis*, Butterworth-Heinemann Publishers, Boston.
11. Wilemski, G. and Fixman, M. (1973) *J. Chem. Phys.* **58**, 4009; (1974) *J. Chem. Phys.* **60**, 866, 878.
12. Doi, M. (1975) *Chem. Phys.* **9**, 455; **11**, 107, 115.
13. de Gennes, P. G. (1982) *J. Chem. Phys.* **76**, 3316, 3322.
14. Friedman, B. and O'Shaughnessy, B. (1993) *Macromolecules* **26**, 4888, 5726; O'Shaughnessy, B. (1994) *Phys. Rev. Lett.* **73**, 1723.
15. Doi, M. and Edwards, S. F. (1986) *The Theory of Polymer Dynamics*, Oxford University Press, New York.
16. O'Shaughnessy, B. and Sawhney, U. (1996) *Phys. Rev. Lett.* **76**, 3444.
17. Fredrickson, G. H. (1996) *Phys. Rev. Lett.* **76**, 3440.
18. Helfand, E. and Tagami, Y. (1972) *J. Chem. Phys.* **56**, 3592.
19. This question was raised by Professor Marrucci during the lectures.

Subject Index

MIX
Papier aus verantwortungsvollen Quellen
Paper from responsible sources
FSC® C105338

If you have any concerns about our products,
you can contact us on
ProductSafety@springernature.com

In case Publisher is established outside the EU,
the EU authorized representative is:
Springer Nature Customer Service Center GmbH
Europaplatz 3, 69115 Heidelberg, Germany

Printed by Libri Plureos GmbH
in Hamburg, Germany